Statistics for Industry and Technology

Advances in Data Analysis

Theory and Applications to Reliability and Inference, Data Mining, Bioinformatics, Lifetime Data, and Neural Networks

Christos H. Skiadas
Editor

Birkhäuser
Boston • Basel • Berlin

Editor
Christos H. Skiadas
Technical University of Crete
Data Analysis and Forecasting Laboratory
73100 Chania, Crete
Greece
skiadas1@otenet.gr

ISBN 978-0-8176-4798-8 e-ISBN 978-0-8176-4799-5
DOI 10.1007/978-0-8176-4799-5

Library of Congress Control Number: 2009939133

Mathematics Subject Classification (2000): 03E72, 05A10, 05C80, 11B65, 11K45, 37A50, 37E25, 37N40, 58E17, 60A10, 60B12, 60E05, 60E07, 60F05, 60F17, 60G05, 60G15, 60G17, 60G50, 60G60, 60H05, 60H10, 60H30, 60J10, 60J22, 60J27, 60J65, 60J80, 60J85, 60K10, 60K15, 62-07, 62-09, 62C10, 62F03, 62F15, 62F30, 62F40, 62G05, 62G08, 62G10, 62G30, 62G32, 62H15, 62H25, 62H30, 62J02, 62J05, 62J07, 62J12, 62M10, 62M40, 62N05, 62P20, 62Q05, 65C30, 65C40, 65D10, 68P15, 68P20, 68P30, 68U35, 74E30, 74F20, 76M25, 78A70, 82B41, 82C41, 90B60, 90C35, 90C70, 91A43, 91A90, 91B24, 91B26, 91B30, 91B32, 91B38, 91B40, 91B60, 91B62, 91B70, 91B84, 91C20, 91E10, 92B20, 92C15, 93C42, 93C55, 93C57

Printed on acid-free paper

Birkhäuser Boston is a part of Springer Science+Business Media (www.birkhauser.com)

Contents

Part IV Bioinformatics and Markov Chains

Part V Life Table Data, Survival Analysis, and Risk in Household Insurance

Preface

This book contains the main part of the invited papers presented at the Twelfth International Conference on Applied Stochastic Models and Data Analysis (ASMDA), which took place in Chania, Crete, Greece, May 29–June 1, 2007. ASMDA, since 1981, aims at serving as the interface between stochastic modeling and data analysis and their real-life applications.

We include both theoretical and practical papers, presenting new results having the potential for solving real-life problems. An important objective was to select material presenting new methods for solving these problems by analyzing relevant data and leading to the advancement of related fields.

This book contains chapters on various important topics of data analysis such as: Data Mining and Text Mining, Asymptotic Behaviour of Stochastic Processes and Random Fields, Bioinformatics and Markov Chains, Life Table Data, Survival Analysis and Risk in Household Insurance, Neural Networks and Self-Organizing Maps, Parametric and Nonparametric Statistics, and Statistical Theory and Methods.

We thank all the contributors for the success of the Twelfth ASMDA 2007 Conference, the reviewers, and especially the authors of this volume. Special thanks go to the conference secretary, Dr. Anthi Katsirikou, for her work and assistance. We also acknowledge the valuable support of Professor N. Balakrishnan, Mrs. Debbie Iscoe for her assistance in compiling the manuscript, and Mr. Tom Grasso and Mrs. Regina Gorenshteyn for their assistance with the production of the book.

The book is dedicated to 25 years of ASMDA meetings and especially to Professor Jacques Janssen, founder of ASMDA.

Christos H. Skiadas
Technical University of Crete
Chania, Crete
Greece

List of Contributors

Afanas'eva, L. G.
Department of Mathematics and Mechanics, Moscow State University, Russia

Afanasyev, V. I.
Department of Discrete Mathematics, Steklov Institute, Moscow, Russia

Alvarez-Esteban, R.
Department of Economics and Statistics, University of León, Spain

Arató, M.
Department of Probability Theory and Statistics, Eötvös Loránd University, Budapest, Hungary

Atsalakis, G.
Department of Production Engineering and Management, Technical University of Crete, Chania, Crete, Greece

Balbi, S.
Dipartimento di Matematica e Statistica, Università di Napoli "Federico II", Italy

Barnes, A.
GE Global Research, 1 Research Circle, Niskayuna, NY, USA

Bécue-Bertaut, M.
Department of Statistics and Operations Research, Universitat Politècnica de Catalunya, Barcelona Spain

Bertrand, D.
UR1268, Biopolymères Interactions Assemblages, INRA, Nantes, France

Beutner, E.
Department of Quantitative Economics, Maastricht University, Tongersestract 53, NL-6200 MD Maastricht, The Netherlands

Boutalis, Y. S.
Department of Electrical and Computer Engineering, Democritus University of Thrace, 67100 Xanthi, Greece

Bulinskaya, E. V.
Department of Mathematics and Mechanics, Moscow State University, Russia

Bulinski, A.
Department of Mathematics and Mechanics, Moscow State University, Russia

Camproux, A. C.
Equipe de Bioinformatique Gènomique et Moléculaire, INSERM UMR-S726/ Université Denis Diderot Paris 7, Paris F-75005, France
MTI, Inserm UMR-S 973; Université Denis Diderot Paris 7, Paris F-75205, France

Cariou, V.
Sensometrics and Chemometrics Laboratory, INRA–ENITIAA, Nantes, France

Christodoulou, M. A.
Department of Electronic and Computer Engineering, Technical University of Crete, 73100 Chania, Greece

Corbellini, A.
Department of Economics and Social Sciences, Università Cattolica del Sacro Cuore, 29100 Piacenza, Italy

Crosato, L.
Statistics Department, Università di Milano Bicocca, 20126 Milano, Italy

Fernández-Aguirre, K.
Facultad de CC. Económicas y Empresariales, Universidad del País Vasco/Euskal Herriko Unibertsitatea (UPV/EHU), Bilbao, Spain

Freeman, J. M.
Manchester Business School, University of Manchester, Manchester, UK

Ganugi, P.
Department of Economics and Social Sciences, Università Cattolica del Sacro Cuore, Piacenza, Italy

Haferkamp, L.
Universität zu Köln, Köln, Germany

Karagrigoriou, A.
University of Cyprus, Nicosia, Cyprus

Kipouridis, I.
Technological Institution of West Macedonia, Department of General Sciences, Koila Kozanis, Greece

Koleva, E. G.
Institute of Electronics, Bulgarian Academy of Sciences, Sofia, Bulgaria

Kottas, T. L.
Department of Electrical and Computer Engineering, Democritus University of Thrace, 67100 Xanthi, Greece

Lhote, L.
GREYC, CNRS UMR 6072, Université de Caen Basse-Normandie, Caen, France

Márkus, L.
Department of Probability Theory and Statistics, Eötvös Loránd University, Budapest, Hungary

Martin, J.
Unité Mathématique Informatique et Génome UR1077, INRA, Jouy-en-Josas F-78350, France
Equipe de Bioinformatique Génomique et Moléculaire, INSERM UMR-S726/ Université Denis Diderot Paris 7, Paris F-75005, France
Université de Lyon, Lyon, France; Université Lyon 1; IFR 128; CNRS, UMR 5086; IBCP, Institut de Biologie et Chime des Protéines, 7 passage du Vercors, Lyon F-69367, France

Mattheou, K.
University of Cyprus, Nicosia, Cyprus

Mazzoli, M.
Department of Economics and Social Sciences, Università Cattolica del Sacro Cuore, 29100 Piacenza, Italy

Migliorati, S.
Department of Statistics, University of Milano-Bicocca, Milano, Italy

Misuraca, M.
Dipartimento di Economia e Statistica, Università della Calabria, Italy

Modroño-Herrán, J. I.
Facultad de CC. Económicas y Empresariales, Universidad del País Vasco/Euskal Herriko Unibertsitatea (UPV/EHU), Bilbao, Spain

Mosler, K.
Universität zu Köln, Köln, Germany

Nezis, D.
Department of Production Engineering and Management, Technical University of Crete, Chania, Crete, Greece

Nuel, G.
CNRS, Paris, France; MAP5 UMR CNRS 8145, Laboratory of Applied Mathematics, Department of Mathematics and Computer Science, Université Paris Descartes, Paris F-75006, France

Ongaro, A.
Department of Statistics, University of Milano-Bicocca, Milano, Italy

Papaioannou, T.
Department of Statistics & Insurance Science, University of Piraeus, Greece

Pardo, M. C.
Department of Statistics and O.R (I), Complutense University of Madrid, Spain

Paul, S.
GE Global Research, John F. Welch Technology Center, EPIP Phase 2, Bangalore, India

Prokaj, V.
Department of Probability Theory and Statistics, Eötvös Loránd University, Budapest, Hungary

Regad, L.
Equipe de Bioinformatique Génomique et Moléculaire, INSERM UMR-S726/Université Denis Diderot Paris 7, Paris F-75005, France
MTI, Inserm UMR-S 973; Université Denis Diderot Paris 7, Paris F-75205, France

Sachlas, A. P.
Department of Statistics & Insurance Science, University of Piraeus, Greece

Shashkin, A.
Department of Mathematics and Mechanics, Moscow State University, Russia

Skiadas, C.
Hanover College, Indiana, USA

Skiadas, C. H.
Technical University of Crete, Chania, Greece

Tsaklidis, G.
Department of Mathematics, Aristotle University of Thessaloniki, Thessaloniki, Greece

Valencia, O.
Department of Applied Economics, University of Burgos, Burgos, Spain

Vuchkov, I. N.
European Quality Center, University of Chemical Technology and Metallurgy, Sofia, Bulgaria

Yarovaya, E.
Department of Mathematics and Mechanics, Moscow State University, Russia

Zografos, K.
Department of Mathematics University of Ioannina, 45110 Ioannina, Greece

Zopounidis, C.
Department of Production Engineering and Management, Technical University of Crete, Chania, Crete, Greece

List of Tables

List of Figures

Part I

Data Mining and Text Mining

1

Assessing the Stability of Supplementary Elements on Principal Axes Maps Through Bootstrap Resampling. Contribution to Interpretation in Textual Analysis

Ramón Alvarez-Esteban[1], Olga Valencia[2], and Mónica Bécue-Bertaut[3]

[1] Department of Economics and Statistics, University of León, Spain (e-mail: ramon.alvarez@unileon.es)
[2] Department of Applied Economics, University of Burgos, Burgos, Spain (e-mail: oval@ubu.es)
[3] Department of Statistics and Operations Research, Universitat Politècnica de Catalunya, Barcelona, Spain (e-mail: monica.becue@upc.edu)

Abstract: Bootstrap resampling is commonly used to assess the stability of the configurations issued from principal axes methods. In the case of textual analysis, the interpretation is usually supported by the characteristics of the individuals, used as supplementary variables. To assess the stability of these variables gives information about the global structure stability.

An example issued from a wine guide illustrates the interest of computing confidence regions for supplementary categorical or quantitative variables in correspondence analysis applied to lexical tables.

Keywords and phrases: Correspondence analysis, bootstrap, textual analysis, free-text comments

1.1 Introduction

Bootstrap resampling has shown its potentiality to assess the stability of the configurations issued from principal axes methods. It allows for computing confidence regions for the elements represented on the principal subspaces (Efron and Tibshirani, 1993; Lebart et al., 2006). In many cases, the supplementary rows and/or columns provide essential information to interpret the results. In textual studies, when correspondence analysis (CA) is performed on a lexical table crossing individuals and words, the interpretation is usually supported by the characteristics of the individuals, used as supplementary variables. To assess the stability of these variables gives information about global structure stability.

Section 1.2 presents the data. Section 1.3 reviews the principles of bootstrap, and Section 1.4 offers some results obtained in the example data. Finally, Section 1.5 concludes with some remarks.

C.H. Skiadas (ed.), *Advances in Data Analysis*, Statistics for Industry and Technology, DOI 10.1007/978-0-8176-4799-5_1,

1.2 Data

Wine tasting is becoming an increasing domain for textual data analysis. The wine guide *El Mundo* (El Mundo, 2005) has analysed 522 wines from 'Castile and Leon'. This region (94.273 km^2) is located in the northwest of Spain and comprises five AOC designations (*Bierzo*, *Cigales*, *Ribera del Duero*, *Rueda*, and *Toro*).

Here, we only focus on the 364 red wines. Every wine is described by free-text tasting notes and complementary information such as score (between 70 and 97), price, type of grape, vintage, etc. (ten Kleij and Musters, 2003) (Table 1.1).

Table 1.1. Free-text tasting notes example

— ***Wine 30 Tares P3-2001 premium. Score = 91.***
A lot of 'terroir' stands out in this great red wine bouquet; hint of minerals, silex, slate, warm roasted character with a contrast of damp soil and much ripe fruit, concentrated, fatty finish on the palate, impressive viscosity on the tongue, again, flavours of damp soil and minerals in the lengthy end.

A lemmatization step has been carried out (Labbé, 1990; Muller, 1977–1992). Then the nouns, adjectives, verbs, and adverbs have been selected and, among these categories, only the words used at least eight times in the whole of the tasting notes are kept. Thus, the resulting lexical table crosses 364 wines (rows) and 222 words (columns).

1.3 Methodology

To assess the stability of the configurations issued from CA, partial or total bootstrap can be considered. In the former case, the principal subspace issued from the analysis performed on the original table is considered as a reference space and the rows or columns of the replicated tables are considered as supplementary elements. In the latter case, a new analysis is performed on every replicated table and the resulting configurations are compared (Lebart, 2004). In this work, we only focus on partial bootstrap.

In the following, we use the terms of the example. Thus, the statistical units (rows) refer to the wines, the active columns represent the words, and the supplementary columns correspond to the characteristics of the individuals (quantitative or categorical).

One basic principle of bootstrap consists in reproducing the process that is used to extract the random sample from the population, but using the distribution of the observed sample as an approximate distribution of the parent population (Lebart, 2006). In our case, the wine sample selection does not follow any random method, but is explicitly chosen by the expert owing to its qualities. Thus, no actual sampling error exists. Nevertheless, bootstrap resampling can be performed, by means of drawing with replacement a sample of size 364 out of the initial wine sample. It allows for studying the stability of the results facing perturbations in the wine choices by the expert.

The replicated tables have the same columns (words) as the original table (although the word frequencies can be different) and 364 bootstrapped rows. For a particular replication, some wines may not appear whereas others may be present more than once. This step is repeated B times (in our case $B = 500$); from every B bootstrap sample, a replicated wines $\times$ words table is built up. At every stage, the margins can differ from the original table margins. Nevertheless, as usual in CA, the latter are used as reference to compute the coordinates of rows and columns of the replicated table, considered as supplementary elements.

Depending on the replication, the coordinates of the wines remain constant, but the coordinates of the columns vary. We can compute these coordinates for the active and supplementary columns (frequency, quantitative or categorical) and the confidence regions (Lebart, 2006; Beran and Srivastava, 1985).

1.4 Results

1.4.1 CA results

Table 1.2 shows the highest five eigenvalues as well as the proportion of total inertia that they explain.

Table 1.2. Eigenvalues and proportion of inertia

Axes	Eigenvalue	Proportion of inertia	Cumulative proportion
1	0.22929	0.02046	0.02046
2	0.19946	0.01780	0.03826
3	0.17162	0.01531	0.05357
4	0.17034	0.01520	0.06877
5	0.16495	0.01472	0.08349

As usual in a sparse table, the first eigenvalues of the CA express a very small part of the total inertia (Lebart et al., 1998, pp. 120–126).

Despite the low percentage of inertia explained by the first axis (2.046%), the high correlation between the initial score and the first axis of CA (0.70) shows that the main dimension induced by the words expresses the score, at least for a large amount. Thus, we interpret the first axis as a score level axis (Figure 1.1).

The wines with the highest scores have positive coordinates while the wines with lowest scores have negative coordinates. On Figure 1.1, the wines located on the right have a higher score than 88 whereas the wines located on the left of the vertical show a lower score than 82.

Furthermore, to make the relationship precise between the first axis (and eventually, the second axis) the values of the score are grouped into six categories (or score levels) and projected onto the first principal plane (Figure 1.2). Except for the lower score level (level 1), the categories follow the natural order along the first axis.

The information given by the relationship between the score and the first axis allows for disclosing the meaning of the words in the context of a wine guide. For example,

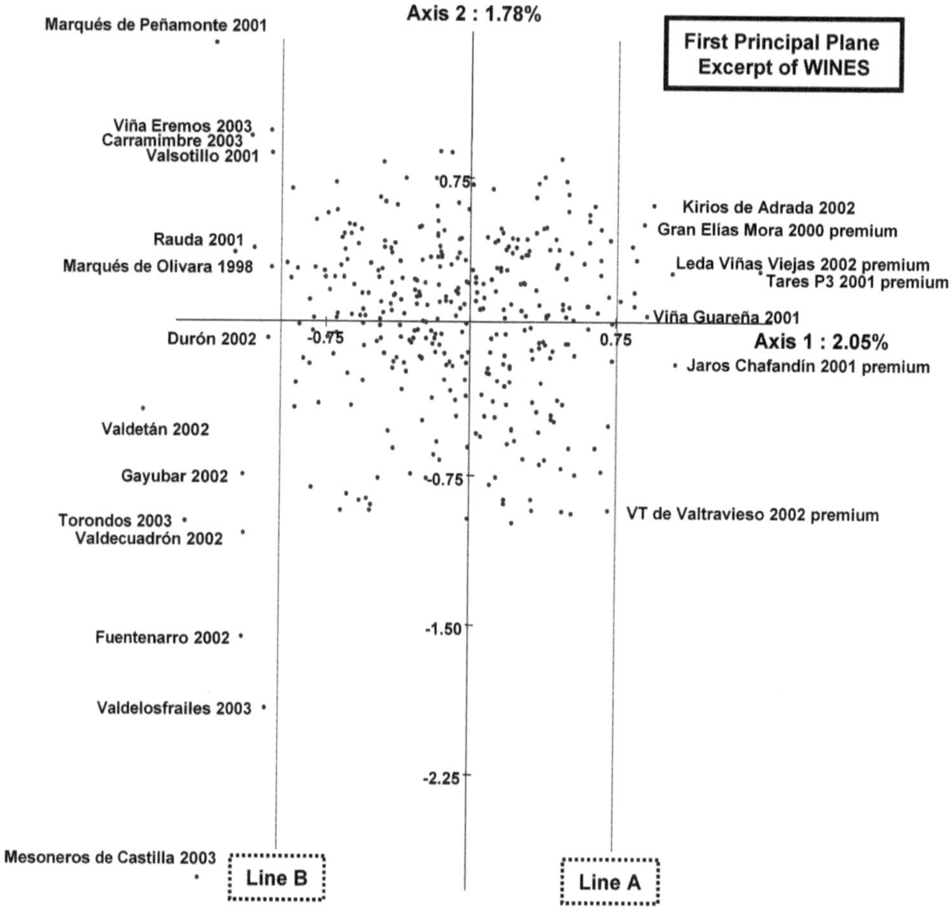

Figure 1.1. First principal plane. Excerpt of the wines

concerning the words related to hedonic features, the first axis contrasts words such as *impressive*, *fatty*, *nutty*, *gun powder*, and *modern* on the right, with *amiable*, *easy*, *traditional*, *consistency*, and *young* on the left (Figure 1.3). The latter words, albeit positive in current language, present here a negative reading. We are able to assert this remark thanks to the relationship between the score and the first axis.

1.4.2 Stability

As the interpretation mainly relies on the supplementary columns, we have to combine the study of the stability of the words and the supplementary variables by means of the bootstrap procedure. Here, we favor the latter. To address this problem, 500 bootstrap resamplings on the 364 wines have been performed. For each replicated table, the coordinates of each score category are computed using the CA transition formula.

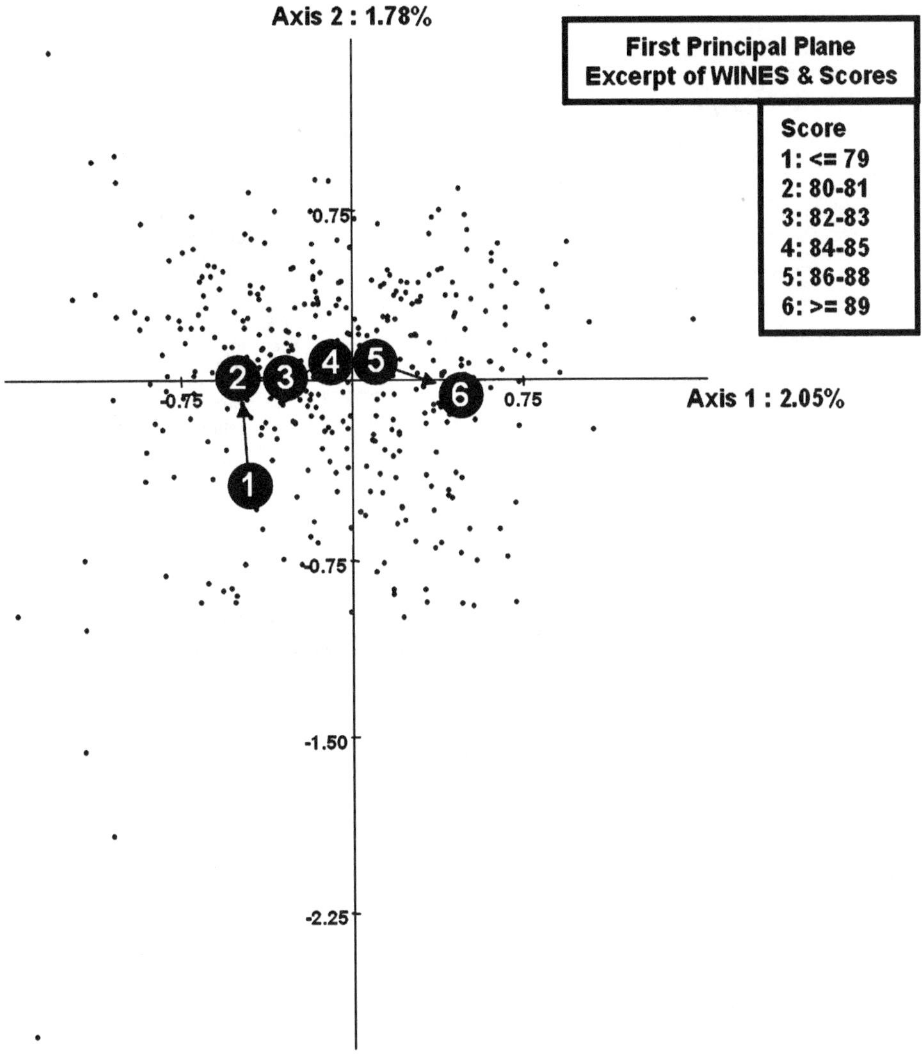

Figure 1.2. First principal plane. Score levels projected as supplementary categories

Table 1.3 shows the means and the standard deviations of the score levels. A high value of the standard deviation of the coordinates of the lower category is observed (only five wines with the lowest scores)

Figure 1.4 shows the confidence regions of every score level. The highest score levels (6:score $\geq$ 89, and 5:score 86–88) present confidence regions that do not overlap with the others. On the contrary, the confidence region, as well as the high standard deviation of the lower score level on the first principal plane, shows that the first category does not hold any relationship with the first two axes.

Referring to the score as a quantitative variable, Table 1.4 shows that its correlation with the first original axis varies between 0.63 and 0.78 among the replicated tables, presenting a low deviation standard. The interpretation of the first axis as a score level axis is stable (Figure 1.5).

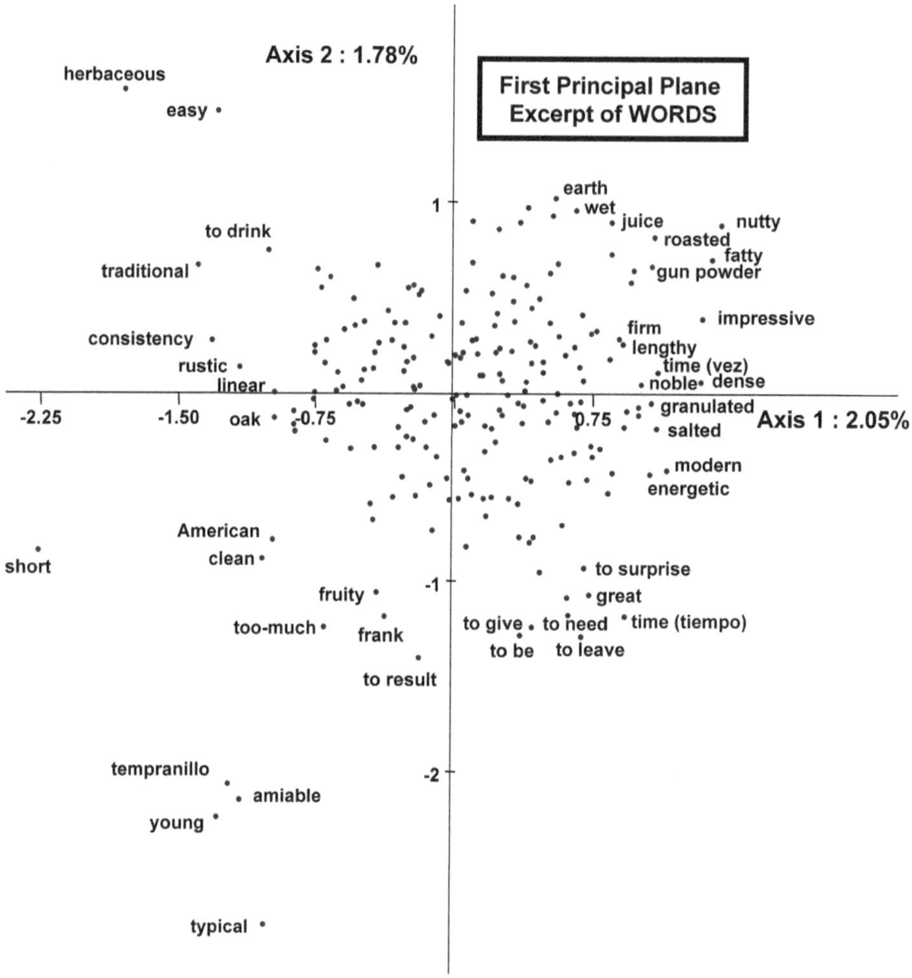

Figure 1.3. First principal plane. Excerpt of the words

Table 1.3. Mean and standard deviation of bootstrap coordinates of the score levels

Score levels	Count	Original coordinates Axis 1	Mean of coordinates Axis 1	Standard deviation Axis 1	Original coordinates Axis 2	Mean of coordinates Axis 2	Standard deviation Axis 2
1	5	0.4976	0.5108	**0.24806**	−0.3900	−0.3909	**0.16586**
2	77	0.5395	0.5401	0.04206	0.0290	0.0261	0.06591
3	61	0.2702	0.2719	0.04626	0.0214	0.0181	0.04568
4	57	0.1330	0.1339	0.04863	0.0928	0.0948	0.05581
5	85	−0.1358	−0.1373	0.03733	0.0367	0.0402	0.04920
6	79	−0.4252	−0.4272	0.03292	−0.1019	−0.1008	0.05093

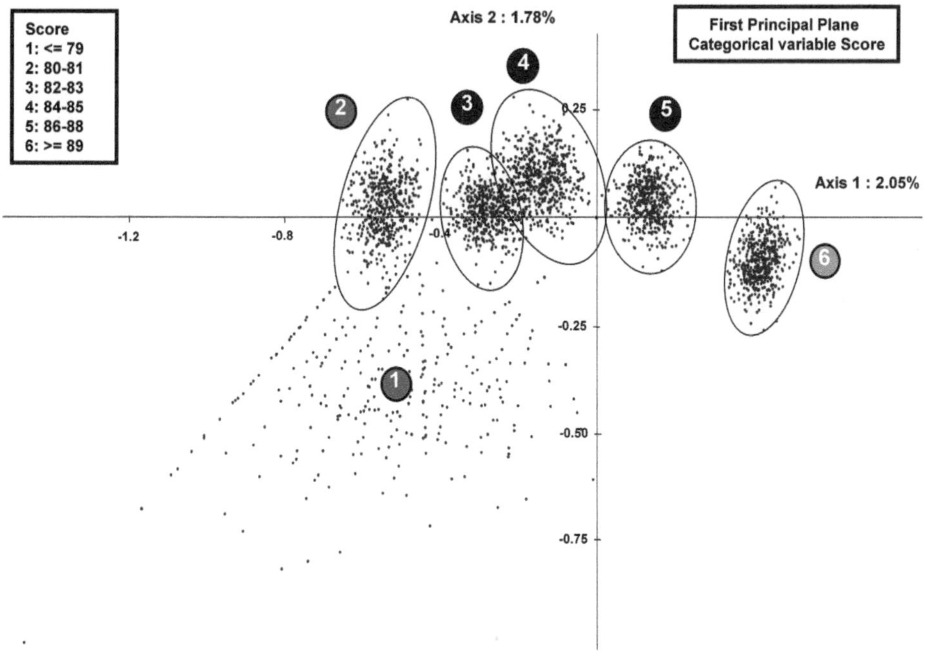

Figure 1.4. Bootstrapped regions of score levels

Table 1.4. Descriptive values of bootstrapped correlations between the score and the first principal coordinate vector

	Original correlation	Minimum bootstrap correlation	Maximum bootstrap correlation	Mean bootstrap correlation	Standard deviation correlation
Correlation F1-Score	0.7013	0.6271	0.7760	0.7027	0.0230
Correlation F2-Score	−0.0596	−0.2623	0.1308	−0.0566	0.0632

1.5 Conclusion

Using the external variable 'score' as a supplementary variable, the bootstrap resampling proves the stability of the relationship between the first principal coordinate vector and the wine score. The latter has been considered as a quantitative variable but also as a categorical variable, through grouping the values into categories.

The analysis of a lexical table through CA benefits from the validation of the structure by using the bootstrap procedure on both active and supplementary columns.

Software note

Bootstrap simulations as well as statistical computations have been carried out by means of specific software developed by the authors called SIMTEXT. This software run under Windows and can be downloaded free from:

http://www3.unileon.es/personal/wwdderae/simtext/publish.htm

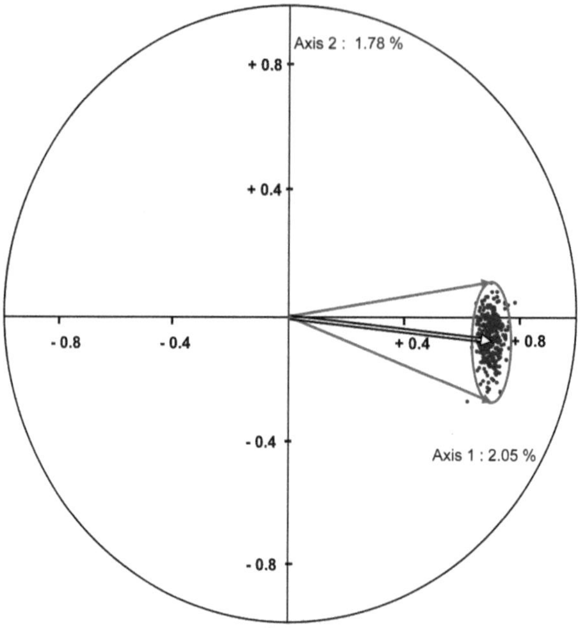

Figure 1.5. Replicated correlations between the score and the first principal coordinate vector

Acknowledgements. This work has been partially supported by the Spanish Ministry of Education and Science, FEDER (Grant SEJ2005-00741/ECON) as well as the Catalan Commission for the Universities DURSI (Grant SGR 00004/2005) and Junta of Castile and León (E-107/2004). We acknowledge the *Wine Guide* editor Mr. Fernando Lázaro for providing us information.

References

Beran, R.B, and Srivastava, M.S. (1985). Bootstrap tests and confidence regions for functions of a covariance matrix. *Ann. Statist.*, 13, 95–115.

Efron, B. (1979). Bootstrap methods: Another look at the jackknife. *Ann. Statist.*, 7, 1–26.

Efron, B., and Tibshirani, R.J. (1993). *Introduction to the Bootstrap.* London: Chapman and Hall.

El Mundo (2005). *Guía de catas 2005. Vinos de Castilla y León.* Biblioteca la Posada.

Labbé, D. (1990). Normes de saisie et de dépouillement des textes politiques. CERAT. Cahier No.7. http://web.upmf-grenoble.fr/cerat/Recherche/PagesPerso/LabbeNormes.pdf.

Lebart, L. (2004). Validation techniques in text mining. In S. Sirmakessis, Ed., *Text Mining and Its Applications*, 169–178. New York: Springer.

Lebart, L. (2006). Validation in multiple correspondence analysis. In M.J. Greenacre and J. Blasius, Eds., *Multiple Correspondence Analysis and Related Methods.* London: Chapman and Hall.

Lebart, L., Piron, M., and Morineau, A. (2006). *Statistique exploratoire multidimensionnelle. Validation et inférence en fouilles de données.* Paris: Dunod.

Lebart. L., Salem, A., and Berry, L. (1998). *Exploring Textual Data.* Norwell, MA: Kluwer.

Muller, Ch. (1977–1992). *Principes et méthodes de statistique lexicale.* Paris: Larousse [Reprint Genève: Champion-Slatkine].

ten Kleij, F., and Musters, P.A.D. (2003). Text analysis of open-ended survey responses: A complementary method to preference mapping. *Food Qual. Prefer.*, 14, 43–52.

2

A Doubly Projected Analysis for Lexical Tables

Simona Balbi[1] and Michelangelo Misuraca[2]

[1] Dipartimento di Matematica e Statistica, Università di Napoli "Federico II," Italy
[2] Dipartimento di Economia e Statistica, Università della Calabria, Italy

Abstract: This chapter aims to show how external information contributes in analysing a lexical table by enriching the readability of factorial maps. The theoretical frame is given by principal component analysis onto a reference subspace, a method based on the orthogonal projection of a correlation structure on the space spanned by an external set of explanatory variables. In previous papers the idea of a projected lexical analysis has been introduced by using a single reference space for terms. Here we consider a double projection strategy by involving external informative structures both on documents and terms, i.e., on rows and columns of a lexical table.

Keywords and phrases: External information, orthogonal projectors, factorial maps

2.1 Introduction

The necessity of introducing additional information in exploring multivariate structures, by means of Principal Component Analysis (PCA) and related techniques, has been a recurring topic in the literature since C. R. Rao introduced a set of q instrumental (explanatory) numerical variables in PCA (Rao, 1964).

In textual data analysis this state is particularly pressing, due to the nature of the data. A preprocessing step is always necessary in analysing a document collection from a statistical standpoint. This process often allows one to reduce the linguistic variability in the data, considered in terms of noise for analysis purposes, but on the other hand it leads to a loss of information for comprehension of the phenomenon.

We often have information dealing with the document categorisation process and the context in which terms have been used. This external information (also known as *metadata*) can be used both in an informal way to aid subjective interpretations of the results and in a formal way by incorporating it in the data.

It is possible to consider two different kinds of information in a textual analysis:

- *Intratextual information*, usually quantitative and corpus-driven, that takes into account the relationships between terms and documents

C.H. Skiadas (ed.), *Advances in Data Analysis*, Statistics for Industry and Technology, DOI 10.1007/978-0-8176-4799-5_2,

- *Extratextual information*, usually qualitative, involving all those aspects strictly linked to the context in which the documents are produced not directly readable from the dataset

The introduction of additional information both on individuals and variables, by considering two external informative matrices, was proposed by Takane and Shibayama (1991) combining features of regression analysis and PCA. The method was developed afterwards by Takane in the so-called Constrained Principal Component Analysis (CPCA) (Takane, 1997).

Sharing the data structure and having in mind CPCA properties and metrics considerations, in the following we choose as our methodological starting point another method, i.e., Principal Component Analysis onto a Reference Subspace (PCAR) (D'Ambra and Lauro, 1982), in order to emphasise the geometrical approach in terms of orthogonal projectors. After reviewing the main issues on the topic, in the following we propose a double projection strategy, simultaneously using orthogonal projectors on the spaces spanned by the additional variables related to documents and to terms.

The effectiveness of the proposed strategy is shown by analysing the educational offerings of the Italian University.

2.2 Some methodological recall

In geometry an *orthogonal projection* of a k-dimensional object onto a d-dimensional subspace spanned by the d columns linearly independent of a matrix $\mathbf{P}(n,d)$, is obtained by considering a projection operator $\mathbf{P}(\mathbf{P}'\mathbf{P})^{-1}\mathbf{P}'$, symmetric and idempotent. From a statistical viewpoint projecting a data structure onto a reference subspace means to analyse the relations between the rows and the columns in the frame of the information listed in $\mathbf{P}$.

2.2.1 Constrained principal component analysis

CPCA data structure is given by an individual-by-variable matrix $\mathbf{Z}$, and two external information matrices, $\mathbf{G}$ (on individuals) and $\mathbf{H}$ (on variables). According to Takane a wide variety of multivariate statistical analyses different from PCA are considered as interesting special cases of CPCA, including, e.g., correspondence analysis.

It has been thought of as a comprehensive method. For that reason there are no prescriptions in terms of distributional assumptions, preprocessing, or metric choices. The individual empirical interests suggest the proper behaviour to researchers. CPCA consists in two main analytical steps. In the first one, the so-called *external analysis*, $\mathbf{Z}$ is orthogonally projected onto the spaces spanned by $\mathbf{G}$ and $\mathbf{H}$, in order to decompose the influence of the "external" variables into the sum of four terms: the first one pertains to what can be jointly explained by $\mathbf{G}$ and $\mathbf{H}$, the second one and the third one, respectively, pertain to what can be explained by $\mathbf{G}$ and $\mathbf{H}$, while the fourth one is a matrix of residuals. This solution is achieved in a least square estimation framework by minimising the residual matrix. In the second step the *internal analysis* is performed on the decomposition matrices by means of one or more PCA.

2.2.2 Principal component analysis onto a reference subspace

PCAR data structure is given by two individuals-by-variables matrices **Z** and **X**. PCAR aims at visualising, in a proper geometrical framework, the dependence of **Z** on **X**.

Namely PCAR looks for the principal components of the orthogonal projection of **Z** on the space spanned by the columns of **X**. It can be seen as a special case of a CPCA internal analysis, when only the first term of the decomposition is considered and we want to introduce external information only on variables. Moreover the variables in **Z** are centered and frequently standardised. In this sense it is a proper PCA. The advantages of PCAR are strictly connected to graphical aspects and interpretation. In fact factorial maps show both the correlations within the same set of variables and the correlations between the two sets.

2.3 Basic concepts and data structure

External information on both documents and terms can help in explaining the use of some keywords under defined conditions. We can focus our attention on the residual uses in order to enhance peculiarity in the terms used in single documents, not connected to the main interpretation keys.

Suppose we are considering two indicator matrices **I** (n,I) and **J** (n,J), representing two categorical variables observed on the same set of individuals. Let **N** (I,J) be the contingency table cross-classifying the variables in I and J. In the frame of a textual statistics viewpoint $\mathbf{N} = \mathbf{I}'\mathbf{J}$ is a lexical table having I documents in rows and J terms in columns. Correspondence Analysis (CA) is usually performed on this table (Lebart et al., 1997) in order to analyse and graphically represent the latent lexical relationships between documents and terms.

Frequently in analysing document collections we dispose of additional information concerning a possible categorisation scheme for documents and about the context in which terms are used. Let us consider therefore an indicator matrix **Y** (I,K), assigning each document to a category k $(k = 1, \ldots, K)$. It is possible to perform a CA on the so-called *aggregated lexical table* **T** (K,J) obtained as the dot product of **Y** and **N**, in order better to read the relationships between groups of similar documents and the terms.

In previous papers, e.g., Balbi and Giordano (2000) and Balbi et al. (2002), the introduction of external information on documents and terms has already been performed for emphasising the different role played in the analysis. Here we focus mainly on an internal analysis in the sense of CPCA, stressing the geometrical features proper of PCAR. In other terms, by means of orthogonal projections we want to visualise on factorial maps the association structure in **N** due to the external information.

Additionally to the introduced matrices **Y** (containing information on documents) and **N** (the lexical table), let us consider a matrix **X** (J,L) containing information on the vocabulary of the corpus we want to analyse. By using the residual part a context-independent representation is obtained. In Figure 2.1 the complete data structure is shown.

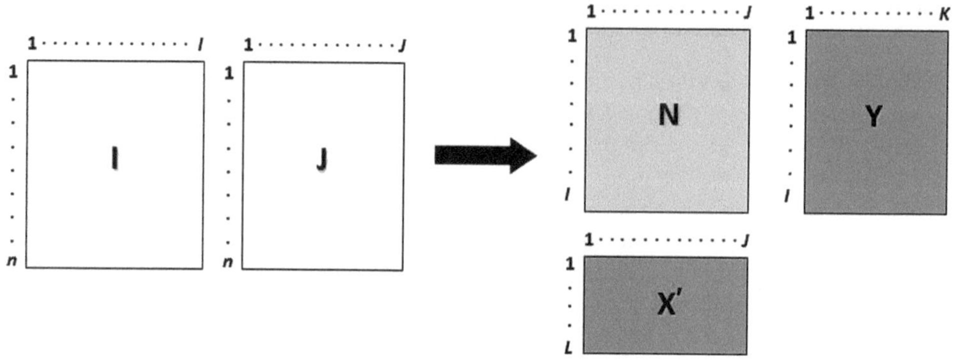

Figure 2.1. Data structure

2.4 A doubly projected analysis

Let $\mathbf{P_Y} = \mathbf{Y}(\mathbf{Y}'\mathbf{Y})^{-1}\mathbf{Y}'$ and $\mathbf{P_X} = \mathbf{X}(\mathbf{X}'\mathbf{X})^{-1}\mathbf{X}'$, the orthogonal projectors considered both in CPCA and in PCAR. Our proposal consists in analysing the residual matrix $\mathbf{A} = (\mathbf{I_y} - \mathbf{P_Y})'\mathbf{N}(\mathbf{I_x} - \mathbf{P_X})$.

If we consider the matrices $\mathbf{I}'$ and $\mathbf{J}$ cross-tabulated in $\mathbf{N}$ it could be possible to split the matrix $\mathbf{A}$ in a first term $(\mathbf{X}'\mathbf{X})^{-1}\mathbf{X}'\mathbf{I}'$ and in a second one given by $\mathbf{J}\mathbf{Y}(\mathbf{Y}'\mathbf{Y})^{-1}$, the profile matrices containing the conditional distribution of I on X and the conditional distribution of J on Y, respectively. Those two matrices are the basic matrices of PCAR or more properly in its version developed for categorical variables, known as NonSymmetrical Correspondence Analysis (NSCA) (Lauro and D'Ambra, 1984).

In other terms, we jointly study the residual part of dependence of the terms' distribution, with respect to the external information in $\mathbf{Y}$, and the residual part of dependence of the documents' distribution, with respect to the external information in $\mathbf{X}$. Dealing with elementary elements (i.e., each single occurrence in the corpus) it is worth noting that the dimensions of the two matrices are very huge, therefore this approach is infeasible in practice but would be useful in understanding results. In a similar way we can decide to carry out our analysis on any of the matrices considered by Takane for internal analysis.

The reference to PCAR is very useful in graphically visualising the results, because it makes it possible to represent on the factorial maps all the elements taken into account, the association structure in $\mathbf{N}$ together with the distribution of the terms conditioned to the information on $\mathbf{Y}$ and the distribution of documents conditioned to the information on $\mathbf{X}$.

2.5 The Italian academic programs: A study on skills and competences supply

In the frame of the European Union harmonisation policies the Italian university system has been reformed in 2000/2001 by introducing a new academic organisation. Two different kinds of degree have been introduced, a first-level three-year course (*Laurea Triennale*) followed by a specialising two-year course (*Laurea Specialistica*). All the courses are classified in 47 and 109 categories, respectively, divided among four general areas (*humanities*, *social*, *scientific*, and *medical issues*). The Ministry for Research and University has prepared for each course an explanatory declaration with the main contents and matters, used as a model by the University for planning their specific academic programs.

According to a request of the National Board for University System Evaluation, an official advisory body named by the ministry, the entire collection of academic programs has been analysed for investigating the correspondence between the competences and the skills supplied in the different courses and the employing outlets.

In this work we focus on the first-level academic programs. We have extracted 2812 declarations involving the different courses and in particular the fragments related to the competences offered have been selected. In this way the corpus of interest contains about 800,000 occurrences with a 17,200-term vocabulary. The lexical richness is not very wide, because in many cases the universities have decided to follow exactly the original declarations in drawing up the course programs.

The external information we introduce is the 47 ministerial course classes for the declarations (documents) and an eight-class competences categorisation obtained by previous analyses (terms).

Results of the PCA are strictly dependent on the choices of preprocessing procedures (centering, standardising, etc.) on the metrics used for computing distances and on the weighting systems. In this case we want to read the nontrivial uses of terms for describing competences in a context-independent framework, so that in the decomposition we assume an Euclidean metric and unitary weights. By introducing in the factorial coordinates a weighted Euclidean metric it is possible to recover the comparability in a common scale.

In Figure 2.2 we try to examine the peculiarities in the single courses' descriptions, independent of both the kind of competences offered and on the courses' nature. It is interesting to note how the first factorial axis ($\approx$11%) opposes general and abstract nouns (*attività*, *competenze*, *formazione*) on the right side, to terms describing more practical activities (*predisposizione di progetti*, *ambiti differenziati*, *azioni di pianificazione*, *uso di tecnologie*). This can be seen as a proper characterisation of the different university programs, in terms of a teaching frame oriented to "technical knowhow" or to a "way of thinking", which is a much-discussed topic in Italy. A deeper insight on this question is given by the second axis ($\approx$7%), where we can see an opposition, from the top to the bottom, between basic and professional/technical competences.

The interpretation of this map is coherent with the European debate about the new role of the university, in which it is necessary to include in the academic programs not only theoretic knowledge (*to know*) but more and more technical skills (*to know how*). It can be possible to perform the same analysis also on the specialising courses and to compare the two configurations of points in order to evaluate the internal coherence of the formative projects offered by the universities in terms of general and specific competences.

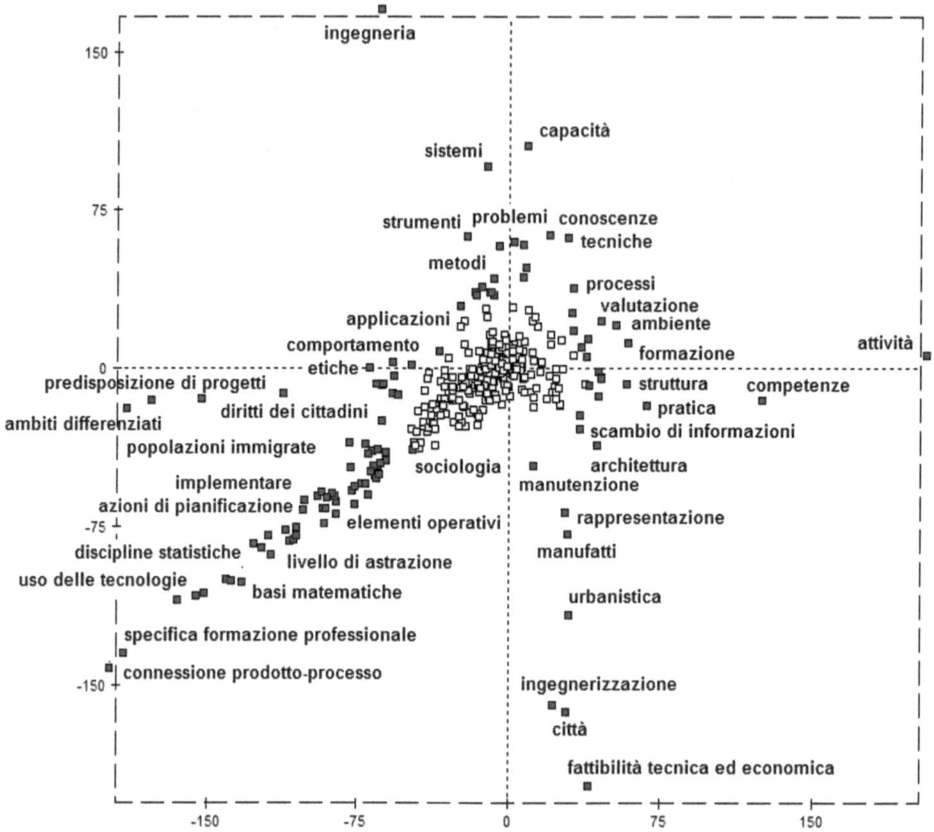

Figure 2.2. Factorial representation on the first two axes (≈18% of explained inertia)

References

Balbi, S. and Giordano, G. (2000). A factorial technique for analysing textual data with external information. In S. Borra et al., Eds., *Advances in Classification and Data Analysis*, pp. 169–176. Springer, Heidelberg.

Balbi, S., Bolasco, S., and Verde, R. (2002) Text mining on elementary forms in complex lexical structures. In A. Morin and P. Sébillot, Eds., *Actes des 6es Journées internationales d'Analyse statistique des Données Textuelles*, volume 1, pp. 89–100. IRISA-INRIA, Paris.

D'Ambra, L. and Lauro, N.C. (1982). Analisi in componenti principali in rapporto ad un sottospazio di riferimento. *Rivista di Statistica Applicata*, volume 15, number 1, pp. 51–67.

Lauro, N.C. and D'Ambra, L. (1984). L'analyse non symétrique des correspondances. In E. Diday et al., Eds., *Data Analysis and Informatics*, volume III, pp. 433–446. North Holland, Amsterdam.

Lebart, L., Salem, A., and Berry, L. (1997). *Exploring Textual Data.* Kluwer Academic, Boston.

Rao, C.R. (1964). The use and interpretation of principal component analysis in applied research. *Sankhya*, volume 26, series A, pp. 329–358.

Takane, Y. (1997). CPCA: A comprehensive theory. *Proceedings of the 1997 IEEE-SMC International Conference*, volume 1, pp. 35–40.

Takane, Y. and Shibayama, T. (1991). Principal component analysis with external information on both subjects and variables. *Psychometrika*, volume 56, number 1, pp. 97–120.

3

Analysis of a Mixture of Closed and Open-Ended Questions in the Case of a Multilingual Survey

Mónica Bécue-Bertaut[1], Karmele Fernández-Aguirre[2], and Juan I. Modroño-Herrán[2]

[1] Universitat Politècnica de Catalunya, Departament EIO, Barcelona, Spain
[2] Facultad de CC. Económicas y Empresariales, Universidad del País Vasco/Euskal Herriko Unibertsitatea (UPV/EHU), Bilbao, Spain

Abstract: Results obtained from surveys are often a mixture of quantitative, categorical, and textual data that leads to a mixed multiple table. Multiple factor analysis, extended to consider textual variables, can be applied to this kind of table. When survey questionnaires are filled in two (or more) languages, an additional difficulty arises. The aim of this work is to adapt the extended multiple factor analysis to these multilingual data. The methodology is applied to the analysis of a survey including both closed and open-ended questions in two languages, Basque and Spanish.

Keywords and phrases: Open-ended questions, free answers, multilingual texts, multiple mixed tables, multiple factor analysis, clustering

3.1 Introduction

Survey questionnaires frequently include both closed and open-ended questions about the same topic. The closed questions lead to quantitative or categorical individuals × variables tables and the open-ended questions to individuals × words frequency tables, also called lexical tables. To simultaneously take into account the closed and open-ended questions as active questions in a principal axes method, Bécue and Pagès (2008) have proposed an extension of multiple factor analysis. When different languages are used in the free answers a new problem arises which is the target of this work. Section 3.2 presents the application data and objectives. Section 3.3 sets out the notation and Section 3.4 our methodological proposal. Section 3.5 offers some of the results obtained in the application. As a conclusion (Section 3.6), we point out some perspectives.

3.2 Data and objectives

The data are extracted from a survey carried out at the University of the Basque Country (UPV/EHU) to know its members' perceptions about the institution better. Thus, the success of a shop selling objects exhibiting the university logo could be forecast.

C.H. Skiadas (ed.), *Advances in Data Analysis*, Statistics for Industry and Technology, DOI 10.1007/978-0-8176-4799-5_3,

The Basque Country is a cultural area in southwestern Europe with an extension of 20,864 km^2 and 2.9 million inhabitants, divided between France and Spain. The UPV/EHU answers the demand for higher education in the Autonomous Community of the Basque Country. This political entity has an extension of 7542 km^2 and 2.1 million inhabitants and is divided into three provinces (Bizkaia, Gipuzkoa, and Araba; see Figure 3.1), each with its own campus. About 30% of the inhabitants are speakers of Basque, a non-Indoeuropean agglutinative language. Spanish, Romanic languages, and Basque show very different structures.

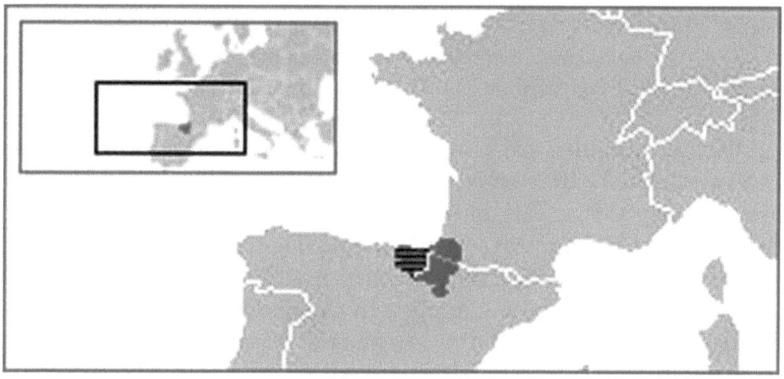

Figure 3.1. The Basque Country (*grey*) and the Autonomous Community of the Basque Country (*striped*)

The survey was performed during one month in 2005 on a sample of university members. The sample has been extracted by using a proportional stratified method, considering three strata (teaching and research staff, administrative staff, and students). The questionnaire was available at an Internet website, only accessible by e-mail invitation. The response rate has been, respectively, 50.3%, 40.0%, and 23.9% in the three strata, producing 1742 effective answers.

The questionnaire was mostly composed of closed multiple choice questions concerning:

- Different aspects of the institution
- The interest about buying 26 corporate products
- Classical demographic aspects

This questionnaire also included two open-ended questions. Only the second, the propensity to buy corporate products with the university logo, has been included in this analysis. Respondents answered either in Basque or in Spanish at their own convenience. Table 3.1 presents the questions used in this work. Only the respondents having answered the open-ended question are kept: 304 Basque-speaking and 1243 Spanish-speaking.

The main objective is to identify the different patterns with respect to the university corporate products. We propose to determine these patterns by building homogeneous clusters of individuals:

- In such a way that the closed and the open-ended questions relative to this topic are both taken into account
- But providing that the process resolves the different language problem in an automatic and transparent way

Table 3.1. Questions, possible answers, and type of variables

Question	Possible answer	Type
I am satisfied about being a member of this university	1 = completely unsatisfied, rather unsatisfied or neither safisfied or unsatisfied 2 = rather satisfied 3 = very satisfied	categorical, active
Would you be interested in buying a product featuring the university logo?	1 = yes 2 = no	categorical, active
⟶ Could you state why?	*(open answer)*	text, active
Gender	male, female	cat., sup.
Age	1=17–22, 2=23–29, 3=30–44, 4=+44	cat., sup.
Campus	Araba, Bizkaia, Gipuzkoa	cat., sup.
Link	Student, Admin., Teach.-Research	cat., sup.

3.3 Notation

The closed categorical questions lead to an Individuals × Indicator Variables table. For every language, a lexical table, that is, an Individuals × Words table is built up by counting up the occurrences of every word in every answer. Both lexical tables have as many rows as the total amount of individuals but the cells corresponding to the respondents who did not use the corresponding language are filled with zeroes.

There are I_1 Basque-speaking and I_2 Spanish speaking; $I_1 + I_2 = I$. There are three subtables corresponding to, respectively, the categorical set ($j = 1; K_1$ indicator-columns) and to the two lexical tables (Basque words, $j = 2, K_2$ word-columns; Spanish words, $j = 3, K_3$ word-columns). Figure 3.2 shows the structure of the resulting multiple mixed table.

	Categorical variables	**Open-ended question in Basque**	**Open-ended question in Spanish**
	Indicator Vars.	Word frequency	Word frequency
Individuals	z_{ik1}	f_{ik2}	0
		0	f_{ik3}
	K_1 cat.-columns	K_2 word-columns	K_3 word-columns

Figure 3.2. Multiple table resulting from juxtaposing the indicator and the lexical tables

In the case of the individuals × indicator variables table, z_{ik1} indicates if respondent i ($i = 1, \ldots, I$) presents ($z_{ik1} = 1$) or not ($z_{ik1} = 0$) the k column-indicator, $k = 1, \ldots, K_1$. In the case of the lexical tables f_{ikj} indicates the frequency with which individual i ($i = 1, \ldots, I$), answering in Basque ($j = 2$) or Spanish ($j = 3$) uses the column-word k, $k = 1, \ldots, K_j$ in his or her answer. This frequency is relative to the corresponding table grand total. Thus,

$$\sum_{i \in I} \sum_{k \in K_1} f_{ikj} = 1; \qquad j = 2, 3.$$

We also consider the row-margins of every separate lexical table:

$$f_{i.2} = \sum_{k=1}^{K_2} f_{ik2} \qquad f_{i.3} = \sum_{k=1}^{K_3} f_{ik3}.$$

We note that the global row-margin of the juxtaposed table, $f_{i..} = \sum_{j=2}^{3} \sum_{k=1}^{K_j} f_{ikj}$ is equal either to $f_{i.2}$ (for $i = 1, \ldots, I_1$) or to $f_{i.3}$ (for $i = I_1 + 1, \ldots, I_1 + I_2$).

3.4 Methodology

We want to identify the various patterns with respect to attitudes and opinions about the university corporate products. Thus, we aim at clustering the individuals starting with the answers given to the corresponding closed and open-ended questions. However, a principal axes method is used as a preprocessing step and clustering starts from the first principal coordinate vectors (Lebart, 1994; Lebart et al., 2006, pp. 295–311).

The principal axis method has to simultaneously include categorical and frequency sets (one frequency table = one set). Extended MFA (Bécue and Pagès, 2008) allows for this property while maintaining a MCA-like approach to the categorical sets as well as a CA-like approach to the frequency sets and balancing the influence of the different sets (Escofier and Pagès, 1998).

Clustering is performed over the principal axes, which are in fact quantitative variables, and then a suitable algorithm can be used. The elimination of the last axes, far from being a drawback, acts as a filter for the random fluctuations that could mask important features (Lebart, 1994; Lebart et al., 2006).

Hereafter, we comment on the main points of the extended multiple factor analysis.

3.4.1 Principle of multiple factor analysis

Multiple factor analysis (MFA), proposed by Escofier and Pagès (1998), deals with mixed data while keeping a PCA-like and a MCA-like approach to, respectively, the quantitative and the categorical sets. MFA balances the influence of the different sets by dividing the weight of its columns by the first eigenvalue of its separate analysis. Thus, the inertia of every set on the first principal axis is standardized to 1.

3.4.2 Integrating categorical sets in MFA

To integrate categorical sets into MFA (Pagès, 2002), in particular when the individuals present nonuniform weights, the equivalence between MCA and a nonstandardized weighted PCA is used (Bécue and Pagès, 2008). In that case the weight for the individual i is denoted by p_i. MCA can be performed as PCA:

- Applied to the table with the general term $(z_{ik1} - w_{k1})/w_{k1}$, where $z_{ik1} = 1$ if i belongs to the category k and 0 if it does not, and $w_{k1} = \sum_{i \in I} p_i \cdot z_{ik1}$ ($\sum_k w_{k1} = Q_1$) the number of categorical variables in set 1)
- Giving the weight w_{k1}/Q_1 to column j of the categorical set
- Giving the weight p_i to row i

3.4.3 Integrating frequency tables in MFA

Abdessemed and Escofier (1996) have extended MFA to include one frequency table (or several frequency tables but with the same margins). Bécue and Pagès (2008) have generalized this extension to several frequency tables with different margins (one frequency table = one set), keeping as far as possible the CA characteristics for these tables.

Extended MFA relies on the equivalence between CA and a particular PCA (Escofier and Pagès, 1998, pp. 95–96). As in classical MFA, the overweighting is obtained by dividing the initial weight of the columns by the first eigenvalue issued from the CA applied to the lexical table, always smaller than 1.

The proposal by Bécue and Pagès can be applied as shown in the presentation of the global process hereinafter.

3.4.4 Extended MFA performed as a weighted PCA

The MFA applied to a table juxtaposing one categorical set (with K_1 columns-indicator) and the two lexical tables corresponding to the answers in the two languages (with, respectively, K_2 and K_3 columns-word) is equivalent to perform a nonstandardized weighted PCA on the multiple table presented in Figure 3.3, using:

Units	Indicator variable k (=category) in categorical set (set=1)	Word k in Basque lexical table (set=2)	Word k in Spanish lexical table (set=3)
1 i I_1	$\frac{z_{ik1} - w_{k1}}{w_{k1}}$	$\frac{f_{ik2} - f_{i.2}f_{.k2}/f_{..2}}{p_i f_{.k2}}$	0
I_1+1 i $I = I_1 + I_2$	$\frac{z_{ik1} - w_{k1}}{w_{k1}}$	0	$\frac{f_{ik3} - f_{i.3}f_{.k3}/f_{..3}}{p_i f_{.k3}}$

Figure 3.3. Mixed multiple table issued from the original table by convenient transformations

- $\{p_i \; i = 1, \ldots, I\}$ as row-unit weights (and as metric in the column space)
- The initial weights of the columns but divided by the first eigenvalues of the separate analysis of every table as column weights (and as metric in the row space), that is, $\left((w_{kj}/Q_j)/\lambda_1^j\right)$ in the case of a categorical set, $\left(f_{.kj}/\lambda_1^j\right)$ in the case of a frequency set). λ_1^j denotes the first eigenvalue issued from the separate PCA of subtable j

To choose the row weights p_i, different criteria can be used. To adopt those imposed by CA (either $\{f_{i.2}; i = 1, \ldots, I_1; f_{i.3}; i = I_1 + 1, \ldots, I_1 + I_2\}$ or $\{f_{i.2}/2f_{..2}, i = 1, \ldots, I_1; f_{i.3}/2f_{..3}, i = I_1 + 1, \ldots, I_1 + I_2\}$) favors the respondents with long answers, generally with a richer vocabulary. Other options could be considered, such as uniform weights, provided that the rows keep the same weight through all the tables. The former option is considered in the application, thus giving the same importance to every sample.

MFA provides the classical results of any PCA, mainly the principal components that can be viewed as good compromises between those obtained in the separate analyses (CA for the frequency table and MCA for the categorical table).

3.5 Results

3.5.1 Clustering from closed questions only

First, we cluster the respondents from only their answers to the closed questions. We follow the strategy "principal axes method-clustering step."

A multiple correspondence analysis is performed on the categorical table. The first principal axis opposes the respondents intending to buy corporate products and very satisfied about their UPV membership to those with the opposite categories. Thus we find a tight link between interest in buying these items and satisfaction about the university, as previously shown in Fernández-Aguirre et al. (2008). This axis also opposes, as supplementary categories, "over 44" and "administrative staff" to "23–29" and "students", showing that the interest is more intense in the former categories than in the latter ones.

The two supplementary categories Basque-speaking (B) and Spanish-speaking (S) lie close to the centroid. A classical test (Lebart, 1994) allows for assessing that the coordinates are not significantly different from null value on the three first factor axes (p-value < 0.05). To complete this test, we also assess the stability of these categories on the first principal plane through partial bootstrap (Lebart, 2006). In this case, the coordinates of the supplementary categories are computed from the replicated bootstrap samples (1000 bootstrap replications in our case). Means and standard deviations of these coordinates are shown in Table 3.2 and the 95% confidence ellipses of both categories are displayed in Figure 3.4. Both Basque and Spanish-speaker centroids are not significantly different from the global centroid and thus they have the same behavior with respect to buying intention as well as satisfaction about the institution. Then, clustering is performed starting from the three first factors (Lebart, 1994) leading to

Table 3.2. Mean and standard deviations of bootstrap replicates of Basque and Spanish respondents' coordinates

	Basque			Spanish		
	Axis 1	Axis 2	Axis 3	Axis 1	Axis 2	Axis 3
Original	0.0254	−0.0560	0.0049	−0.0257	0.0565	−0.0050
Mean	−0.0216	−0.0036	0.0601	−0.0230	−0.0094	0.0629
Standard deviation	0.0869	0.0859	0.0877	0.0410	0.0442	0.0436

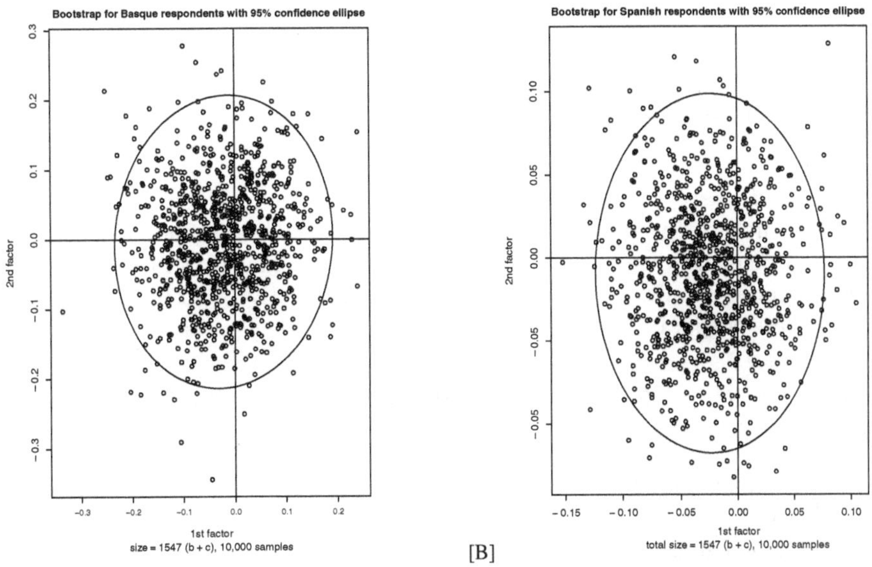

[A] [B]

Figure 3.4. Bootstrap of language categories projected on the main plane as supplementary categories with 95% confidence ellipse: (**A**) projections of Basque replicates; (**B**) projections of Spanish replicates

four clusters. The clustering method combines, first, a K-means algorithm followed by a generalized Ward's hierarchical method to reduce the number of clusters and, finally, a consolidation of the partition by reassignment of the individuals to the closest clusters.

For every cluster, the significantly over- and underrepresented categories (Lebart et al., 1998) are selected and shown in Table 3.3.

3.5.2 Clustering from closed and open-ended questions

In order to cluster the individuals from closed and open-ended questions, we apply the sequence "extended MFA-clustering" to the table presented in Figure 3.2.

Thus we consider three active sets: the categorical variables used in the former section (set 1), the Basque lexical table (set 2), and the Spanish lexical table (set 3). The variables listed in Table 3.1 form a supplementary set.

Table 3.3. Clusters formed over main factors and their description

Cluster no.	Size (%)	Active categories	Cat./Grp. (%)	Cat./All (%)	Supplementary categories
1	27.29	Rather satisfied	99.63	42.36	Teaching-research,
		Would buy	99.37	59.63	male
2	15.40	Rather satisfied	98.47	42.36	Gipuzkoa campus
		Would not buy	99.28	39.35	
3	33.63	Very satisfied	97.94	32.94	Female,
		Would buy	72.07	59.63	age 30–44
4	23.67	Little or not satisfied	98.64	23.35	Students, age 23–29,
		Would not buy	64.22	39.35	age 17–22

Language preprocessing

We have considered as equivalent the words with the same root (for instance, *prefer* and *preferred* correspond to the same word *prefer*). We have decided to keep not only the words but also the repeated segments (Lebart et al., 1998, pp. 35–38), and to use the same frequency threshold for both lexical units, equal to 4 in the Basque corpus and to 15 in the Spanish corpus. Thus we obtain, respectively, 223 and 247 total lexical units in these corpora.

Extended MFA performed on the global table including the frequency tables

We perform an extended MFA on the table shown in Figure 3.2. Partial first eigenvalues are equal to, respectively, 0.6451 (closed questions), 0.1858 (Basque free answers), and 0.1937 (Spanish free answers). The largest influence, in terms of inertia, that could have the closed questions set, is corrected by MFA overweighting.

The three sets contribute to the first global axis with, respectively, 53.4%, 24.3%, and 22.3% of the total inertia. This dispersion direction is common to the three sets. The global first eigenvalue (1.49), far from its maximum (equal to 3 in this case), has a value similar to those that are usually observed in the case of mixed tables including textual data (Bécue and Pagès, 2008).

The first set loses importance in the second global axis (the textual groups contribute more to the inertia of this axis) but recovers importance on the third global axis. The correlations between canonical variables and the global axes, on the one hand, and the contributions of the partial axes to the global axes, corroborate this result.

Clustering is performed from the coordinates on the three principal axes of this global analysis. Selecting a larger number of axes' results would lead to a considerable instability favoring one of the textual groups to the detriment of the other. By means of a mixed cluster algorithm (similar to the one used above) we obtain a well-balanced partition solution in four clusters whose interpretation (Table 3.4) is very close to the interpretation of the former partition in Table 3.3.

Characteristic words and answers of the four clusters

The over – and underrepresented words of each cluster are identified by using a statistical criterion (Lebart et al., 1998, pp. 130–136). The modal answers are also extracted

Table 3.4. Clusters: closed and open-ended questions, Basque and Spanish respondents' global analysis

Cluster no.	Size (%)	Active categories	Cat./Grp. (%)	Cat./All (%)	Supplementary categories
1	29.63	Rather satisfied	97.39	42.36	Teaching-Research,
		Would buy	92.17	59.63	Male
2	30.51	Very satisfied	79.30	32.94	Age +44,
		Would buy	97.25	59.63	Female
3	16.23	Rather satisfied	67.74	42.36	Students,
		Would not buy	94.18	39.35	Gipuzkoa Campus
4	23.63	Little or not satisfied	57.06	23.35	Age 23–29,
		Would not buy	90.57	39.35	Students

Table 3.5. Overrepresented words in cluster 4 (Little or not satisfied, Would not buy) with both internal and global frequencies

Basque			**Spanish**		
Word segments	Int. freq.	Glob. freq.	Word segments	Int. freq.	Glob. freq.
ez dut uste – I don't think	11	15	no – not	418	737
asko – much	8	13	no me gusta – I don't like	91	96
jende – people	6	13	ningún – none	47	59
diru – money	5	10	propaganda – advertising	41	56
behar – need	14	47	con logotipos – with logo	22	24
ez nuke erosiko – wouldn't buy	3	5	nada – at all	29	37
logotipoa duen – with logo	3	5	pagar – pay	20	22
zalea – keen on	4	9	hacer publicidad – to advertize	14	22
logo – logo	23	100	no me parece – I don't think	25	38
erosi – buy	21	100	comprar – buy	118	271
kontsumismo – consumerism	3	5			

Table 3.6. Overrepresented words in cluster 2 (Very satisfied, Would buy) with both internal and global frequencies

Basque			**Spanish**		
Word segments	Int. freq.	Glob. freq.	Word segments	Int. freq.	Glob. freq.
ikasi – learn, teach	23	37	pertenecer – belong, membership	68	124
arro – proud	5	5	de pertenecer – about belonging	37	61
euskal – basque	6	7	a la upv – to the upv/ehu	46	84
publikoa – public	5	6	pertenecer a la upv – be upv member	19	32
egon – be	23	37	para regalar – to make a gift	48	100
bat – one	36	93	orgulloso – proud	37	63
bertako – from here (w. pride)	11	13	profesores – teachers	21	37
unibertsitatea – university	27	70	visitar – visit	14	17
herriko – of the ... country	6	8	a conocer – (to make) known	31	51
prezio – price	7	11	universidad – university	117	267
detaile – gift	4	5	tambien – too	31	51
zergatik ez – why not	10	17	detalle – gift	22	36
polita – beautiful	12	25	imagen – image	49	100
zerbait – something	11	22	ehu – (upv/ehu)	60	132

(Lebart et al., 1998, pp. 137–141) for showing the actual context in which these words are used.

We present (Tables 3.5 and 3.6) the clusters formed at both extremes of the principal axis.

The cluster including the individuals satisfied with the university and intending to buy corporate products is very similar to the corresponding cluster obtained in the former clustering. The characteristic words and segments show that some expressions located close to the centroid in the partial CA applied to the lexical table do appear, for instance, *visit* (a mention of visiting professors), *Why not* or *to make a gift* (Table 3.6). We can see that many terms appear simultaneously in both languages, related to being very satisfied and proud of their university.

In the case of the cluster including the individuals not favorable to buying corporate products, we observe negative terms reflecting the ideas of not buying or not paying or not liking, supposedly referring to these products.

Table 3.7 shows some actual full responses featuring some of the most frequent lexical terms, cited above. We then conclude that we have automatically got two opposite clusters internally equivalent with respect to both languages.

Table 3.7. Some modal sentences in extreme clusters; Basque and Spanish answers

Cluster	Basque	Spanish
Little or not satisfied, Would not buy	ez dut uste erabiliko nuenik (I don't think I'd use it)	no me gustan (I don't like)
	ez dut holakorik erosten (I don't buy such things)	porque no me llama la atención (because it doesn't draw my attention)
	ez dut nire dirua holako gauzetan xahutu nahi (I don't waste money in such things)	No acostumbro (I usually don't)
Very satisfied, Would buy	ikasten dudan lekuaz arro nagoelako (because I am proud of learning here)	porque me siento orgullosa de pertenecer a la UPV/EHU (because I am proud of my UPV/EHU membership)
	karrera ikasten egon naizeneko detailetxo bat oroigarri gisa (as a gift from where I have studied)	porque se favorece la expansión de la UPV (because it favours UPV expansion)
	ongi deritzot bertako logotipodun zerbait erabiltzea (I like to wear something with logo from here – with pride)	por sentir a la UPV/EHU como nuestra (to feel the UPV/EHU as ours – with pride)

3.6 Conclusion

We propose a methodology that allows for clustering individuals according to a battery of questions on a specific topic, formulated as either closed or open-ended questions, the latter being eventually answered in different languages.

We apply this methodology to a survey in which two different languages, Basque and Spanish, are used.

Thus, we offer a new way for the automatic treatment of open-ended questions in multilingual surveys, which is of prime interest for international and/or multilingual area surveys.

Acknowledgements. This work has been partially supported by Spanish Ministry of Education and Science and FEDER (grant SEJ2005-00741/ECON). Financial support from Grupo de investigación del sistema universitario vasco IT-321-07 is gratefully acknowledged.

References

Abdessemed, L. and Escofier, B. (1996). Analyse factorielle multiple de tableaux de fréquences ; comparaison avec l'analyse canonique des correspondances. *Journal de la Société de statistique de Paris*, 137(2):3–18.

Bécue, M. and Pagès, J. (2008). Multiple factor analysis and clustering of a mixture of quantitative, categorical and frequency data. *Journal of Computational Statistics & Data Analysis*, 52:3255–3268.

Escofier, B. and Pagès, J. (1998). *Analyses factorielles simples et multiples: Objectifs, méthodes et interpretation*, 3rd edition. Dunod, Paris.

Fernández-Aguirre, K., Landaluce, M., Martín, A., and Modroño, J.I. (2008). Data mining of an on-line survey. A market research application. In Preisach, C., Burkhardt, H., Schmidt-Thieme, L., and Decker, R. (Eds.), *Data Analysis, Machine Learning and Applications.* Proceedings of the 31st Annual Conference of the Gesellschaft für Klassifikation. Springer, New York.

Lebart, L. (1994). Complementary use of correspondence analysis and cluster analysis. In Greenacre, M.J., and Blasius, J. (Eds.), *Correspondence Analysis in the Social Sciences.* Academic Press, San Diego.

Lebart, L. (2006). Validation techniques in multiple correspondence analysis. In Greenacre, M.J., and Blasius, J. (Eds.), *Multiple Correspondence Analysis and Related Methods*, Academic Press, London, 179–195.

Lebart, L., Piron, M., and Morineau, A. (2006). *Statistique Exploratoire Multidimensionnelle*, 4th edition. Dunod, Paris.

Lebart, L., Salem, A., and Berry, L. (1998). *Exploring Textual Data.* Kluwer Academic, New York.

Pagès, J. (2002). Analyse factorielle multiple appliquée aux variables qualitatives et aux données mixtes. *Rev. de Statist. Appl.* 50(4):5–37.

4

Number of Frequent Patterns in Random Databases

Loïck Lhote

GREYC, CNRS UMR 6072, Université de Caen Basse-Normandie, F-14032 Caen, France

Abstract: In a tabular database, patterns that occur over a frequency threshold are called frequent patterns. They are central in numerous data processes and various efficient algorithms were recently designed for mining them. Unfortunately, very little is known about the real difficulty of this mining, which is closely related to the number of such patterns. The worst case analysis always leads to an exponential number of frequent patterns, but experimentation shows that algorithms become efficient for reasonable frequency thresholds. In order to explain this behaviour, we perform here a probabilistic analysis of the number of frequent patterns. We first introduce a general model of random databases that encompasses all the previous classical models. In this model, the rows of the database are seen as independent words generated by the same probabilistic source (i.e., a random process that emits symbols). Under natural conditions on the source, the average number of frequent patterns is studied for various frequency thresholds. Note that the source may be nonexplicit since the conditions deal with the words. Then, we exhibit a large class of sources, the class of dynamical sources, which is proven to satisfy our general conditions. This finally shows that our results hold in a quite general context of random databases.

Keywords and phrases: Data mining, models of databases, frequent patterns, probabilistic analysis, dynamical sources

4.1 Introduction

Data mining, which applies to various fields (astronomy, fraud detection, marketing, biology, ...), aims at extracting new knowledge from large databases. We consider here tabular databases where knowledge is represented by a collection of pairs $(column, value)$, also called a pattern.

Patterns that occur frequently at the same time in several rows are of great interest since they indicate a correlation between the columns that compose the pattern. A pattern is said to be frequent if it occurs over a frequency threshold, which is defined by users. Frequent patterns intervene in numerous data processes such as classification or

C.H. Skiadas (ed.), *Advances in Data Analysis*, Statistics for Industry and Technology, DOI 10.1007/978-0-8176-4799-5_4,

clustering (Goethals, 2003). They are also essential (Agrawal et al., 1993) for generating the well-known association rules that apply in bioinformatic, physics, or geography.

The frequent pattern mining problem was first described in Agrawal et al. (1993) and during the last decade, several algorithms have been designed to solve it (Agrawal et al., 1996; Savasere et al., 1995; Toivonen, 1996; Han et al., 2000; Zaki, 2000). Their complexities are closely related to the number of frequent patterns which is, in the worst case, always exponential with respect to the number of columns. But the actual behaviour appears to be quite different. The algorithms fail when the frequency threshold is too small, which suggests an exponential behaviour and, they become efficient for reasonable frequency thresholds, which suggests a polynomial behaviour. There already exist bounds for the number of frequent patterns in Geerts et al. (2001), but they are involved and do not elucidate the influence of the frequency threshold on the number of frequent patterns.

In this chapter we perform a probabilistic analysis that elucidates the real behaviour of the number of frequent patterns. There exist such analyses dealing with the maximal size of the frequent patterns (Agrawal et al., 1996), or the fail rate of the APRIORI algorithm (Purdom et al., 2004). But the previous analyses dealt with a model based on column independence, whereas the algorithms are precisely designed for searching correlations between columns. We introduce a general model of random databases which avoids this contradiction. In our model, the rows of the database are independent words generated by the same source. A source is a probabilistic process that emits a symbol at each unit time, and the complete process builds a word. Since successive emitted symbols may be correlated, the columns are no longer independent. Under natural conditions on words produced by the source (Conditions 1 and 2-γ), we obtain two main results (Theorems 1 and 2) on the number of frequent patterns in two main cases: the first one is related to a fixed frequency threshold, whereas the second one deals with a linear frequency threshold (with respect to the number of rows). We then describe a large class of sources, called *dynamical sources*, which are proven to satisfy Conditions 1 and 2-γ (Theorem 3). This class contains all the classical sources (memoryless sources and Markov chains), but also many other sources which may possess a higher degree of correlations. It then follows that Theorem 1 and Theorem 2 apply to various models of databases (classical or not).

4.2 Model of databases

4.2.1 Frequent pattern mining

Frequent pattern mining is often described in the framework of *market basket analysis*, but we adopt here the more general framework of the *multiple-choice questionnaire*. In this context, a set of persons (of cardinality n) answers to a number m of multiple-choice questions. The set $\mathcal{E}$ of possible answers to each question is the same, and is called the alphabet. The word $\mathcal{E}^m$ formed by the answers of one person to all the questions is called a transaction. A natural data structure for storing all the transactions is an $n \times m$ matrix over $\mathcal{E}$.

	Questions						
persons	q_1	q_2	q_3	q_4	q_5	q_6	q_7
p_1	2	1	2	1	2	1	1
p_2	1	2	2	1	2	1	3
p_3	2	3	2	1	2	1	1
p_4	2	1	3	2	1	2	1

Pattern	Support	Frequency
$(q_1, 2), (q_3, 2)$	p_1, p_3	2
$(q_4, 1), (q_7, 3)$	p_2	1
$(q_5, 2)$	p_1, p_2, p_3	3

Figure 4.1. On the *left*, an instance of a database with seven questions and four persons whose answers to the questionnaire belong to $\mathcal{E} = \{1, 2, 3\}$. On the *right*, instances of patterns with the associated support and frequency

A pattern is a set of pairs $(\texttt{question}, \texttt{answer})$ where each question appears at most once. A person p supports a pattern X if for all pairs *(q,a)* in X, the answer of p to the question q is the answer a. We also say that the transaction contains the pattern X.

The support of a pattern X is the set of persons that support X, and the frequency of X is the size of its support. Figure 4.1 gives instances of patterns with their support and frequency. A pattern is said to be γ-frequent in a database $\mathcal{B}$, with a frequency threshold $\gamma \geq 1$, if the cardinality of its support is greater than γ. In the table of Figure 4.1, the pattern $(q_5, 2)$ is 1; 2-, or 3-frequent since its frequency is 3.

When the database contains at least γ copies of each possible transaction (this means that $n \geq \gamma \cdot |\mathcal{E}|^m$), all possible patterns are γ-frequent. In this case, the number of frequent patterns equals $(1 + |\mathcal{E}|)^m - 1$ (for m questions). Now, if the matrix coefficients are all equal to v, all the patterns that contain pairs only of the form $(\texttt{question}, v)$ are frequent (for any frequency threshold less than n). In this case, the number of frequent patterns equals $2^m - 1$. In particular, the worst case is always at least exponential.

However, the previous situations are rare and do not correspond to real applications. This is why we propose a probabilistic analysis of frequent patterns under a general model that we now introduce.

4.2.2 Model of random databases

Our model considers all the transactions as different words produced by the same probabilistic source defined on the alphabet $\mathcal{E}$. For instance, the word associated with the first transaction (or row or person) in Figure 4.1 is 2121211. Since frequent patterns aim at describing correlations between questions, we always suppose that the transactions are independent, even if the persons themselves may not be independent. Finally, we are interested in asymptotics when the databases become large, with a number of persons and a number of questions which are polynomially related. The next definition summarises these three hypotheses.

Definition 1. *We call a* random database *a probabilistic database that satisfies the three following conditions.*

(i) Each transaction is a word produced by the same probabilistic source over a finite alphabet $\mathcal{E}$.
(ii) The transactions form a family of independent random variables.
(iii) The number n of persons and the number m of questions are polynomially related, namely of the form $\log n = \Theta(\log m)$.

4.3 Main results

We study the average number of frequent patterns in random databases (Definition 1) for two types of frequency thresholds: the linear frequency threshold and the constant threshold (with respect to n). A general result which would hold for all existing sources is certainly unexpected. This is why we introduce a condition on the source for each frequency threshold. Both conditions only concern the words produced and not the way they are produced. Hence, the source may be nonexplicit. In addition, we show in Section 4.4 that both conditions are natural since they are verified by a large class of sources, classical or not. In the whole section, m and n, respectively, denote the number of questions and persons in a *random* database $\mathcal{B}$.

4.3.1 Linear frequency threshold

A frequency threshold γ is said to be linear if it satisfies, $\gamma \sim r \cdot n$ (for some $r \in]0,1[$) as n tends to infinity. The probability that a person, or equivalently a word, supports the pattern X is noted p_X. The quantities p_X are essential in our different conditions. One has clearly $p_Y \leq p_X$ as soon as $X \subseteq Y$. The next condition considers sources whose pattern probability is exponentially decreasing with the size of the pattern.

Condition 1 *There exist $M > 0$ and $\theta \in]0,1[$ such that for any pattern X, the probability p_X satisfies:*

$$p_X \leq M \cdot \theta^{|X|}.$$

In practice, Condition 1 implies that questions discriminate persons.

Theorem 1. *Let $\mathcal{B}$ be an $n \times m$ random database generated by a probabilistic source that satisfies Condition 1 with parameters M and θ. For a linear frequency threshold $\gamma \sim r \cdot n$, the average number $F_{\gamma,m,n}$ of γ-frequent patterns is polynomial with respect to the number m of questions,*

$$F_{\gamma,m,n} = O\left(m^{j_0}\right) \quad \textit{with} \quad j_0 = \max\{j \geq 0 \mid M\theta^j \geq r\}.$$

This polynomial behaviour explains the efficiency of the algorithms for reasonable frequency thresholds. It is also possible to obtain an estimate of $F_{\gamma,m,n}$ under the weaker condition$(1-\theta) \cdot \min(m,n) \to \infty$, but, in this case, the asymptotic behaviour is no longer polynomial with respect to m.

4.3.2 Constant frequency threshold

Here, the frequency threshold γ is now constant and does not evolve with the number of persons. Given γ random transactions over m questions, the probability that the γ transactions support X is p_X^γ. Hence, the average number of patterns supported by the γ transactions is

$$\Sigma_{\gamma,m} = \sum_X p_X^\gamma.$$

The sum $\Sigma_{\gamma,m}$ is proven to be greater than 1 and it admits a closed form for various (classical) sources (see Sections 4.4 and 4.5). The next condition implies that, for γ constant, $\Sigma_{\gamma,m}$ is exponential with respect to the number m of questions.

Condition 2-γ *There exists $\theta_\gamma > 1$ such that, for large m,*

$$\Sigma_{\gamma,m} > \theta_\gamma^m \cdot \Sigma_{\gamma+1,m}.$$

With Condition 2-γ, we prove our second main result.

Theorem 2. *Fix $\gamma \in \mathbb{N}^\star$ and consider an $n \times m$ random database $\mathcal{B}$ generated by a probabilistic source that satisfies Condition 2-γ with parameter θ_γ. The mean number of γ-frequent patterns verifies*

$$F_{\gamma,m,n} = \binom{n}{\gamma} \Sigma_{\gamma,m} \cdot \left[1 + n \cdot O\left(\frac{1}{\theta_\gamma^m}\cdot\right)\right].$$

In other words, for a constant frequency threshold, the number of frequent patterns is exponential with respect to the number m of questions, and polynomial with respect to the number n of persons. This result explains why the algorithms fail for small frequency thresholds.

4.3.3 Sketch of proofs

For a given frequency threshold γ, the average number of frequent patterns is the sum over all possible patterns X and all possible supports S, of the probability that X has support S. Now, the size of the support of X follows a binomial law with parameter p_X, so that

$$F_{\gamma,m,n} = \sum_X F_{\gamma,m,n,X} \quad \text{with} \quad F_{\gamma,m,n,X} := \sum_{k=\gamma}^{n} \binom{n}{k} p_X^k (1-p_X)^{n-k}.$$

The fundamental step transforms $F_{\gamma,m,n,X}$ into an integral. Developing $(1-p_X)^k$, doing a change of variable, inverting two signs sum and using a recurrence lead to the alternative formula

$$F_{\gamma,m,n,X} = \gamma \binom{n}{\gamma} \int_0^{p_X} t^{\gamma-1}(1-t)^{n-\gamma} dt.$$

The proofs for constant and linear thresholds separate here.

For a constant threshold, we use the bounds $1-(n-\gamma)t < (1-t)^{n-\gamma} < 1$ and get a lower bound of $F_{\gamma,m,n}$ that involves the sums $\Sigma_{m,\gamma}$ and $\Sigma_{m,\gamma+1}$, whereas the upper bound only involves $\Sigma_{m,\gamma}$. Condition 2-γ is then used to conclude.

For a linear threshold $\gamma \sim r \cdot n$, we prove that $F_{\gamma,m,n,X}$ tends to 0 if $p_X < r - \epsilon$ for some positive ϵ (with an explicit error term). Otherwise, it is bounded by 1. Hence, the sum $F_{\gamma,m,n}$ only involves patterns with probability greater than $r-\epsilon$ and Condition 1 ensures that the number of such patterns is at most polynomial.

4.4 Dynamical databases

The results of the previous section are valid for any database generated by any source, provided that it satisfies Conditions 1 and 2-γ. In this chapter, we prove that a large class of sources satisfies these conditions. We now present this class, formed by a large subset of dynamical sources introduced by Brigitte Vallée (2001), and further used in Clément et al. (2001), Bourdon (2001), and Bourdon et al. (2001). The model of dynamical sources gathers classical sources such as the Bernoulli sources or the Markov chains, as well as more correlated ones. It is sufficiently general and can yet be precisely analysed. This class is then a good candidate for generating general databases that we call *dynamical databases.* We prove the following.

Theorem 3. *A (Markovian and irreducible) dynamical source satisfies Condition 1 and, for all $\gamma \geq 1$, Condition 2-γ. Moreover, $\Sigma_{\gamma,m}$ is of the form:*

$$\Sigma_{m,\gamma} = \kappa_\gamma \cdot \lambda_\gamma^m (1 + O(\theta_\gamma^{-m})), \qquad \kappa_\gamma > 0, \quad \lambda_\gamma > 1, \quad \theta_\gamma > 1 \quad \text{and} \quad \lambda_\gamma > \lambda_{\gamma+1}.$$

In particular, Theorems 1 *and* 2 *hold for (Markovian and irreducible) dynamical databases.*

4.4.1 Dynamical sources

A dynamical source is defined by six elements:

(i) An interval I
(ii) An alphabet $\mathcal{E}$
(iii) A topological partition $(I_\alpha)_{\alpha\in\mathcal{E}}$ of I (i.e., $\alpha \neq \beta \Rightarrow I_\alpha \cap I_\beta = \emptyset$ and $\cup_\alpha \overline{I}_\alpha = I$)
(iv) A coding function $\sigma : I \to \mathcal{E}$ such that $\sigma(I_\alpha) = \alpha$
(v) A shift function T on I, of class C^2 and strictly monotone on each interval I_α, and strictly expansive (namely there exists ρ with $0 < \rho < 1$ and $|T'| > \rho^{-1} > 1$)
(vi) An initial density f_0 on I

Figure 4.2 describes some instances of dynamical sources. A dynamical source emits symbols in the following way. (i) First a random real x is chosen in I according to the initial density f_0, (ii) then, the emitted symbol at the ith step is the symbol associated with the interval that contains the ith iterate of x $[\alpha_i = \sigma(T^i x)]$, so that the (infinite) word $\mathcal{M}(x)$ produced by the source is $\mathcal{M}(x) := \alpha_1\alpha_2\alpha_3\cdots$.

A dynamical source is then similar to a pseudorandom generator, where a probabilistic seed is used to initialise the process, and after this random choice, the process is completely deterministic.

There exist several types of dynamical sources according to the geometric or analytic properties of T. The simplest family occurs when T is affine and surjective on each interval of the partition. Such sources model the classical memoryless sources that emit symbols independently from the previous ones, but always following the same probabilistic law. When such a source is used for generating a database, the questions are not correlated. Figure 4.2 gives an example of a Bernoulli source.

In order to introduce some correlations between questions, we first consider sources with *bounded memory,* such as Markov chains. A Markov chain emits a new symbol according to a constant probabilistic law that depends on a bounded number of previous symbols. Used to generate databases, it entails that close questions are correlated.

A Markov chain is a particular dynamical source. In this case, T is piecewise affine and the image of an interval of the partition is the union of intervals of the partition. Figure 4.2 gives an instance of a Markov chain.

In this chapter, we deal with more general sources, called Markovian dynamical sources. A Markovian dynamical source has the same geometry as a Markov chain (the image by T of the interval I_α is a union of such intervals), but the shift function is not necessary affine. Moreover, we suppose that the process is irreducible; i.e., the matrix $M = (m_{\alpha,\beta})$ with $m_{\alpha,\beta} = 1$ if $T(I_\beta) \cap I_\alpha \neq \emptyset$, and $m_{\alpha,\beta} = 0$ elsewhere, satisfies M^k has strictly positive coefficients for some positive integer k. Figure 4.2 presents a Markovian source. More general dynamical sources are not used in this chapter.

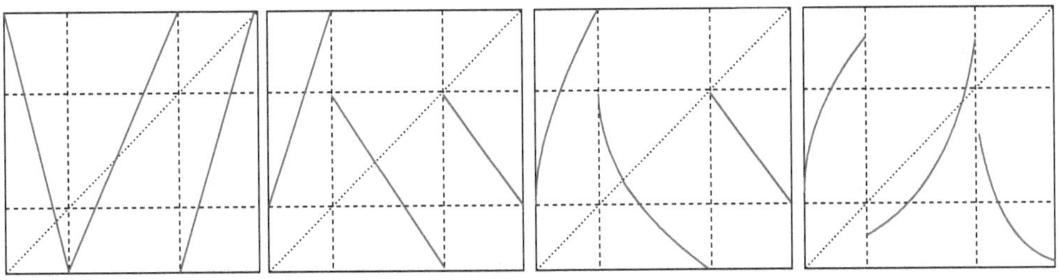

Figure 4.2. Instances of dynamical sources (without the initial density). From *left to right*: a Bernoulli source, a Markov chain, a Markovian dynamical source, a general dynamical source

4.4.2 Main tools

Fix $w = \alpha_1 \cdots \alpha_m$ a word of length m over $\mathcal{E}$. The set of real $x \in I$ such that the word $\mathcal{M}(x)$ begins with w forms an interval noted I_w. The image by the mth iterate T^m of I_w is also an interval J_w and the inverse function $h_w : J_w \to I_w$ of T^m is called an inverse branch of depth m. Note that h_w admits the alternative formula

$$h_w = h_{\alpha_1} \circ h_{\alpha_2} \circ \cdots \circ h_{\alpha_m}.$$

The main tools for analysing a dynamical source are functional operators. As does the transition matrix for Markov chains, the density transformer $\mathbf{G}$ describes the evolution of the density under iteration of the shift. If f_0 is the initial density of the source, the new density after m iterations of T is $f_m = \mathbf{G}^{\mathrm{m}}[\mathrm{f}_0]$, with

$$\mathbf{G}^m = \sum_{w \in \mathcal{E}^m} \mathbf{G}_w \quad \text{and} \quad \mathbf{G}_w[f](x) = |h'_w(x)|\, f \circ h_w(x) \mathbf{1}_{J_w}(x).$$

where $\mathbf{1}_E(x)$ equals 1 if x belongs to E, and is zero elsewhere. The operators $\mathbf{G}_{\mathrm{w}}$ are fundamental since the probability p_w that a word begins by w satisfies

$$p_w = \int_{I_w} f_0(t)dt = \int_I \mathbf{G}_{\mathrm{w}}[\mathrm{f}_0](\mathrm{t})\mathrm{dt}.$$

In particular, the probability $p_{w_1 \cdot w_2}$ (here $\cdot$ is the concatenation) expresses with the operator $\mathbf{G}_{\mathrm{w}_1 \cdot \mathrm{w}_2} = \mathbf{G}_{\mathrm{w}_2} \circ \mathbf{G}_{\mathrm{w}_1}$. Note that this functional relation replaces the well-known properties $p_{w_1 \cdot w_2} = p_{w_1} \cdot p_{w_2}$ valid only for memoryless sources.

Now, fix W_1 and W_2 two sets of words. By additivity, the probability that a word belongs to $W_1 \cdot W_2$ [words $w_1 \cdot w_2$ with $(w_1, w_2) \in W_1 \times W_2$] is related to the operator

$$\mathbf{G}_{\mathrm{W}_1 \cdot \mathrm{W}_2} = \mathbf{G}_{\mathrm{W}_2} \circ \mathbf{G}_{\mathrm{W}_1} \quad \text{with} \quad \mathbf{G}_{\mathrm{W}} = \sum_{\mathrm{w} \in \mathrm{W}} \mathbf{G}_{\mathrm{w}}. \tag{4.1}$$

Moreover, if W_1 and W_2 are disjoint, the probability of belonging to $W_1 \cup W_2$ involves the operator

$$\mathbf{G}_{W_1 \cup W_2} = \mathbf{G}_{\mathrm{W}_1} + \mathbf{G}_{\mathrm{W}_2}. \tag{4.2}$$

Equations (4.1) and (4.2) are just the "dynamical source version" of the following equalities valid for memoryless sources

$$p_{W_1 \cdot W_2} = p_{W_1} \cdot p_{W_2} \quad \text{and} \quad p_{W_1 \cup W_2} = p_{W_1} + p_{W_2} \quad [\text{disjoint union}].$$

The previous operators are all relative to one random transaction (or word). Various generalisations exist, but in the sequel we only deal with operators that generate the sum $\Sigma_{\gamma,m}$. Precisely, a γ-word of length m, noted $\overline{w}$, is a γ-tuple of words of same length m. If $\overline{w} = (w_1, \dots, w_\gamma)$, we define the applications

$$H_{\overline{w}}(t_1, \cdots, t_\gamma) = h_{w_1}(t_1) \cdots h_{w_\gamma}(t_\gamma) \quad \text{and} \quad \mathbf{1}_{J_{\overline{w}}}(t_1, \dots, t_\gamma) = \mathbf{1}_{J_{w_1}}(t_1) \cdots \mathbf{1}_{J_{w_\gamma}}(t_\gamma).$$

For instance, the probability $p_{\overline{w}}$ that γ independent and identical dynamical sources emit simultaneously the words $w_1, \dots, w_\gamma$ satisfies

$$p_{\overline{w}} = \int_{I^\gamma} \mathbb{G}_{\overline{w}}[F](t_1, \dots, t_\gamma) dt_1 \cdots dt_\gamma \quad \text{with} \quad \mathbb{G}_{\overline{w}}[F] = \mathrm{Jac}[H_{\overline{w}}]\ F \circ H_{\overline{w}}\ \mathbf{1}_{J_{\overline{w}}},$$

and $F(t_1, \dots, t_\gamma) = f_0(t_1) \cdots f_0(t_\gamma)$. Of course, the previous properties with the operators $\mathbf{G}$ extend to the operators $\mathbb{G}$ and one has

$$\mathbb{G}_{\overline{W}} := \sum_{\overline{w} \in \overline{W}} \mathbb{G}_{\overline{w}}, \quad \mathbb{G}_{\overline{W}_1 \cdot \overline{W}_2} = \mathbb{G}_{\overline{W}_2} \circ \mathbb{G}_{\overline{W}_1}, \quad \mathbb{G}_{\overline{W}_1 \cup \overline{W}_2} = \mathbb{G}_{\overline{W}_1} + \mathbb{G}_{\overline{W}_2}. \tag{4.3}$$

For a fixed γ-word $\overline{w} = (w_1, \dots, w_\gamma)$, a natural question is how many patterns are contained at the same time in the words $w_1, \dots, w_\gamma$? We define the cost function $c_\gamma(\overline{w})$ by the number of positions where the γ words of $\overline{w}$ have exactly the same value. Then, the number of patterns contained at the same time in $w_1, \dots, w_\gamma$ is $2^{c_\gamma(\overline{w})} - 1$. If $\overline{w}_1$ and $\overline{w}_2$ denote two γ-words, the following additive property is satisfied

$$c_\gamma(\overline{w}_1 \cdot \overline{w}_2) = c_\gamma(\overline{w}_1) + c_\gamma(\overline{w}_2).$$

Now, the sum $\Sigma_{\gamma,m}$ admits the alternative form

$$\begin{aligned} \Sigma_{\gamma,m} + 1 &= \sum_{\overline{w} \in (\mathcal{E}^m)^\gamma} 2^{c_\gamma(\overline{w})} p_{\overline{w}} \\ &= \int_{I^\gamma} \sum_{\overline{w} \in (\mathcal{E}^m)^\gamma} 2^{c_\gamma(\overline{w})} \mathbb{G}_{\overline{w}}[F](t_1, \dots, t_\gamma) dt_1 \cdots dt_\gamma \end{aligned}$$

with $F = (t_1, \ldots, t_\gamma) = f_0(t_1) \cdots f_0(t_\gamma)$. The additive property of cost c_γ and relations (4.3) entail a new expression of the sum in the integral that involves the operator $\mathbb{G}_\gamma$ defined by

$$\mathbb{G}_\gamma := \sum_{\overline{w} \in \mathcal{E}^\gamma} 2^{c_\gamma(\overline{w})} \mathbb{G}_{\overline{w}}.$$

Precisely, the sum $\Sigma_{\gamma,m}$ satisfies

$$\Sigma_{\gamma,m} = \int_{I^\gamma} \mathbb{G}_\gamma^m [F](t_1, \ldots, t_\gamma) dt_1 \cdots dt_\gamma \tag{4.4}$$

with $F = (t_1, \ldots, t_\gamma) = f_0(t_1) \cdots f_0(t_\gamma)$.

4.4.3 Proof of Theorem 3

Condition 1. The set of words (or transactions) that support a pattern X is of the form $\mathcal{E}_1 \cdot \mathcal{E}_2 \cdot \ldots \cdot \mathcal{E}_m$ with $\mathcal{E}_i := \mathcal{E}$ if the ith question q_i is not a question in X and $\mathcal{E}_i := \{\alpha, (q_i, \alpha) \in X\}$ in the other case. Then with Relation (4.1), the probability p_X satisfies

$$p_X = \int_I \mathbf{G}_{\mathcal{E}_m} \circ \cdots \circ \mathbf{G}_{\mathcal{E}_1}[f_0](t) dt.$$

Note that if $\mathcal{E}_i$ is replaced by the whole alphabet $\mathcal{E}$, one obtains a upper bound of p_X. To prove Condition 1, it is then sufficient to carefully replace some of the subalphabets.

On a convenient functional space (for instance, the Banach space of Lipschitz functions), the density transformer admits a unique dominant eigenvalue 1, separated from the remainder of the spectrum by a spectral gap. In addition, the dominant eigenvector g is strictly positive on I. This spectral property entails the decomposition

$$\mathbf{G}^k[f] := \left(\int_I f(t) dt \right) \cdot g + \mathbf{N}^k[f]$$

with $\mathbf{N}[g] = 0$ and $\mathbf{N}$ an operator with spectral radius strictly less than 1. In particular, for all positive ϵ_1, there exist k_0 such that on I, $\mathbf{G}^{k_0}[f] \leq (1+\epsilon_1)g$. Replacing the first k_0 subalphabets in the expression of p_X by the whole alphabet $\mathcal{E}$ yields the inequality

$$p_X \leq (1+\epsilon_1) \int_I \mathbf{G}_{\mathcal{E}_m} \circ \cdots \circ \mathbf{G}_{\mathcal{E}_{k_0+1}}[g](t) dt = (1+\epsilon_1) \int_I \mathbf{G}_{\mathcal{E}_m} \circ \cdots \circ \mathbf{G}_{\mathcal{E}_{k_1}}[g](t) dt$$

with $k_1 = \min\{k \geq k_0 + 1 \ : \ \mathcal{E}_k \neq \mathcal{E}\}$. Note that k_1 exists as soon as $|X| > k_0$.

There exist $\epsilon_2 = \rho_{\min} \cdot \inf_I g$ with $\rho_{\min} = \min_{m \in \mathcal{E}} \inf_{J_m} |h'_m|$ such that for all subalphabets $\mathcal{E}'$, the function $\mathbf{G}_{\mathcal{E}'}[g]$ satisfies

$$\mathbf{G}_{\mathcal{E}'}[g] \leq g - \epsilon_2 \quad \text{on} \quad E_{\mathcal{E}'} := \bigcup_{\alpha \notin \mathcal{E}'} J_\alpha$$

and $\mathbf{G}_{\mathcal{E}'}[g] \leq g$ elsewhere. For all positive integers k, one also has

$$\mathbf{G}^k \circ \mathbf{G}_{\mathcal{E}'}[g] = \sum_{w : I_w \subset E_{\mathcal{E}'}} \mathbf{G}_w[g] + \sum_{w : I_w \not\subset E_{\mathcal{E}'}} \mathbf{G}_w[g] \leq g - \epsilon_2 \cdot \left(\min_{w \in \mathcal{E}^k} \min_{J_w} |h'_w| \right) \cdot \sum_{w : I_w \subset E'_{\mathcal{E}}} 1.$$

Since the source is irreducible, there exist k_2 such that the last sum is strictly positive and then, we have found $\theta < 1$ such that on I,

$$\mathbf{G}^{k_2} \circ \mathbf{G}_{\mathcal{E}'}[g] \leq g - \epsilon_2 \cdot \left(\min_{w \in \mathcal{E}^k} \min_{J_w} |h'_w| \right) \leq \theta \cdot g.$$

Applying this relation recursively to the expression of p_X, we prove that Condition 1 is satisfied.

Condition 2-γ. The sum $\Sigma_{m,\gamma}$ is the average number of patterns supported by γ random transactions. We recall the alternative expression for $\Sigma_{m,\gamma}$, namely

$$\Sigma_{m,\gamma} = \int_{I^\gamma} \mathbb{G}_\gamma^m[(f_0, \dots, f_0)](t_1, \dots, t_\gamma) dt_1 \cdots dt_\gamma.$$

Here, $\mathbb{G}_\gamma$ is a multidimensional functional operator that admits a unique dominant eigenvalue $\lambda(\gamma)$, separated from the remainder of the spectrum by a spectral gap, and $\lambda(\gamma) > \lambda(\gamma+1)$. This spectral property entails a decomposition of $\mathbb{G}_\gamma$ of the form

$$\mathbb{G}_\gamma^m[F] = \lambda(\gamma)^m \mathbb{P}[F] \left(1 + O\left(r_\gamma^m\right)\right),$$

with $\mathbb{P}$ a projector and $r_\gamma < 1$. Theorem 3 follows with $\theta_\gamma^{-1} = \max(r_\gamma, r_{\gamma+1}, ((\lambda(\gamma+1))/\lambda(\gamma)))$.

4.5 Improved memoryless model of databases

All the existing databases are not a particular case of dynamical databases. Consider, for instance, a quite simple one, which is called the improved memoryless model in Lhote et al. (2005). Persons and questions are independent, and each question has its own probabilistic behaviour. More precisely, the answer to the ith question follows a Bernoulli law $B_i = (p_{i,\alpha})_{\alpha \in \mathcal{E}}$ over the alphabet $\mathcal{E}_i$, where the B_is and the $\mathcal{E}_i$s may depend on the index i. In the "simple" memoryless model, used for the classical probabilistic analyses, the B_is were the same.

Let p denote the maximum of all the probabilities $p_{i,v}$. When $p < 1$, the relation $p_X \leq p^{|X|}$ holds and ensures Condition 1. Moreover, the sum $\Sigma_{\gamma,m}$ admits the closed formula

$$\Sigma_{\gamma,m} = \prod_{j=1}^{m} \left(1 + \sum_{v \in \mathcal{E}} p_{j,v}^\gamma\right)$$

and Condition 2-γ is clearly satisfied with $\theta_\gamma = (1 + |\mathcal{E}|^{-\gamma})/(1 + \mathrm{p}|\mathcal{E}|^{-\gamma})$.

4.6 Experiments

This section presents some experiments realised with classical databases of the FIMI website (Frequent Itemset Mining Implementations). In Figure 4.3, the plain line in the graphics represents the number of frequent patterns in the function of the frequency threshold for two real databases (Chess.dat and Mushroom.dat) and a synthetic one (T10I4D100K.dat). The dotted (resp., dashed) line represents the average number of frequent patterns of the simple (resp., improved) Bernoulli model naturally associated with the real database.

In the graphics, the improved model gives very good estimations whereas the simple model is quite bad. This result is not surprising for synthetic data since they have, by construction, few correlations. However, such closed results were unexpected for real-life databases.

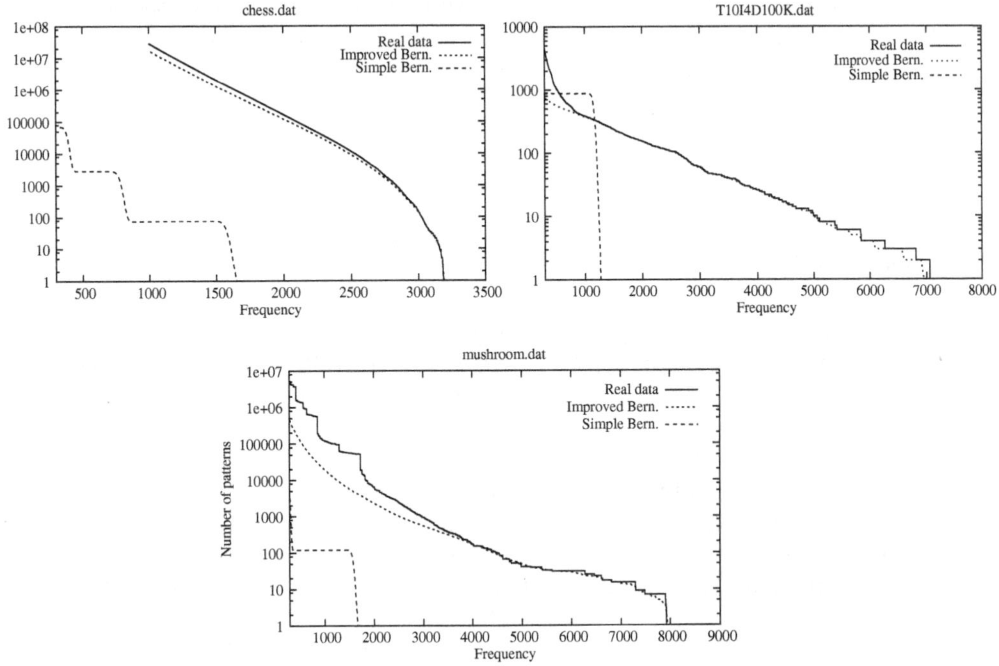

Figure 4.3. Number of frequent patterns in the function of the frequency threshold in the real database (*plain line*), in the associated simple Bernoulli model (*dashed*), and in the associated improved Bernoulli model (*dotted*)

4.7 Conclusion

Frequent pattern mining is a fundamental task in data mining but its complexity is closely related to the parameters (frequency threshold, size of the database, etc.). In this chapter, we have introduced a general model of random databases that are generated by sources. Under natural conditions on the words produced by the sources, we elucidate (for the first time) two different behaviours for the average number of frequent patterns in large databases. First, if the frequency threshold is constant, the mean number of frequent patterns is exponential with respect to the number of columns. That explains why in practice the algorithms fail for small frequency thresholds. On the other hand, for linear frequency thresholds, the mean number of frequent patterns is polynomial with respect to the number of columns. That explains the efficiency of the algorithms for reasonable frequency thresholds.

In the second part, we introduced the large class of dynamical sources that encompass various classical sources (simple Bernoulli sources, Markov chains). We prove that they satisfy the natural conditions which implies that our results apply in a quite general context.

However, frequent patterns are not the only patterns useful in data mining. One can cite, for instance, the closed patterns, the minimal nonfrequent patterns, the general constrained patterns, and so on. Future work will consist in the analysis of such patterns. The analysis of the associated algorithms will also be of great interest since it is not rare that several authors claim that their algorithms are the most efficient.

References

R. Agrawal, T. Imielinski, and A. Swami. Mining association rules between sets of items in large databases. In *Proc. of the 1993 ACM SIGMOD International Conference on Management of Data, Washington, USA*, pp. 207–216, 1993.

R. Agrawal, H. Mannila, R. Srikant, H. Toivonen, and A. Verkamo. Fast discovery of association rules. *Advances in Knowledge Discovery and Data Mining*, pp. 307–328, AAAI/MIT Press, Cambridge, MA 1996.

J. Bourdon. Size and path-length of Patricia tries: Dynamical sources context. *Random Structures and Algorithms*, 19:289–315, 2001.

J. Bourdon, M. Nebel, and B. Vallée. On the stack-size of general tries. *Theoretical Informatics and Applications*, 35:163–185, 2001.

J. Clément, P. Flajolet, and B. Vallée. Dynamical sources in information theory: A general analysis of trie structures. *Algorithmica*, 29(1):307–369, 2001.

F. Geerts, B. Goethals, and J. Van den Bussche. A tight upper bound on the number of candidate patterns. In *IEEE International Conference on Data Mining (ICDM'01), San Jose, USA*, pp. 155–162, 2001.

B. Goethals. Survey on frequent pattern mining, *Helsinki, Institute for Information Technology, Technical report*, 2003.

J. Han, J. Pei, and Y. Yin. Mining frequent patterns without candidate generation: A frequent-pattern tree approach. In *ACM SIGMOD International Conference on Management of Data (SIGMOD'00), Dallas, USA*, pp. 1–12, 2000.

L. Lhote, F. Rioult, and A. Soulet. Average number of frequent (closed) patterns in Bernouilli and Markovian databases. In *IEEE International Conference on Data Mining (ICDM'05), Houston, USA*, pp. 713–716, 2005.

P.W. Purdom, D. Van Gucht, and D.P. Groth. Average-case performance of the Apriori algorithm. *SIAM Journal on Computing*, 33(5):1223–1260, 2004.

A. Savasere, E. Omiecinski, and S. Navathe. An efficient algorithm for mining association rules in large databases. In *VLDB'95*, 1995.

H. Toivonen. Sampling large databases for association rules. In *International Conference on Very Large Data Bases (VLDB'96), Mumbai, India*, pp. 134–145. Morgan Kaufman, San Francisco, 1996.

B. Vallée. Dynamical sources in information theory: Fundamental intervals and word prefixes. *Algorithmica*, 29:262–306, 2001.

M.J. Zaki. Scalable algorithms for association mining. *IEEE Transactions or Knowledge and Data Engineering*, 12(2):372–390, 2000.

Part II

Information Theory and Statistical Applications

5

Introduction

Koustautiuos Zografos

Department of Mathematics, University of Ioannina, 45110 Ioannina, Greece

5.1 Introduction

Part II is based upon three papers that were presented in the Special Session *Information Theory and Statistical Applications* of the 12th International Conference on Applied Stochastic Models and Data Analysis (ASMDA 2007) which was held in May 2007 in Chania, Greece. Information theory includes research dealing, among others with statistical inference, association, prediction, and modelling of statistical data. The last two or three decades are characterised by a vigorous growth in the use of information-theoretic ideas and methods in statistics. The reason is that Statistical Information Theory (SIT) provides a number of measures which obey nice probabilistic and statistical properties and moreover can be used to formulate and solve a great variety of statistical problems. In this sense SIT contributes to the advancement in probability theory and statistics in particular, and progress in almost all areas of science and engineering.

Entropy and divergence measures are the main constitutive elements of statistical information theory (cf. Vajda, 1989). The concept of entropy, although introduced in the context of classical thermodynamics, has become a major tool in information theory and in almost every branch of science and engineering after the seminal work of Shannon in the mathematical theory of communication (Shannon, 1948). One of the most powerful techniques, which is based on Shannon entropy, is the maximum entropy principle, a method of constrained inference that generalises the principle of insufficient reason. It has been introduced by Jaynes (1957) and provides the unknown probabilistic model which describes a set of statistical data. On the other hand the concept of divergence between two probability distributions and the respective measures of divergence (cf. Csiszar, 1963; Ali and Silvey, 1966) are used for the development of information-theoretic ideas and methods in problems of estimation and testing statistical hypotheses. In this context, a measure of divergence between the empirical and the true model can serve as a test statistic for testing goodness-of-fit. In a similar manner, the minimization, in respect to the unknown parameter, of a divergence measure between the true and the empirical model leads to nice alternatives of the maximum likelihood estimators. For a comprehensive discussion on the development of information-theoretic methods in statistics the reader is referred to the monographs by Read and Cressie (1988) and Pardo (2006), among others.

C.H. Skiadas (ed.), *Advances in Data Analysis*, Statistics for Industry and Technology, DOI 10.1007/978-0-8176-4799-5_5,

The section starts with two chapters with direct applications of divergence measures in statistics while the third chapter focuses on applications of divergences to actuarial science and in particular to the problem of graduating mortality rates. All the papers include methodological developments and relevant applications. In particular, in the chapter by Prof. A. Karagrigoriou and Dr. K. Mattheou, a model selection criterion, the Divergence Information Criterion (DIC), is proposed by means of a generalised family of divergence measures. Furthermore, a lower bound for the mean squared error of prediction is furnished. This criterion generalises, in a sense, the well-known Akaike Information Criterion (AIC) (cf. Akaike, 1973). A simulation study is used to illustrate that the proposed model selection criteria are at least comparable to the well-known and used model selection criteria such as AIC, BIC, and so on. The second chapter, by Prof. M. C. Pardo, deals with the influence or leverage of points in the fitted value of a generalized linear model for ordinal multinomial data. For this purpose the family of the phi-divergence measure is used. The minimum phi-divergence estimator is applied and a generalized Hessian (hat) matrix is introduced as a diagnostic tool for measuring the influence. Residuals based on the phi-divergence measure are introduced for detecting outliers. The third chapter, by Dr. A. Sachlas and Prof. T. Papaioannou discusses extensions of the classical divergence measures so as to be useful and applicable for the study of graduating mortality rates. The properties of the proposed divergences are studied and moreover used to formulate the problem of graduating mortality rates via Lagrangian duality theory. Numerical examples are also given to illustrate the theoretic results.

References

Akaike, H. (1973). Information theory and an extension of the maximum likelihood principle. *Proc. of the 2nd Intern. Symposium on Information Theory*, (Petrov B. N. and Csaki F., Eds.), Akademiai Kaido, Budapest.

Ali, S. M. and Silvey, S. D. (1966). A general class of coefficients of divergence of one distribution from another. *J. R. Statist. Soc.* B, **28**, 131–142.

Csiszar, I. (1963). Eine Informationstheoretische Ungleichung und ihre Anwendung auf den Bewis der Ergodizitat on Markhoffschen Ketten. *Publ Mathe Inst Hungarian Acad Sci*, **8**, 84–108.

Jaynes E. T. (1957). Information theory and statistical mechanics. *Phys. Rev.*, **106**, 620–630.

Pardo, L. (2006). *Statistical Inference Based on Divergence Measures.* Chapman and Hall, London.

Read, T. R. C. and Cressie, N. A. C. (1988). *Goodness-of-Fit Statistics for Discrete Multivariate Data.* Springer-Verlag, New York.

Shannon, C. E. (1948). The mathematical theory of communication. *Bell Syst. Tech. J.*, **27**, 379–423.

Vajda, I. (1989). *Theory of Statistical Inference and Information.* Kluwer Academic, Dordrecht.

6

Measures of Divergence in Model Selection

Alex Karagrigoriou and Kyriacos Mattheou

University of Cyprus, Nicosia, Cyprus

Abstract: In this chapter a number of measures of divergence are presented and the way model selection criteria are constructed via measures of divergence is discussed. The construction of the divergence information criterion based on a new family of measures of divergence is presented and the lower bound of the mean squared error of prediction is established. Some illustrative simulation results are also given.

Keywords and phrases: Measures of divergence, model selection, AIC, BIC, DIC, MSE of prediction

6.1 Introduction

The measures of divergence are powerful tools used in measuring the distance or affinity between two probability distributions for the purpose of association, clustering, or classification of the distributions involved. The measures of divergence together with the measures of information have a very long history since the fundamental work of Fisher, Shannon, and Kullback. Such measures have numerous applications in statistical inference and can be applied in various fields such as signal processing and pattern recognition, analysis of contingency tables and tests of fit, model selection, finance and economics, approximations of probability distributions, etc.

The aim of this chapter is to present some recent developments on measures of information in connection with model selection criteria. In particular, in Section 6.2 we present a number of measures of divergence while in Section 6.3 we review the model selection criteria. In Section 6.4 we discuss the construction of the Divergence Information Criterion (DIC) and in Section 6.5 we discuss some of its asymptotic properties. Finally, in Section 6.6 we use an illustrative example for the use of DIC.

C.H. Skiadas (ed.), *Advances in Data Analysis*, Statistics for Industry and Technology, DOI 10.1007/978-0-8176-4799-5_6,

6.2 Measures of divergence

In what follows assume that $f(\cdot)$ and $g(\cdot)$ are probability density functions (pdfs) corresponding to some random variable X which may or may not depend on some unknown parameter of finite dimension. The divergence is a functional which expresses the dissimilarity between two functions and it should be such that its value increases if the two functions are considered to be "less similar." The most popular measure of (directed) divergence is considered to be the Kullback–Leibler divergence (Kullback and Leibler, 1951) known also as a relative measure of information which is based on the likelihood ratio and is given by

$$I_X^{KL}(f,g) = \int f(z) \ln(f(z)/g(z))dz.$$

Observe that the measure is not symmetric as opposed to the J-divergence (Jeffreys, 1946) given by $I_X^J(f,g) = I_X^{KL}(f,g) + I_X^{KL}(g,f)$.

A general divergence-type measure is the family of "convex likelihood-ratio expectations" introduced by Csiszar, Ali, and Silvey (Csiszar, 1963; Ali and Silvey, 1966) which is also known as φ-divergence, often denoted by CAS (Csiszar–Ali–Silvey) and based on a convex function φ that unifies the above-mentioned as well as many other measures of divergence. The CAS measure is defined by

$$I_X^C(f,g) = \int \varphi(f(z)/g(z))g(z)dz,$$

where φ is a convex function in $[0,\infty)$ such that $0\varphi(0/0) = 0$, $\varphi(u) \underset{u\to 0}{\to} 0$ and $0\varphi(u/0) = u\varphi_\infty$ with $\varphi_\infty = \lim_{u\to\infty}[\varphi(u)/u]$.

Observe that the CAS measure reduces to Kullback–Liebler divergence if $\varphi(u) = u \ln u$. If $\varphi(u) = (1-u)^2$ or $\varphi(u) = \mathrm{sgn}(a-1)u^a$, $a > 0$, $a \neq 1$ CAS's measure yields the Kagan (Pearson's X^2, Kagan, 1963) and Renyi's divergence, respectively (Renyi, 1961). Another member of the CAS family of divergence measures with $\varphi(u) = (u^{\lambda+1} - u - \lambda(u-1))/(\lambda(\lambda+1))$, $\lambda \neq 0, -1$, is the power divergence introduced by Cressie and Read (1984) which is given by

$$I_X^{CR}(f,g) = \frac{1}{\lambda(\lambda+1)} \int f(z) \left[\left(\frac{f(z)}{g(z)} \right)^\lambda - 1 \right] dz, \qquad \lambda \in R,$$

where for $\lambda = 0, -1$ is defined by continuity. Note that the Kullback–Leibler divergence is obtained for $\lambda \downarrow 0$.

A recently proposed measure of divergence is the Basu, Harris, Hjort, and Jones (BHHJ) power divergence between f and g (Basu et al., 1998) which is indexed by a positive parameter a, and defined as

$$I_X^a(g,f) = \int \left\{ f^{1+a}(z) - \left(1 + \frac{1}{a}\right) g(z) f^a(z) + \frac{1}{a} g^{1+a}(z) \right\} dz, \qquad a > 0. \quad (6.1)$$

This family of measures was proposed by Basu et al. (1998) for the development of a minimum divergence estimating method for robust parameter estimation. Indeed, if $\{\mathcal{F}_\theta\}$ a parametric family of probability density functions f_θ indexed by an unknown parameter θ, then the estimation method consists of choosing parameter values in a hypothesized set Θ that minimize $I_X^a(g,f)$. The index a controls the trade-off between robustness and asymptotic efficiency of the parameter estimators which are the quantities that minimize (6.1). It should be also noted that the BHHJ family reduces to the Kullback–Leibler divergence for $a \downarrow 0$ and as it can be easily seen, to the square of the standard L_2 distance between f and g for $a = 1$. As a result, for $a \downarrow 0$ the family, as an estimating method, reduces to the traditional maximum likelihood estimation while for $a = 1$ it becomes the mean squared error estimation. In the former case the resulting estimator is efficient but not robust while in the latter the method results in a robust but inefficient estimator. The authors observed that for values of a close to 0 the resulting estimators have strong robust features without a big loss in efficiency relative to the maximum likelihood estimating method. As a result one is interested in small values of $a > 0$, say between zero and one, although values larger than one are also allowed. One should be aware though of the fact that the estimating method becomes less and less efficient as the index a increases.

6.3 Model selection criteria

Since the measures of divergence are used as indices of similarity or dissimilarity between populations and for measuring mutual information concerning two variables they can be used for the construction of model selection criteria. Irrespectively of the strategy used to select the best model among a set of candidate models, the true and the model ultimately fitted differ in a number of aspects. This lack of fit can be measured by some measure of divergence. The resulting discrepancy between the two models is known as the expected overall discrepancy (EOD) which is a random variable. The distributional characteristics of the EOD are playing the key role in comparing different fitting strategies. Since one should prefer a strategy which on the average, results in a low EOD, it is natural to judge a strategy by its *mean expected overall discrepancy.* In many cases some of the terms of the mean EOD are omitted since they do not involve the fitted model. Since this quantity depends on the true model an appropriate estimator is required. The estimator of the mean expected overall discrepancy (or the essential part of the mean expected overall discrepancy after the irrelevant terms are omitted) is traditionally referred to as a *model selection criterion.* If the value of the criterion is small then the approximated or fitted model is good.

The Kullback–Leibler measure was the divergence used by Akaike (1973) to develop the Akaike Information Criterion (AIC). Let $\mathbf{x} = (x_1, \ldots, x_n)$ a realization of a random vector $\mathbf{X} = (X_1, \ldots, X_n)$ and assume that the X_is are independent and identically distributed each with true unknown density function $g(\cdot, \theta_0)$, with $\theta_0 = (\theta_{01}, \ldots, \theta_{0p})'$ the true but unknown value of the p-dimensional parameter of the distribution. Consider a candidate model $f_\theta(\cdot)$, the log-likelihood $l(\theta; x)$, and let $\hat{\theta}$ be the maximum likelihood estimator (MLE) of θ_0 in some hypothesized set Θ, i.e.,

$$l(\hat{\theta}; \mathbf{x}) = \sum_{i=1}^{n} \log(f_{\hat{\theta}}(x_i)) = \max_{\theta \in \Theta} l(\theta; \mathbf{x})$$

so that $f_{\hat{\theta}}(\cdot)$ is an estimate of $g(\cdot, \theta_0)$. The divergence between the estimate (candidate model) and the true density can be measured by the Kullback–Leibler measure:

$$I_X^{KL}(g, f_{\hat{\theta}}) = \int g(z, \theta_0) \log\Big(\frac{g(z, \theta_0)}{f_{\hat{\theta}}(z)}\Big) dz$$

which is a special case for $a \downarrow 0$ of the BHHJ measure

$$I_X^a\left(g, f_{\hat{\theta}}\right) = \int \left\{ f_{\hat{\theta}}^{1+a}(z) - \left(1 + \frac{1}{a}\right) g\left(z, \theta_0\right) f_{\hat{\theta}}^a(z) + \frac{1}{a} g^{1+a}\left(z, \theta_0\right) \right\} dz. \tag{6.2}$$

Observe that $I_X^{KL}(g, f_{\hat{\theta}})$ can be written in the form

$$I_X^{KL}(g, f_{\hat{\theta}}) = E_g[\log(g(X, \theta_0))] - E_g[\log(f_{\hat{\theta}}(X))].$$

Note that the first term is independent of the candidate model and therefore the divergence can be evaluated using only the second term, usually known as the expected loglikelihood. Akaike proposed the evaluation of the fit of $f_{\hat{\theta}}(\cdot)$ using minus twice the *mean* expected loglikelihood (i.e., the essential part of the mean expected overall discrepancy) given by

$$-2E_g\left[E_g[\log(f_{\hat{\theta}}(X))]\right] = -2\int \cdots \int E_g[\log(f_{\hat{\theta}}(X))] \prod_{i=1}^{n} g(x_i, \theta_0) dx_1 \cdots dx_n \tag{6.3}$$

since the candidate model is close to the true model if the above quantity is small. Furthermore, Akaike provided an unbiased estimator of (6.3) given by

$$[-2l(\hat{\theta}; \mathbf{x}) + 2p]/n$$

so that the resulting AIC is defined to be

$$AIC = -2l(\hat{\theta}; \mathbf{x}) + 2p.$$

A general class of criteria has been introduced by Konishi and Kitagawa (1996) which also estimates the Kullback–Leibler measure where the estimation is not necessarily based on maximum likelihood and the specified family of candidate distributions does not contain the distribution generating the data.

Following the early work of Akaike, other model selection proposals include Bayesian approaches with the Bayesian Information Criterion (BIC; Schwarz, 1978) and the Deviance Information Criterion (DIC; Spiegelhalter et al., 2002; van der Linde, 2005) being the most popular. The BIC criterion has a number of advantages worth mentioning. More specifically, it has been shown to be consistent (Schwarz, 1978; Wei, 1992) which means that it chooses the correct model with probability 1 as n tends to infinity. The second advantage is that the criterion depends on $\log n$ ($BIC = -2l(\hat{\theta}; \mathbf{x}) + p \log n$) and therefore it downweights the effective sample size which in some cases prevents the erroneous rejection of null hypothesis for small sample sizes.

Other model selection proposals inspired by or related to the pioneer work of Akaike include among others approaches based on bootstrapping (Shang and Cavanaugh, 2008) and on Monte Carlo simulations (Shang, 2008), approaches for high-dimensional medical data (Koukouvinos et al., 2008), as well as variations of AIC (Cavanaugh, 2004; Bengtsson and Cavanaugh, 2006).

6.4 The divergence information criterion

Here we apply the same methodology used for AIC to the BHHJ divergence in order to develop a new criterion, the divergence information criterion. Note that the DIC proposed here is not related to the above-mentioned deviance information criterion which is a Bayesian criterion for posterior predictive comparisons.

Consider a random sample $X_1, \ldots, X_n$ from the distribution g (the true model) and a candidate model f_θ from a parametric family of models $\{f_\theta\}$, indexed by an unknown parameter $\theta \in \Theta$, where Θ is a one-dimensional parametric space. To construct the new criterion for goodness-of-fit we consider the quantity:

$$W_\theta = \int \left\{ f_\theta^{1+a}(z) - \left(1 + a^{-1}\right) g(z) f_\theta^{a}(z) \right\} dz, \qquad a > 0, \tag{6.4}$$

which is the same as the BHHJ divergence $I_X^a(g, f_\theta)$ given in (6.1) without the last term that remains constant irrespectively of the model f_θ used. Observe that (6.4) can also be written as

$$W_\theta = E_{f_\theta}\left(f_\theta^{a}(Z)\right) - \left(1 + a^{-1}\right) E_g\left(f_\theta^{a}(Z)\right), \qquad a > 0. \tag{6.5}$$

The target theoretical quantity that needs to be later approximated by an unbiased estimator is given by

$$EW_{\hat{\theta}} = E\left(W_\theta \middle| \theta = \hat{\theta}\right), \tag{6.6}$$

where $\hat{\theta}$ is any consistent and asymptotically normal estimator of θ. This quantity can be viewed as the average distance between g and f_θ up to a constant and is the essential part of the mean expected overall discrepancy between g and f_θ equivalent to (6.3).

Observe that the mean expected overall discrepancy can be easily evaluated by using a Taylor expansion around θ_0. The necessary derivatives of (6.5) in the case where g belongs to the family $\{f_\theta\}$ are given by (see Mattheou et al., 2009)

$$\frac{\partial W_\theta}{\partial \theta} = (a+1)\left[\int u_\theta(z) f_\theta^{1+a}(z)\, dz - E_g\left(u_\theta(Z) f_\theta^{a}(Z)\right)\right]$$

and

$$\begin{aligned}\frac{\partial^2 W_\theta}{\partial \theta^2} &= (a+1)\left\{(a+1)\int \left[u_\theta(z)\right]^2 f_\theta^{1+a}(z)\, dz - \int i_\theta f_\theta^{1+a}(z)\, dz = 0 \right. \\ &\quad \left. + E_g\left(i_\theta(Z) f_\theta^{a}(Z)\right) - E_g\left(a\left[u_\theta(Z)\right]^2 f_\theta^{a}(Z)\right) = (a+1)J(\theta_0)\right\},\end{aligned}$$

where $u_\theta(z) = (\partial/\partial\theta)(\log(f_\theta(z)))$, $i_\theta(z) = -(\partial^2/\partial\theta^2)(\log(f_\theta(z)))$, θ_0 represents the best fitting value of the parameter, and $J(\theta_0) = \int \left[u_{\theta_0}(z)\right]^2 f_{\theta_0}^{1+a}(z)\, dz$.

Using a Taylor expansion of W_θ around the true point θ_0, we can show that the mean expected overall discrepancy at $\theta = \hat{\theta}$ is given by

$$EW_{\hat{\theta}} = W_{\theta_0} + \frac{(a+1)}{2} E\left[\left(\hat{\theta} - \theta_0\right)^2 J(\theta_0)\right] + ER_n, \tag{6.7}$$

where $R_n = o((\hat{\theta} - \theta_0)^2)$.

As in the case of the AIC criterion we construct now an unbiased estimator of the mean expected overall discrepancy (6.7). First we deal though with the estimation of the unknown density g. An estimator of (6.5) with respect to g is given by replacing $E_g\left(f_\theta^a(Z)\right)$ by its sample analogue

$$Q_\theta = \int f_\theta^{1+a}(z)\,dz - \left(1+\frac{1}{a}\right)\frac{1}{n}\sum_{i=1}^{n} f_\theta^a(X_i), \tag{6.8}$$

with derivatives given by

$$\frac{\partial Q_\theta}{\partial \theta} = (a+1)\left[\int u_\theta(z)\, f_\theta^{1+a}(z)\,dz - \frac{1}{n}\sum_{i=1}^{n} u_\theta(X_i)\, f_\theta^a(X_i)\right]$$

and

$$\frac{\partial^2 Q_\theta}{\partial \theta^2} = (a+1)\left\{(a+1)\int [u_\theta(z)]^2 f_\theta^{1+a}(z)\,dz\right.$$
$$\left. - \int i_\theta f_\theta^{1+a}(z)\,dz + \frac{1}{n}\sum_{i=1}^{n} i_\theta(X_i)\, f_\theta^a(X_i) - \frac{1}{n}\sum_{i=1}^{n} a\,[u_\theta(X_i)]^2 f_\theta^a(X_i)\right\}.$$

It is easy to see that by the weak law of large numbers, as $n \to \infty$, we have:

$$\left[\frac{\partial Q_\theta}{\partial \theta}\right]_{\theta=\theta_0} \xrightarrow{P} \left[\frac{\partial W_\theta}{\partial \theta}\right]_{\theta=\theta_0} \text{ and } \left[\frac{\partial^2 Q_\theta}{\partial \theta^2}\right]_{\theta=\theta_0} \xrightarrow{P} \left[\frac{\partial^2 W_\theta}{\partial \theta^2}\right]_{\theta=\theta_0}. \tag{6.9}$$

The consistency of $\hat{\theta}$, the continuity of $J(\theta)$, expressions (6.8) and (6.9), and a Taylor expansion of Q_θ around the point $\hat{\theta}$ can be used to evaluate the expectation of Q_θ at $\theta = \theta_0$ and W_θ at $\theta = \hat{\theta}$:

$$EQ_{\theta_0} = EQ_{\hat{\theta}} + \frac{a+1}{2}E\left[\left(\theta_0 - \hat{\theta}\right)^2 J(\theta_0)\right] + ER_n \equiv W_{\theta_0}$$

and

$$EW_{\hat{\theta}} = E\left\{Q_{\hat{\theta}} + (a+1)\left(\hat{\theta} - \theta_0\right)^2 J(\theta_0) + R_n\right\} \tag{6.10}$$

so that the quantity within the brackets is an unbiased estimator of the mean expected overall discrepancy. Recall that the estimator $\hat{\theta}$ is a consistent and asymptotically normal estimator of the parameter θ. For such an estimator one could select the value of θ that either maximizes the loglikelihood function ($\hat{\theta}$, MLE method) or minimizes the BHHJ discrepancy or equivalently the quantity W_θ ($\hat{\theta}_a$, Basu method). In the latter case the consistency as well as the asymptotic normality of the estimator $\hat{\theta}_a$ is verified by Basu et al. (1998).

The previous results can be easily extended to the multivariate case. This extension is possible since under certain regularity conditions (Basu et al., 1998) the p-dimensional estimator $\hat{\theta}_a = \left(\hat{\theta}_{a1}, \ldots, \hat{\theta}_{ap}\right)'$ is consistent for $\theta_0 = (\theta_{01}, \ldots, \theta_{0p})'$ and $\sqrt{n}(\hat{\theta}_a - \theta_0)$ is asymptotically multivariate normal with vector mean $\mathbf{0}$ and variance–covariance matrix $J^{-1}(\theta_0)K(\theta_0)J^{-1}(\theta_0)$ where

$$J(\theta_0) = \int u_{\theta_0}(z) u'_{\theta_0}(z) f_{\theta_0}^{1+a}(z)\,dz$$

and

$$K(\theta_0) = \int u_{\theta_0}(z) u'_{\theta_0}(z) f_{\theta_0}^{1+2a}(z)\, dz - \xi\xi', \tag{6.11}$$

$\xi = \int u_{\theta_0}(z) f_{\theta_0}^{1+a}(z)\, dz$, $u_\theta(z) = (\partial/\partial\theta)(\log(f_\theta(z)))$, and ψ' the transpose of the vector ψ.

As a result, for a p-dimensional parameter θ, we can see that the p-dimensional unbiased estimator of the mean expected overall discrepancy takes the form:

$$Q_{\hat{\theta}_a} + (a+1)\left(\hat{\theta}_a - \theta_0\right)' J(\theta_0)\left(\hat{\theta}_a - \theta_0\right) + o(||\hat{\theta}_a - \theta_0||^2) \tag{6.12}$$

which for $p = 1$ reduces to the corresponding univariate estimator (see (6.10)).

Consider now the case that the candidate model f_θ comes from the family of the p-variate normal distribution where θ is the mean vector and $\hat{\theta}_a$ is obtained by minimizing (6.2) (Basu method). Then, it can be shown that (see Basu et al., 1998)

$$J(\theta_0) = (2\pi)^{-(a/2)}(1+a)^{-(1+(p/2))}\Sigma^{-(1+(a/2))}$$

and

$$\mathrm{Var}\left(\hat{\theta}_a\right) = \left(1 + \frac{a^2}{1+2a}\right)^{1+(p/2)} \Sigma$$

so that

$$J(\theta_0) = (2\pi)^{-(a/2)}\left(\frac{1+a}{1+2a}\right)^{1+(p/2)} \Sigma^{-(a/2)}\left[\mathrm{Var}\left(\hat{\theta}_a\right)\right]^{-1},$$

where Σ is the $p \times p$ asymptotic covariance matrix of the maximum likelihood estimator of the p-dimensional parameter θ_0. Taking into consideration the fact that $n \cdot o(||\hat{\theta}_a - \theta_0||^2) = o_P(1)$ since $\sqrt{n}(\hat{\theta}_a - \theta_0)$ is asymptotically normal, we have that

$$n\left(\hat{\theta}_a - \theta_0\right)' \Sigma^{-(a/2)}\left[\mathrm{Var}\left(\hat{\theta}_a\right)\right]^{-1}\left(\hat{\theta}_a - \theta_0\right) \tag{6.13}$$

has approximately a $\mathcal{X}_p^2$ distribution for a small. Then, the divergence information criterion defined as the asymptotically unbiased estimator of the mean expected overall discrepancy is introduced in the theorem below which is due to Mattheou et al. (2009).

Theorem 1. *Assume that the candidate model comes from the family of the p-variate normal distribution with θ the mean vector and $\hat{\theta}_a$ the estimator obtained by minimizing (6.2). An asymptotically unbiased estimator of n-times the mean expected overall discrepancy evaluated at $\hat{\theta}_a$ is given by*

$$DIC = nQ_{\hat{\theta}_a} + (a+1)(2\pi)^{-(a/2)}\left(\frac{1+a}{1+2a}\right)^{1+(p/2)} p. \tag{6.14}$$

The DIC criterion as it has been derived in the above theorem uses as an estimator of the unknown parameter the estimator obtained by minimizing (6.2) (Basu method). As mentioned earlier, the researcher may alternatively choose to use the maximum likelihood method ($\hat{\theta}$, MLE method) in which case the correction term is adjusted accordingly. Indeed, in this case

$$J(\theta_0) = (2\pi)^{-(a/2)} (1+a)^{-(1+(p/2))} \Sigma^{-(1+(a/2))}$$
$$= (2\pi)^{-(a/2)} (1+a)^{-(1+(p/2))} \Sigma^{-(a/2)} \left[\mathrm{Var}\left(\hat{\theta}\right)\right]^{-1}$$

since $\mathrm{Var}\left(\hat{\theta}\right) = \Sigma$ is the covariance matrix of the maximum likelihood estimator. Using (6.12) and the fact that (6.13) follows again approximately a $\mathcal{X}_p^2$ distribution it is easy to see that the adjusted DIC is given by

$$DIC^{MLE} = nQ_{\hat{\theta}} + (2\pi)^{-(a/2)} (1+a)^{-(p/2)} p. \tag{6.15}$$

By comparing the correction terms of DIC and DIC^{MLE} we observe that they are similar in the sense that for small a

$$(1+a)\left(\frac{1+a}{1+2a}\right)^{1+(p/2)} \simeq (1+a)^{-(p/2)} < 1.$$

In order to put into the proposed criterion some extra penalty for too large models (models with large number of parameters) we can replace the above term(s) by a (common) quantity larger than 1. Observe that for small values of a the denominator of the left-hand side of the above expression can be assumed to be close to 1 and therefore it can be disregarded. As a result both of the above terms can be replaced in DIC and DIC^{MLE} by the remaining part of the expression on the left-hand side, namely $(1+a)^{2+(p/2)}$. Observe that the above quantity is now larger than 1 so that the penalty term of the criterion will be larger for large values of p. Both criteria are adjusted accordingly, and in fact now they are both given by the same corrected formula (although $\hat{\theta}$ is obtained by different estimating methods), namely

$$DIC_c = nQ_{\hat{\theta}} + (2\pi)^{-(a/2)} (1+a)^{2+(p/2)} p. \tag{6.16}$$

Both the MLE and the Basu estimating methods have a number of advantages. In particular, in linear models the MLE method is computationally faster than the Basu method. This is due to the fact that the MLE method is given in closed form for such models as opposed to the Basu method which is not in closed form and as a result we rely on a numerical method to obtain the desired estimator. On the other hand the Basu estimating method as mentioned earlier has been proved to work better than the MLE in some contexts due to its robust features. Theoretically speaking both estimating methods result in equally good estimators since both satisfy the standard properties required by such estimators, namely the consistency and the asymptotic normality. The practical implications of these two forms of the DIC criterion become evident in Section 6.6 where simulations are performed.

6.5 Lower bound of the MSE of prediction of DIC

One of the main issues in model selection is the notion of asymptotic efficiency (Shibata, 1980, 1981). The asymptotic efficiency deals with the selection of a model with finitely many variables that provides the best possible approximation of the true model with infinitely many variables with respect to the mean squared error (MSE) of prediction. The issue of asymptotic efficiency is of great interest whenever the object of the analysis is a model selection that yields a good inference. Here we provide a lower bound for the mean squared error of prediction and in particular we show that the MSE of prediction of DIC is never below the so-called Average Mean Squared Error (Average MSE) of prediction. For the evaluation of the MSE the original set of n observations is used for the estimation of the parameters and the one-step-ahead prediction is used for measuring the accuracy of the selection. Following Shibata's assumption (Shibata, 1981) infinitely many independent variables are assumed so that the design matrix $\mathbf{X}$ is an $n \times \infty$ matrix.

Let $\mathbf{X}$ be the design matrix of the model

$$\mathbf{Y} = \mathbf{X}\beta + \varepsilon,$$

where $\beta = (\beta_0, \beta_1, \ldots)'$, the vector of unknown coefficients, $\varepsilon \sim \mathrm{N}(0, \sigma^2 I)$, the error sequence, and I the infinite-dimensional identity matrix.

Let

$$V(j) = \left\{ c(j) \text{, such that } c(j) = \left(c_0, 0, \ldots, c_{j_1}, 0, \ldots, c_{j_{k_j}}, 0, \ldots \right)' \right\}$$

be the subspace that contains the $k_j + 1$ parameters involved in the model and let

$$\beta^{(n)} = \left(\beta_0, 0, \ldots, \beta_{j_1}, 0, \ldots, \beta_{j_{k_j}}, 0, \ldots \right)'$$

be the projection of β on $V(j)$.

The prediction $\hat{\mathbf{Y}} = (\hat{Y}_1, \ldots, \hat{Y}_n)'$ is given by $\hat{\mathbf{Y}} = \mathbf{X_j}\hat{\beta}$, where the estimator of $\beta^{(\mathbf{n})}$ obtained through a set of observations $(X_{ij_1}, \ldots, X_{ij_{k_j}}, Y_i)$, $i = 1, 2, \ldots, n$ and for the model selected by DIC, is denoted by

$$\hat{\beta} = \left(\hat{\beta}_0, 0, \ldots, \hat{\beta}_{j_1}, 0, \ldots, \hat{\beta}_{j_2}, 0, \ldots, \hat{\beta}_{j_{k_j}}, 0, \ldots \right)'.$$

Observe that the design matrix $\mathbf{X_j}$ is an $n \times \infty$ matrix where only the columns $j_1, \ldots, j_{k_j}$ have entries different from zero.

The mean squared error of prediction (up to a constant) and the average MSE of prediction are defined, respectively, by

$$S_n(j) = \mathrm{E}\left[\left(\hat{\mathbf{Y}} - \mathbf{Y} \,|\mathbf{X_j} \right)' \left(\hat{\mathbf{Y}} - \mathbf{Y} \,|\mathbf{X_j} \right) \right] - n\sigma^2$$

and

$$L_n(j) \equiv \mathrm{E}\left(S_n(j) \right).$$

We now prove that the above two quantities take the form given in the following lemma.

It is not difficult to see that

$$E(\hat{\mathbf{Y}} - \mathbf{Y} \,|\mathbf{X_j})'(\hat{\mathbf{Y}} - \mathbf{Y} \,|\mathbf{X_j}) = E\left((\hat{\beta} - \beta)'\mathbf{X_j}'\mathbf{X_j}(\hat{\beta} - \beta) + \varepsilon'\varepsilon - 2\varepsilon'\mathbf{X_j}(\hat{\beta} - \beta)|\mathbf{X_j} \right)$$

so that under the notation and conditions of this section we have that

$$S_n(j) = \left\|\hat{\beta} - \beta\right\|^2_{\mathbf{M(j)}} \quad \text{and} \quad L_n(j) = \mathrm{E}\left\|\hat{\beta} - \beta\right\|^2_{\mathbf{M(j)}},$$

where $\mathbf{M}(\mathbf{j}) = \mathbf{X}_\mathbf{j}'\mathbf{X}_\mathbf{j}$ and $\|A\|^2_R = A'RA$.

The lemma below provides a lower bound for the MSE of prediction. In particular, we show that $S_n(j)$ is asymptotically never below the quantity

$$L_n(j^*) = \min_j L_n(j).$$

Let $L_n(j^*) = \min_j L_n(j)$. Assume also that for $0 < \delta < 1$,

$$\lim_{n\to\infty} \sum_j [(1 - \delta\omega_n(j)) \exp(\delta\omega_n(j))]^{(k_j+1)/2} = 0,$$

where
$$\omega_n(j) = \frac{L_n(j)}{(k_j+1)g(\alpha, k_j+1)\sigma^2}$$

and $g(a,m) = \left(1 + a^2/(1+2a)\right)^{((m/2)+1)}$. Then, for every $0 < \delta < 1$,

$$\lim_{n\to\infty} P\left[\frac{S_n(j)}{L_n(j^*)} > 1 - \delta\right] = 1.$$

Proof. For every $0 < \delta < 1$ and for every j and using the fact that

$$\|\hat{\beta} - \beta\|^2_{\mathbf{M(j)}} = \|\hat{\beta} - \beta^{(n)}\|^2_{\mathbf{M(j)}} + \|\beta^{(n)} - \beta\|^2_{\mathbf{M(j)}}$$

we have

$$\begin{aligned} P\left[\frac{S_n(j)}{L_n(j^*)} \le 1 - \delta\right] &\le P\left[\frac{S_n(j)}{L_n(j)} \le 1 - \delta\right] \le \sum_j P\left[\frac{\left\|\hat{\beta} - \beta\right\|^2_{\mathbf{M(j)}}}{L_n(j)} \le 1 - \delta\right] \\ &= \sum_j P\left[\frac{\left\|\hat{\beta} - \beta^{(n)}\right\|^2_{\mathbf{M(j)}} + \left\|\beta^{(n)} - \beta\right\|^2_{\mathbf{M(j)}}}{L_n(j)} \le 1 - \delta\right] \\ &= \sum_j P\left[\left\|\hat{\beta} - \beta^{(n)}\right\|^2_{\mathbf{M(j)}} \le (1-\delta) L_n(j) - \left\|\beta^{(n)} - \beta\right\|^2_{\mathbf{M(j)}}\right]. \end{aligned} \tag{6.17}$$

The limiting covariance matrix of $n^{1/2}\hat{\theta}$ is a multivariate normal random variable

$$N_p(\theta_0, g(\alpha, p)\Sigma),$$

where
$$g(\alpha, p) = \left(1 + \frac{\alpha^2}{1+2\alpha}\right)^{p/2+1}.$$

Then, in this case we have

$$\begin{aligned} \left\|\hat{\beta} - \beta^{(n)}\right\|^2_{\mathbf{M(j)}} &= \left(\hat{\beta} - \beta^{(n)}\right)' \left\{\sigma^2 g(\alpha, k_j+1)\mathbf{M}(\mathbf{j})\right\}^{-1} \left(\hat{\beta} - \beta^{(n)}\right) \sigma^2 g(\alpha, k_j+1) \\ &\sim \sigma^2 g(\alpha, k_j+1)\, \mathcal{X}^2_{k_j+1} \end{aligned} \tag{6.18}$$

and

$$L_n(j) = E\left\|\hat{\beta} - \beta\right\|^2_{\mathbf{M(j)}} = \left\|\beta - \beta^{(n)}\right\|^2_{\mathbf{M(j)}} + E\left\|\hat{\beta} - \beta^{(n)}\right\|^2_{\mathbf{M(j)}}$$
$$= \left\|\beta - \beta^{(n)}\right\|^2_{\mathbf{M(j)}} + (k_j + 1)\, g\,(\alpha, k_j + 1)\, \sigma^2,$$

where $\mathcal{X}_k^2$ is a chi-square distribution with k degrees of freedom. Using (6.18) we have that (6.17) is bounded by

$$\sum_j P\left[\mathcal{X}^2_{k_j+1} \leq (k_j + 1) - \delta(k_j + 1)\omega_n(j)\right]$$
$$\leq \sum_j \left[\exp\left(\delta\omega_n\left(j\right)\right)\left(1 - \delta\omega_n\left(j\right)\right)\right]^{(k_j+1)/2}$$

where the last inequality follows from the fact that for $k > \delta$ (see Shibata, 1981)

$$P\left[\mathcal{X}_k^2 \leq k - \delta\right] \leq \exp\!\left(\frac{\delta}{2}\right)\left(1 - k^{-1}\delta\right)^{(k/2)} \leq \exp\!\left(\frac{-\delta^2}{4k}\right). \tag{6.19}$$

The result follows immediately.

■

6.6 Simulations

In order to check the performance of the DIC criterion proposed in Section 6.4 we performed a simulation study using (a) the divergence information criterion based on the Basu method, (b) the corrected DIC_c based on the MLE method, (c) the Akaike information criterion, (d) the Bayesian information criterion, (e) the AIC for small sample sizes, and (f) the AIC with the estimator of the variance obtained by the minimization of the BHHJ measure.

The simulation study has the following characteristics. Fifty observations of 4 variables X_1, X_2, X_3, X_4 were independently generated from the normal distributions $N(0,3)$, $N(1,3)$, $N(2,3)$, and $N(3,3)$, correspondingly. Sample correlation coefficients between these variables were less than 0.13 (in absolute values) in all cases and the independence is verified since the test of independence gives a p-value <0.05 in all cases. The first two of these variables were planned to be used to generate values of Y_i, $i = 1, \ldots, 50$ using the following model specification,

$$Y_i = a_0 + a_1 X_{1,i} + a_2 X_{2,i} + \varepsilon_i$$

with $a_0 = a_1 = a_2 = 1$ and $\varepsilon_i \sim N(0,1)$. Due though to contamination of the above model by 10% from the model

$$Y_i = 1 + X_{1,i} + X_{2,i} + \varepsilon_i^*$$

with $\varepsilon_i^* \sim N(5,1)$ the simulated values were generated from the model

$$Y_i = 0.9(a_0 + a_1 X_{1,i} + a_2 X_{2,i} + \varepsilon_i) + 0.1\,(a_0 + a_1 X_{1,i} + a_2 X_{2,i} + \varepsilon_i^*)\,.$$

The reason for introducing contamination into the simulation study was to put into a test the robust features of the BHHJ measure. In other words, we wanted to force the DIC to perform to the fullest extent and activate its prime feature according to which when $a > 0$, observations significantly discrepant with respect to the model get an almost zero weight and therefore their contribution to the final selection is minimal.

With a set of four possible regressors there are $2^4 - 1 = 15$ possible specifications that include at least one regressor. These 15 possible regression specifications constitute the set of candidate models for the experiment. As a result the candidate set consists of the full model (X_1, X_2, X_3, X_4) given by

$$Y = b_0 + b_1 X_1 + b_2 X_2 + b_3 X_3 + b_4 X_4 + \varepsilon$$

as well as all 14 possible subsets of the full model consisting of one (X_{j_1}), two (X_{j_1}, X_{j_2}), and three $(X_{j_1}, X_{j_2}, X_{j_3})$, with $j_i \in \{1,2,3,4\}, i = 1,2,3$ of the four regressors X_1, X_2, X_3, X_4. Fifty such experiments were performed with the intention to select the best model among the available candidate models.

First we consider the standard AIC criterion given by

$$AIC = n \log \hat{\sigma}_p^2 + 2\,(p+2)\,,$$

where n is the sample size, p the number of variables of the model, and $\hat{\sigma}_p^2$ the estimate of the variance of the model with p variables.

We also consider the corrected AIC criterion introduced by Hurvich and Tsai (1989) and used in small sample situations. The corrected AIC is given by

$$AIC_c = n \log \hat{\sigma}_p^2 + \frac{n(n+p+1)}{n-p-3}.$$

Another variant of the AIC criterion used in the simulations is the one given by

$$AIC_a = n \log \hat{\sigma}_{p,a}^2 + 2\,(p+2)\,,$$

where $\hat{\sigma}_{p,a}^2$ is the estimator of the variance $\sigma_{p,a}^2$ of the model with p variables which is obtained by the minimization of the BHHJ measure. AIC_a is evaluated for $a = 0.01, 0.05$, and 0.10.

From the various Bayesian approaches we have chosen to include in the simulations the Bayesian information criterion (Schwarz, 1978) because of its consistency property. The BIC is given by

$$BIC = n \log \hat{\sigma_p}^2 + (p+2) \log n.$$

Finally the DIC is used with both corrected and uncorrected penalty terms and with both estimating methods, namely the Basu and the MLE methods. The original DIC (uncorrected) based on the Basu method (expression (6.14)), is used with index $a = 0.01, 0.05$, and 0.10 and the corrected DIC_c based on the MLE method (expression (6.16)), with $a = 0.01, 0.05, 0.10$, and 0.15. To make the notation precise we use in the sequel, DIC^{BHHJ} for the former and DIC_c^{MLE} for the latter case.

For each of the 50 experiments the value of each of the above model selection criteria was calculated for each of the 15 possible regression specifications under consideration.

As a result, for each of the 50 experiments and for each model selection criterion the 15 candidate models were ranked from 1st to 15th according to the value of the criterion. Recall that the model chosen by a criterion is the one for which the value of the criterion is the lowest among all 15 candidate models. Table 6.1 presents for each selection criterion, the proportion of times each candidate model has been selected by the criterion. Notice that only 4 of the 15 candidate models have been ranked 1st and therefore selected, namely the true model (X_1, X_2), and the "larger" models (X_1, X_2, X_3), (X_1, X_2, X_4), and (X_1, X_2, X_3, X_4). Obviously, all selections contain the correct variables of the model, namely X_1 and X_2.

Table 6.1. Proportion of the selected models by model selection criteria ($n = 50$)

Criteria		Variables	%	Variables	%	Variables	%
AIC	AIC	X_1, X_2	80	X_1, X_2, X_4	20		
	AIC_c	X_1, X_2	88	X_1, X_2, X_4	12		
BIC	BIC	X_1, X_2	96	X_1, X_2, X_4	4		
AIC_a	$AIC_{0.01}$	X_1, X_2	80	X_1, X_2, X_4	16	X_1, X_2, X_3	4
	$AIC_{0.05}$	X_1, X_2	76	X_1, X_2, X_4	16	X_1, X_2, X_3	8
	$AIC_{0.10}$	X_1, X_2	68	X_1, X_2, X_4	16	X_1, X_2, X_3 or X_1, X_2, X_3, X_4	16
DIC^{BHHJ}	$DIC_{0.01}$	X_1, X_2	80	X_1, X_2, X_4	20		
	$DIC_{0.05}$	X_1, X_2	76	X_1, X_2, X_4	20	X_1, X_2, X_3	4
	$DIC_{0.10}$	X_1, X_2	72	X_1, X_2, X_4	16	X_1, X_2, X_3 or X_1, X_2, X_3, X_4	12
DIC_c^{MLE}	$DIC_{0.01}$	X_1, X_2	80	X_1, X_2, X_4	20		
	$DIC_{0.05}$	X_1, X_2	80	X_1, X_2, X_4	20		
	$DIC_{0.10}$	X_1, X_2	88	X_1, X_2, X_4	12		
	$DIC_{0.15}$	X_1, X_2	96	X_1, X_2, X_4	4		

Observe that the DIC^{BHHJ} criterion has the same rate of success as the AIC criterion, namely 80%. The AIC_c has a higher success rate (88%) which could be attributed to the relative small sample size used ($n = 50$). The AIC criterion with index a has the smaller rate of success (less than 80%). In fact observe that the larger the value of the index a the worse the performance of the resulting criterion.

On the other hand both BIC and DIC_c^{MLE} with $a = 0.15$ have the best selection rate (96%) among all competing selection criteria. It should be noted that for DIC^{BHHJ} the selection rate improves as a tends to 0 while for DIC_c^{MLE} the rate improves as a increases up to a maximum value. This behavior is due to the different form of the correction term. Indeed, DIC^{BHHJ} decreases as a function of the index a while DIC_c^{MLE} is an increasing function of a. As a result and as a (and p) increases, the DIC_c^{MLE} criterion puts a heavier penalty in large models (in models where the dimension p of the parameter is large) and therefore for too large values of a (and p) we end up underestimating the true model. Recall that for a tending to zero the BHHJ measure on which the DIC is based reduces to the Kullback–Leibler measure on which the AIC criterion is based. As a result, it is natural that AIC, DIC^{BHHJ} for $a = 0.01$ and DIC_c^{MLE} for $a = 0.01$ produce similar results.

The performance of DIC_c^{MLE} seems to be superior to that of DIC^{BHHJ} not only because of its higher rate of success but also because it is based on the MLE method which is computationally faster than the Basu method since the former is provided in closed form while the latter relies on a numerical method for obtaining the required estimator.

In conclusion, the DIC^{BHHJ} expresses a good medium sample size performance which is comparable to the traditional AIC criterion while the DIC_c^{MLE} is very powerful and comparable to BIC.

References

Akaike, H. (1973). Information theory and an extension of the maximum likelihood principle. *Proc. of the* 2nd *Intern. Symposium on Information Theory* (Petrov, B. N. and Csaki, F., Eds.), Akademiai Kaid'o, Budapest.

Ali, S. M. and Silvey, S. D. (1966). A general class of coefficients of divergence of one distribution from another. *J. R. Statist. Soc. B*, **28**, 131–142.

Basu, A., Harris, I. R., Hjort, N. L., and Jones, M. C. (1998). Robust and efficient estimation by minimising a density power divergence. *Biometrika*, **85**, 549–559.

Bengtsson, T. and Cavanaugh. J. E. (2006). An improved Akaike information criterion for state-space model selection. *Comput. Statist. Data Anal.*, **50**, 2635–2654.

Cavanaugh, J. E. (2004). Criteria for linear model selection based on Kullback's symmetric divergence, *Austral. N. Zeal. J. Statist.*, **46**, 257–274.

Cressie, N. and Read, T. R. C. (1984). Multinomial goodness-of-fit tests. *J. R. Statist. Soc.*, **5**, 440–454.

Csiszar, I. (1963). Eine informationstheoretische ungleichung and ihre anwendung auf den beweis der ergodizitat von markoffischen ketten. *Magyar Tud. Akad. Mat. Kutato Int. Kozl.*, **8**, 85–108.

Hurvich, C. M. and Tsai, C. L. (1989). Regression and time series model selection in small samples. *Biometrika*, **76**, 297–307.

Jeffreys, H. (1946). An invariant form for the prior probability in estimation problems. *Proc. Roy. Soc. A*, **186**, 453–561.

Kagan, A. M. (1963). On the theory of Fisher's amount of information (in Russian). *Doklady Academii Nauk SSSR*, **151**, 277–278.

Konishi, S. and Kitagawa, G. (1996). Generalised information criteria in model selection. *Biometrika*, **83**, 875–890.

Koukouvinos, C., Mylona, K., and Vonta, F. (2008). A comparative study of variable selection procedures applied in high dimensional medical problems. *J. Appl. Prob. Statist.*, **3 (2)**, 195–209.

Kullback, S. and Leibler, R. (1951). On information and sufficiency. *Ann. Math. Statist.*, **22**, 79–86.

Mattheou, K., Lee, S., and Karagrigoriou, A. (2009). A model selection criterion based on the BHHJ measure of divergence. *J. Statist. Plan. Infer.* **139**, 128–135.

Renyi, A. (1961). On measures of entropy and information. *Proc.* 4th *Berkeley Symp. on Math. Statist. Prob.*, **1**, 547–561, University of California Press, Berkeley.

Schwarz, G. (1978). Estimating the dimension of a model. *Ann. Statist.*, **6**, 461–464.
Shang, J. (2008). Selection criteria based on Monte Carlo simulation and cross validation in mixed models. *Far East J. Theor. Statist.*, **25**, 51–72.
Shang, J. and Cavanaugh, J. E. (2008). Bootstrap variants of the Akaike information criterion for mixed model selection. *Comput. Statist. Data Anal.*, **52**, 2004–2021.
Spiegelhalter, D. J., Best, N. G., Carlin, B. P., and van der Linde, A. (2002). Bayesian measures of model complexity and fit. *J. R. Statist. Soc. B*, **64**, 583–639.
Shibata, R. (1980). Asymptotically efficient selection of the order of the model for estimating parameters of linear process. *Ann. Statist.*, **8**, 147–164.
Shibata, R. (1981). An optimal selection of regression variables. *Biometrika*, **68**, 45–54.
van der Linde, A. (2005). DIC in variable selection. *Statist. Neerlandica*, **59**, 45–56.
Wei, C. Z. (1992). On predictive least squares principles. *Ann. Statist.*, **20**, 1–42.

7

High Leverage Points and Outliers in Generalized Linear Models for Ordinal Data

M.C. Pardo

Department of Statistics and O.R (I), Complutense University of Madrid, Spain

Abstract: A generalized hat matrix based on ϕ-divergences is proposed to determine how much influence or leverage each data value can have on each fitted value to a generalized linear model for ordinal data. After studying for evidence of points where the data value has high leverage on the fitted value, if such influential points are present, we must still determine whether they have had any adverse effects on the fit. To evaluate it we propose a new family of residuals based on ϕ-divergences. All the diagnostic measures are illustrated through the analysis of real data.

Keywords and phrases: Generalized linear models, ordinal multinomial data, minimum ϕ-divergence estimation, hat matrix, leverage points, outliers

7.1 Introduction

Generalized linear models for ordinal multinomial data (Green, 1984; McCullagh, 1980; McCullagh and Nelder, 1989; Fahrmeir and Tutz, 2001; Liu and Agresti, 2005) are a powerful technique for relating a dependent ordered categorical variable to both categorical and continuous independent variables. In practice, however, the model-building process can be highly influenced by peculiarities in the data. For univariate generalized models, Cook (1986), Cook and Weisberg (1982), McCullagh and Nelder (1989), Thomas and Cook (1990), Pregibon (1981), and Williams (1987) have discussed diagnostic tools for detecting outliers and leverage points. Lesaffre and Albert (1989) have extended Pregibon's regression diagnostics to the case where several groups are envisaged. A wider extension was made by Fahrmeir and Tutz (2001) for multivariate extensions of generalized linear models.

There has been extensive development of diagnostic measures for models fitted by maximum likelihood. Bedrick and Tsai (1993) used the family of power divergence statistics (Cressie and Read, 1984) to construct a diagnostic for dose-response models. In this chapter we focus on ordinal multicategorical response variables and multinomial models. We show that maximum likelihood and deviance-based diagnostics for multivariate extensions of generalized linear models (Fahrmeir and Tutz, 2001) extend naturally to the ϕ-divergence family defined by Ali and Silvey (1966).

C.H. Skiadas (ed.), *Advances in Data Analysis*, Statistics for Industry and Technology, DOI 10.1007/978-0-8176-4799-5_7,

In Section 7.2, we introduce the generalized linear models for ordinal multinomial data (GLM), the relevant notation, and an alternative to the maximum likelihood which is the usual method of fitting these models. Measures for detecting leverage points are presented in Section 7.3, and residuals in Section 7.4, both based on the ϕ-divergence measure. For illustration, the diagnostics are applied to a dataset in Section 7.5.

7.2 Background and notation for GLM

Let $\boldsymbol{Y}$ be the response variable with J possible values, which for simplicity are labeled $1, ..., J$, which is observed together with m explanatory variables $\boldsymbol{x}^T = (x_1, ..., x_m) \in \mathbb{R}^m$. Given $\boldsymbol{x}$, $\boldsymbol{Y}$ is a multinomial random variable with probability vector $\pi^T = (\pi_1, ..., \pi_{J-1})$ and $\pi_r = P(\boldsymbol{Y} = r \mid \boldsymbol{x}^T)$, $r = 1, ..., J-1$.

Suppose that the $\boldsymbol{x}_i^T$ takes N different values,

$$\boldsymbol{x}_i^T = (x_{i1}, ..., x_{im}), \qquad i = 1, ..., N,$$

the multinomial generalized linear model supposes that $\mu_i = E\left[\boldsymbol{Y} \mid \boldsymbol{x}_i^T\right]$ is related to the linear predictor

$$\eta_i = \boldsymbol{Z}_i^T \beta$$

by

$$\mu_i = \mathbf{h}(\eta_i) = \mathbf{h}\left(\boldsymbol{Z}_i^T \beta\right), \qquad i = 1, ..., N,$$

where $\mathbf{h}$ is a vectorial response function, $\boldsymbol{Z}_i$ is a $p \times (J-1)$-design matrix obtained from $\boldsymbol{x}_i$ and β is a p-dimensional vector of unknown parameters.

Let $n(\boldsymbol{x}_i)$ be the number of observations considered when the explanatory variable $\boldsymbol{x}^T$ has the value $\boldsymbol{x}_i^T$, in such a way that if $\boldsymbol{x}^T$ is fixed at $\boldsymbol{x}_i^T$ we have a multinomial distribution with parameters $\left(n(\boldsymbol{x}_i); \pi_1\left(\boldsymbol{Z}_i^T\beta\right), ..., \pi_{J-1}\left(\boldsymbol{Z}_i^T\beta\right)\right)$.

Different models are obtained when we specify the response function and the design matrix such as cumulative models and sequential models among others (see Fahrmeir and Tutz, 2001).

Suppose we observe the sample $\boldsymbol{Y}_1 = \boldsymbol{y}_1, ..., \boldsymbol{Y}_N = \boldsymbol{y}_N$ jointly with the explanatory variables $\boldsymbol{x}_1, ..., \boldsymbol{x}_N$; the usual way to estimate the vector β of unknown parameters is using the maximum likelihood estimator (MLE). To obtain it, we maximize the log-likelihood function which is equivalent to minimizing the Kullback–Leibler divergence, $D_{Kullback}$, between

$$\widehat{\boldsymbol{p}} = \left(\frac{y_{11}}{n}, ..., \frac{y_{J1}}{n}, \frac{y_{12}}{n}, ..., \frac{y_{J2}}{n}, ..., \frac{y_{1N}}{n}, ..., \frac{y_{JN}}{n}\right)^T,$$

with y_{si} the number of observations that takes value s, $s = 1, ..., J$ given the explanatory variable $\boldsymbol{x}_i^T$, $i = 1, ..., N$, $n = n(\boldsymbol{x}_1) + \cdots + n(\boldsymbol{x}_N)$ and

$$\boldsymbol{p}(\beta) = \left(\frac{n(\boldsymbol{x}_1)}{n}\widetilde{\pi}\left(\boldsymbol{Z}_1^T\beta\right)^T, ..., \frac{n(\boldsymbol{x}_N)}{n}\widetilde{\pi}\left(\boldsymbol{Z}_N^T\beta\right)^T\right)^T$$

being $\widetilde{\pi}\left(\boldsymbol{Z}_i^T\beta\right)^T = \left(\pi_1\left(\boldsymbol{Z}_i^T\beta\right), ..., \pi_J\left(\boldsymbol{Z}_i^T\beta\right)\right)$. Therefore, the MLE, $\widehat{\beta}$, can be rewritten as

$$\widehat{\beta} = \arg\min_{\beta \in \boldsymbol{\Theta}} D_{Kullback}\left(\widehat{\mathbf{p}}, \mathbf{p}\left(\beta\right)\right) \tag{7.1}$$

with

$$\Theta = \left\{\beta^T = \left(\beta_1, ..., \beta_p\right) : \beta_s \in \mathbb{R},\ s = 1, .., p\right\}$$

and

$$D_{Kullback}\left(\widehat{\mathbf{p}}, \mathbf{p}\left(\beta\right)\right) = \sum_{l=1}^{J}\sum_{i=1}^{N} \frac{y_{li}}{n} \log \frac{\frac{y_{li}}{n}}{\pi_l\left(\boldsymbol{Z}_i^T\beta\right)\frac{n(\boldsymbol{x}_i)}{n}}.$$

A wide generalization of the Kullback–Leibler divergence measure is the ϕ-divergence measure defined by Ali and Silvey (1966),

$$D_\phi(\widehat{\mathbf{p}}, \mathbf{p}(\beta)) = \sum_{l=1}^{J}\sum_{i=1}^{N} \pi_l\left(\boldsymbol{Z}_i^T\beta\right)\frac{n(\boldsymbol{x}_i)}{n}\phi\left(\frac{y_{li}/n}{\pi_l\left(\boldsymbol{Z}_i^T\beta\right)n(\boldsymbol{x}_i)/n}\right) \tag{7.2}$$

where $\phi \in \Phi$ and Φ is the class of all convex functions $\phi\left(x\right)$, $x > 0$, such that at $x = 1, \phi\left(1\right) = \phi'\left(1\right) = 0, \phi''\left(1\right) > 0$, and at $x = 0$, $0\phi\left(0/0\right) = 0$ and $0\phi\left(p/0\right) = p\lim_{u\to\infty}\phi\left(u\right)/u$. For more details about ϕ-divergences see Vajda (1989) and Pardo (2006). As a particular case, the Kullback–Leibler divergence is obtained for $\phi\left(x\right) = x\log x - x + 1$.

Therefore, as a natural extension of the MLE we define now the minimum ϕ-divergence estimator replacing the Kullback–Leibler divergence in (7.1) by the ϕ-divergence measure; that is to say

$$\widehat{\beta}_\phi = \arg\min_{\beta \in \boldsymbol{\Theta}} D_\phi\left(\widehat{\mathbf{p}}, \mathbf{p}\left(\beta\right)\right). \tag{7.3}$$

Pardo (2008), under mild regularity conditions and assuming that $n\left(\boldsymbol{x}_i\right) \to \infty$, $i = 1, ..., N$ such that $n\left(\boldsymbol{x}_i\right)/n \to \lambda_i > 0$, $i = 1, ..., N$, proved that

$$\sqrt{n}\left(\widehat{\beta}_\phi - \beta^0\right) \underset{n\to\infty}{\overset{L}{\to}} \mathcal{N}\left(\mathbf{0}, \boldsymbol{I}_{F,\lambda}\left(\beta^0\right)^{-1}\right)$$

where β^0 is the true value of the parameter β, $\boldsymbol{I}_{F,\lambda}\left(\beta\right) = \lim_{n\to\infty}\boldsymbol{I}_{F,n}\left(\beta\right)$ being

$$\boldsymbol{I}_{F,n}\left(\beta\right) = \boldsymbol{Z}\mathbf{V}_n\left(\beta\right)\boldsymbol{Z}^T$$

where $\boldsymbol{Z} = \left(\boldsymbol{Z}_1, ..., \boldsymbol{Z}_N\right)$ and $\boldsymbol{V}_n\left(\beta\right) = \text{Diag}\left(\boldsymbol{V}_{n,1}\left(\beta\right), ..., \boldsymbol{V}_{n,N}\left(\beta\right)\right)$ being

$$\boldsymbol{V}_{n,i}\left(\beta\right) = \frac{n\left(\boldsymbol{x}_i\right)}{n}\frac{\partial\pi\left(\boldsymbol{Z}_i^T\beta\right)}{\partial\left(\boldsymbol{Z}_i^T\beta\right)}\mathbf{\Sigma}_i^{-1}\left(\beta\right)\frac{\partial\pi\left(\boldsymbol{Z}_i^T\beta\right)}{\partial\left(\boldsymbol{Z}_i^T\beta\right)}^T \tag{7.4}$$

and $\mathbf{\Sigma}_i^{-1}\left(\beta\right) = \left(v_{sr}\left(\beta\right)\right)_{s,r=1,...,J-1}$ with

$$v_{sr}\left(\beta\right) = \begin{cases} \dfrac{1}{\pi_r\left(\boldsymbol{Z}_i^T\beta\right)} + \dfrac{1}{\pi_J\left(\boldsymbol{Z}_i^T\beta\right)} & r = s \\ \dfrac{1}{\pi_J\left(\boldsymbol{Z}_i^T\beta\right)} & r \neq s \end{cases}.$$

7.3 The hat matrix: Properties

In what follows, we assume that we have fitted a GLM by minimum ϕ-divergence estimation. After fitting the model and prior to drawing inferences from it, it is very useful to determine how much leverage each datum can have. In this section we consider a generalized form of the hat matrix based on minimum ϕ-divergence estimation. The hat matrix yields a measure for the leverage of data.

Definition 1. *The GLM hat matrix, where the parameters are estimated by using the minimum ϕ_2-divergence estimator defined in* (7.3) *with $\phi = \phi_2 \in \Phi$, is given by*

$$\boldsymbol{H}\left(\widehat{\beta}_{\phi_2}\right) = \boldsymbol{V}_n\left(\widehat{\beta}_{\phi_2}\right)^{1/2} \boldsymbol{Z}^T \boldsymbol{I}_{F,n}\left(\widehat{\beta}_{\phi_2}\right)^{-1} \boldsymbol{Z}\boldsymbol{V}_n\left(\widehat{\beta}_{\phi_2}\right)^{1/2}.$$

The square matrices $\boldsymbol{H}$ and $\boldsymbol{M} = \boldsymbol{I} - \boldsymbol{H}$ are projection block matrices, where each block $\boldsymbol{H}^{ij}\left(\widehat{\beta}_{\phi_2}\right)$ and $\boldsymbol{M}^{ij}\left(\widehat{\beta}_{\phi_2}\right)$ $(i, j = 1, ..., N)$ is $(J-1)$-dimensional. Some properties of the GLM hat matrix are given in the following proposition.

Proposition 1. *It holds that*

(i)

$$0 \leq \operatorname{Det}\left(\boldsymbol{M}^{ii}\left(\widehat{\beta}_{\phi_2}\right)\right) < 1, \qquad i = 1, ..., N$$

(ii)

$$\operatorname{Trace}\left(\boldsymbol{H}\left(\widehat{\beta}_{\phi_2}\right)\right) = p$$

Proof. (i) We denote by $\boldsymbol{Z}_{(i)}$ the matrix,

$$\boldsymbol{Z}_{(i)}^T = \left(\boldsymbol{Z}_1^T, ..., \boldsymbol{Z}_{i-1}^T, \boldsymbol{Z}_{i+1}^T, ..., \boldsymbol{Z}_N^T\right)_{(J-1)(N-1)\times p},$$

and by $\widetilde{\boldsymbol{V}}^{(i)}(\beta)$ the matrix

$$\operatorname{Diag}\left(\widetilde{\boldsymbol{V}}_1(\beta), ..., \widetilde{\boldsymbol{V}}_{i-1}(\beta), \widetilde{\boldsymbol{V}}_{i+1}(\beta), ..., \widetilde{\boldsymbol{V}}_N(\beta)\right)$$

with

$$\widetilde{\boldsymbol{V}}_j(\beta) = n(\boldsymbol{x}_j) \frac{\partial \pi\left(\boldsymbol{Z}_j^T \beta\right)}{\partial\left(\boldsymbol{Z}_j^T \beta\right)} \boldsymbol{\Sigma}_j^{-1}(\beta) \frac{\partial \pi\left(\boldsymbol{Z}_j^T \beta\right)}{\partial\left(\boldsymbol{Z}_j^T \beta\right)^T}.$$

The dimension of $\widetilde{\boldsymbol{V}}^{(i)}(\beta)$ is $(N-1)(J-1) \times (N-1)(J-1)$. Then $\boldsymbol{Z}_{(i)}$ and $\widetilde{\boldsymbol{V}}^{(i)}\left(\widehat{\beta}_{\phi_2}\right)$ are the matrices obtained from $\boldsymbol{Z}$ and

$$\widetilde{\boldsymbol{V}}\left(\widehat{\beta}_{\phi_2}\right) = \operatorname{Diag}\left(\widetilde{\boldsymbol{V}}_1\left(\widehat{\beta}_{\phi_2}\right), ..., \widetilde{\boldsymbol{V}}_N\left(\widehat{\beta}_{\phi_2}\right)\right)$$

by deleting the parts $\boldsymbol{Z}_i$ and $\widetilde{\boldsymbol{V}}_i\left(\widehat{\beta}_{\phi_2}\right)$ corresponding to the ith observation. Since

$$\boldsymbol{Z}_{(i)}\widetilde{\boldsymbol{V}}^{(i)}\left(\widehat{\beta}_{\phi_2}\right)\boldsymbol{Z}_{(i)}^T=\sum_{\substack{j=1\\j\neq i}}^{N}\boldsymbol{Z}_j\widetilde{\boldsymbol{V}}_j\left(\widehat{\beta}_{\phi_2}\right)\boldsymbol{Z}_j^T$$
$$=\boldsymbol{Z}\widetilde{\boldsymbol{V}}\left(\widehat{\beta}^{\phi_2}\right)\boldsymbol{Z}^T-\boldsymbol{Z}_i\widetilde{\boldsymbol{V}}_i\left(\widehat{\beta}_{\phi_2}\right)\boldsymbol{Z}_i^T$$

then

$$\begin{aligned}\operatorname{Det}\left(\boldsymbol{Z}_{(i)}\widetilde{\boldsymbol{V}}^{(i)}\left(\widehat{\beta}_{\phi_2}\right)\boldsymbol{Z}_{(i)}^T\right)&=\operatorname{Det}\left(\boldsymbol{Z}\widetilde{\boldsymbol{V}}\left(\widehat{\beta}_{\phi_2}\right)\boldsymbol{Z}^T\right)\\&\quad\times\operatorname{Det}\left(\boldsymbol{I}-\left(\boldsymbol{Z}\widetilde{\boldsymbol{V}}\left(\widehat{\beta}_{\phi_2}\right)\boldsymbol{Z}^T\right)^{-1}\left(\boldsymbol{Z}_i\widetilde{\boldsymbol{V}}_i\left(\widehat{\beta}_{\phi_2}\right)\boldsymbol{Z}_i^T\right)\right)\\&=\operatorname{Det}\left(\boldsymbol{Z}\widetilde{\boldsymbol{V}}\left(\widehat{\beta}_{\phi_2}\right)\boldsymbol{Z}^T\right)Det\left(\boldsymbol{I}-\widetilde{\boldsymbol{V}}_i\left(\widehat{\beta}_{\phi_2}\right)^{1/2}\boldsymbol{Z}_i^T\right.\\&\quad\left.\times\left(\boldsymbol{Z}\widetilde{\boldsymbol{V}}\left(\widehat{\beta}_{\phi_2}\right)\boldsymbol{Z}^T\right)^{-1}\boldsymbol{Z}_i\widetilde{\boldsymbol{V}}_i\left(\widehat{\beta}_{\phi_2}\right)^{1/2}\right)\\&=\operatorname{Det}\left(\boldsymbol{Z}\widetilde{\boldsymbol{V}}\left(\widehat{\beta}_{\phi_2}\right)\boldsymbol{Z}^T\right)Det\left(\boldsymbol{I}-\boldsymbol{H}^{ii}\left(\widehat{\beta}_{\phi_2}\right)\right)\\&=\operatorname{Det}\left(\boldsymbol{Z}\widetilde{\boldsymbol{V}}\left(\widehat{\beta}_{\phi_2}\right)\boldsymbol{Z}^T\right)Det\left(\boldsymbol{M}^{ii}\left(\widehat{\beta}_{\phi_2}\right)\right).\end{aligned}$$

Therefore,

$$\frac{\operatorname{Det}\left(\boldsymbol{Z}_{(i)}\widetilde{\boldsymbol{V}}^{(i)}\left(\widehat{\beta}_{\phi_2}\right)\boldsymbol{Z}_{(i)}^T\right)}{\operatorname{Det}\left(\boldsymbol{Z}\widetilde{\boldsymbol{V}}\left(\widehat{\beta}_{\phi_2}\right)\boldsymbol{Z}^T\right)}=\left(\operatorname{Det}\left(\boldsymbol{M}^{ii}\left(\widehat{\beta}_{\phi_2}\right)\right)_{(J-1)\times(J-1)}\right),\qquad i=1,...,N.$$

The matrix $\boldsymbol{M}^{ii}\left(\widehat{\beta}_{\phi_2}\right)$ is idempotent and hence its eigenvalues are either 0 or 1 (Rao, 1973, p. 72). But this matrix is also symmetric and

$$\operatorname{Diag}(d_1,...,d_{J-1})=\boldsymbol{Q}^T\boldsymbol{M}^{ii}\left(\widehat{\beta}_{\phi_2}\right)\boldsymbol{Q}$$

(Harville, 1997, p. 535), where $\boldsymbol{Q}$ is orthogonal and $d_1,...,d_{J-1}$ are the eigenvalues of $\boldsymbol{M}^{ii}\left(\widehat{\beta}_{\phi_2}\right)$. Then

$$0\leq\operatorname{Det}(\operatorname{Diag}(d_1,...,d_J))=\operatorname{Det}\left(\boldsymbol{Q}^T\boldsymbol{M}^{ii}\left(\widehat{\beta}_{\phi_2}\right)\boldsymbol{Q}\right)=\operatorname{Det}\left(\boldsymbol{M}^{ii}\left(\widehat{\beta}_{\phi_2}\right)\right).$$

The matrix $\boldsymbol{M}^{ii}\left(\widehat{\beta}_{\phi_2}\right)=\left(\boldsymbol{m}_{rs}\left(\widehat{\beta}_{\phi_2}\right)\right)_{r,s=1,\ldots,J-1}$ is idempotent and symmetric, such that

$$\boldsymbol{m}_{jj}\left(\widehat{\beta}_{\phi_2}\right)\left(1-\boldsymbol{m}_{rs}\left(\widehat{\beta}_{\phi_2}\right)\right)=\sum_{i\neq j}\boldsymbol{m}_{ij}\left(\widehat{\beta}_{\phi_2}\right)^2$$

and $0<\boldsymbol{m}_{jj}\left(\widehat{\beta}_{\phi_2}\right)\leq 1$. Then (Rao (1973), page 74) we have

$$\operatorname{Det}\left(\boldsymbol{M}^{ii}\left(\widehat{\beta}_{\phi_2}\right)\right)\leq\prod_{i=1}^{J-1}m_{ii}<1.$$

(ii) The matrix $\boldsymbol{H}\left(\widehat{\beta}_{\phi_2}\right)$ is idempotent since

$$\boldsymbol{Z}\boldsymbol{V}\left(\widehat{\beta}_{\phi_2}\right)^{1/2}\boldsymbol{V}\left(\widehat{\beta}_{\phi_2}\right)^{1/2}\boldsymbol{Z}^T$$

coincides with $\boldsymbol{I}_{F,n}\left(\widehat{\beta}_{\phi_2}\right)$. From a result in Rao (1973, p. 28)

$$\text{Trace}\Big(\boldsymbol{H}\left(\widehat{\beta}_{\phi_2}\right)\Big)=\sum_{i=1}^{N}\text{Trace}\Big(\boldsymbol{H}^{ii}\left(\widehat{\beta}_{\phi_2}\right)\Big)=\text{range}\Big(\boldsymbol{H}\left(\widehat{\beta}_{\phi_2}\right)\Big)=p.$$

The diagonal elements $\boldsymbol{M}^{ii}\left(\widehat{\beta}_{\phi_2}\right)$ can be used to find leverage points. A leverage point is characterized by the fact to increase significantly the variability of the estimations when it is removed from the sample. The variability of $\widehat{\beta}_{\phi_2}$ is given by the volume of the asymptotic confidence ellipsoid for β^0 which (see Cramér, 1946, Section 11.2) is proportional to

$$\left(Det\left(\boldsymbol{Z}\widetilde{\boldsymbol{V}}\left(\widehat{\beta}_{\phi_2}\right)\boldsymbol{Z}^T\right)^{-1}\right)^{1/2}.$$

If the explicative variable $\boldsymbol{x}_i$ is deleted, then the volume of the confidence ellipsoid is proportional to

$$\left(\text{Det}\left(\boldsymbol{Z}_{(i)}\widetilde{\boldsymbol{V}}^{(i)}\left(\widehat{\beta}_{\phi_2}\right)\boldsymbol{Z}_{(i)}^T\right)^{-1}\right)^{1/2}$$

where the subscript (i) indicates that the $\boldsymbol{x}_i$ contribution to the corresponding matrix has been removed. A point with a value near zero of

$$\text{Det}\Big(\boldsymbol{M}^{ii}\left(\widehat{\beta}_{\phi_2}\right)\Big)=\frac{\text{Det}\Big(\boldsymbol{Z}_{(i)}\widetilde{\boldsymbol{V}}^{(i)}\left(\widehat{\beta}_{\phi_2}\right)\boldsymbol{Z}_{(i)}^T\Big)}{\text{Det}\Big(\boldsymbol{Z}\widetilde{\boldsymbol{V}}\left(\widehat{\beta}_{\phi_2}\right)\boldsymbol{Z}^T\Big)}$$

indicates that the deletion of $\boldsymbol{x}_i$ substantially increases the volume. Therefore, a point with a value of

$$\text{Det}\Big(\boldsymbol{M}^{ii}\left(\widehat{\beta}_{\phi_2}\right)\Big)$$

near 0 has a stabilizing effect on the estimated coefficients and it is defined as a leverage point for the GLM.

On the other hand, taking into account the second property, the average size of an element of the diagonal of the hat matrix will be p/N. A reasonable rule of thumb for detecting leverage points is to consider that $\boldsymbol{x}_i$ is a leverage point if

$$\text{Trace}\Big(\boldsymbol{H}^{ii}\left(\widehat{\beta}_{\phi_2}\right)\Big)>2p/N.$$

■

7.4 Outliers

If leverage points are present, we must still determine whether they have any adverse effects on the fit, that it is to say, whether they are outliers. In that case, such a leverage point will be influential. The generalized residuals frequently used are the Pearson residuals defined by Jorgensen (1983) as $\boldsymbol{R}^T = (\boldsymbol{r}_1^T, ..., \boldsymbol{r}_N^T)$ with

$$\boldsymbol{r}_i = n(\boldsymbol{x}_i)^{-1/2} \boldsymbol{\Sigma}_i^{-1/2} \left(\widehat{\beta}\right) \left(\boldsymbol{y}_i - n(\boldsymbol{x}_i)\pi\left(\boldsymbol{Z}_i^T \widehat{\beta}\right)\right)$$

with $\boldsymbol{y}_i = (y_{1i}, ..., y_{J-1i})^T$. A generalization of $\boldsymbol{r}_i$ is obtained substituting the MLE by the minimum ϕ_2-divergence estimator

$$\boldsymbol{r}_i^{\phi_2} = n(\boldsymbol{x}_i)^{-1/2} \boldsymbol{\Sigma}_i^{-1/2} \left(\widehat{\beta}_{\phi_2}\right) \boldsymbol{e}_i^{\phi_2}, \qquad i = 1, ..., N$$

with

$$\boldsymbol{e}_i^{\phi_2} = \left(\boldsymbol{y}_i - n(\boldsymbol{x}_i)\pi\left(\boldsymbol{Z}_i^T \widehat{\beta}_{\phi_2}\right)\right), \qquad i = 1, ..., N.$$

If we denote by

$$\boldsymbol{S}_i\left(\widehat{\beta}_{\phi_2}\right) = \left(n(\boldsymbol{x}_i)\boldsymbol{\Sigma}_i\left(\beta^0\right)\right)^{-1/2} \boldsymbol{e}_i^{\phi_2},$$

where $\boldsymbol{\Sigma}_i(\beta) = (\sigma_{sr}(\beta))_{s,r=1,...,J-1}$ with

$$\sigma_{sr}(\beta) = \begin{cases} \pi_r\left(\boldsymbol{Z}_i^T\beta\right)\left(1 - \pi_r\left(\boldsymbol{Z}_i^T\beta\right)\right) & r = s \\ -\pi_r\left(\boldsymbol{Z}_i^T\beta\right)\pi_s\left(\boldsymbol{Z}_i^T\beta\right) & r \neq s \end{cases} \tag{7.5}$$

being $n(\boldsymbol{x}_i)\boldsymbol{\Sigma}_i(\beta)$ the covariance matrix of $\boldsymbol{Y}_i$ and

$$\boldsymbol{e}_i^{\phi_2} = \boldsymbol{I}_{(J)}\widetilde{\boldsymbol{e}}_i^{\phi_2},$$

being

$$\widetilde{\boldsymbol{e}}_i^{\phi_2} = \left(\widetilde{\boldsymbol{y}}_i - n(\boldsymbol{x}_i)\widetilde{\pi}\left(\boldsymbol{Z}_i^T\widehat{\beta}_{\phi_2}\right)\right)$$

with $\widetilde{\boldsymbol{y}}_i = (y_{1i}, ..., y_{Ji})^T$ and $\boldsymbol{I}_{(J)}$ the matrix obtained from the identity matrix $\boldsymbol{I}_{J\times J}$ by deleting the last row, then we have

$$\begin{aligned}
\text{Cov}\left(\boldsymbol{S}_i\left(\widehat{\beta}_{\phi_2}\right)\right) &= \text{Cov}\left(n(\boldsymbol{x}_i)^{-1/2}\boldsymbol{\Sigma}_i\left(\beta^0\right)^{-1/2} \boldsymbol{I}_{(J)}\,\text{Diag}\left(\widetilde{\pi}\left(\boldsymbol{Z}_i^T\beta^0\right)^{1/2}\right)\right. \\
&\quad \left.\times\ \text{Diag}\left(\widetilde{\pi}\left(\boldsymbol{Z}_i^T\beta^0\right)^{-1/2}\right)\widetilde{\boldsymbol{e}}_i^{\phi_2}\right) \\
&= \text{Cov}\left(\boldsymbol{\Sigma}_i\left(\beta^0\right)^{-1/2} \boldsymbol{I}_{(J)}\,\text{Diag}\left(\widetilde{\pi}\left(\boldsymbol{Z}_i^T\beta^0\right)^{1/2}\right)\widetilde{\boldsymbol{r}}_i^{\phi_2}\right) \\
&\approx \boldsymbol{\Sigma}_i\left(\beta^0\right)^{-1/2} \boldsymbol{I}_{(J)}\,\text{Diag}\left(\widetilde{\pi}\left(\boldsymbol{Z}_i^T\beta^0\right)^{1/2}\right)\left(\text{Diag}\left(\widetilde{\pi}\left(\boldsymbol{Z}_i^T\beta^0\right)^{-1/2}\right)\right. \\
&\quad \times\ \boldsymbol{\Sigma}_{\widetilde{\pi}\left(\boldsymbol{Z}_i^T\beta^0\right)}\text{Diag}\left(\widetilde{\pi}\left(\boldsymbol{Z}_i^T\beta^0\right)^{-1/2}\right) - \left(\boldsymbol{C}_{\lambda,i}^0\right)^T \boldsymbol{Z}_i^T \boldsymbol{I}_{F,\lambda}\left(\beta^0\right)^{-1} \\
&\quad \times\ \left.\boldsymbol{Z}_i\boldsymbol{C}_{\lambda,i}^0\right)\text{Diag}\left(\widetilde{\pi}\left(\boldsymbol{Z}_i^T\beta^0\right)^{1/2}\right)\boldsymbol{I}_{(J)}^T\,\boldsymbol{\Sigma}_i\left(\beta^0\right)^{-1/2}
\end{aligned}$$

$$= \boldsymbol{\Sigma}_i\left(\beta^0\right)^{-1/2} \underbrace{\boldsymbol{I}_{(J)}\, \boldsymbol{\Sigma}_{\widetilde{\pi}\left(\boldsymbol{Z}_i^T\beta^0\right)} \boldsymbol{I}_{(J)}^T}_{\boldsymbol{\Sigma}_i(\beta^0)} \boldsymbol{\Sigma}_i\left(\beta^0\right)^{-1/2}$$

$$-\boldsymbol{\Sigma}_i\left(\beta^0\right)^{-1/2} \underbrace{\boldsymbol{I}_{(J)} \operatorname{Diag}\left(\widetilde{\pi}\left(\boldsymbol{Z}_i^T\beta^0\right)^{1/2}\right)\left(\boldsymbol{C}_{\lambda,i}^0\right)^T}_{\left(\lambda_i^{1/2}\right)\frac{\partial\pi\left(\boldsymbol{Z}_i^T\beta\right)}{\partial\left(\boldsymbol{Z}_i^T\beta\right)^T}} \boldsymbol{Z}_i^T \boldsymbol{I}_{F,\lambda}\left(\beta^0\right)^{-1} \boldsymbol{Z}_i \times$$

$$\underbrace{\boldsymbol{C}_{\lambda,i}^0 \operatorname{Diag}\left(\widetilde{\pi}\left(\boldsymbol{Z}_i^T\beta^0\right)^{1/2}\right) \boldsymbol{I}_{(J)}^T}_{\left(\lambda_i^{1/2}\right)\frac{\partial\pi\left(\boldsymbol{Z}_i^T\beta\right)}{\partial\left(\boldsymbol{Z}_i^T\beta\right)}} \boldsymbol{\Sigma}_i\left(\beta^0\right)^{-1/2}$$

$$= \boldsymbol{I}_{(J-1)\times(J-1)} - \boldsymbol{V}_{\lambda,i}\left(\beta^0\right)^{1/2} \boldsymbol{Z}_i^T \boldsymbol{I}_{F,\lambda}\left(\beta^0\right)^{-1} \boldsymbol{Z}_i \boldsymbol{V}_{\lambda,i}\left(\beta^0\right)^{1/2}$$

where $\boldsymbol{C}_{\lambda,i}^0 = \lim_{n\to\infty} \boldsymbol{C}_{n,i}^0$ with

$$\boldsymbol{C}_{n,i}^0 = \left(\boldsymbol{C}_{n,i}\right)_{\beta=\beta^0} = \left[\left(\frac{n\left(\boldsymbol{x}_i\right)}{n}\right)^{1/2} \frac{\partial\widetilde{\pi}\left(\boldsymbol{Z}_i^T\beta\right)}{\partial\left(\boldsymbol{Z}_i^T\beta\right)^T} \operatorname{Diag}\left(\widetilde{\pi}\left(\boldsymbol{Z}_i^T\beta\right)^{-1/2}\right)\right]_{\beta=\beta^0}, \quad i = 1, \ldots, N,$$

$$\boldsymbol{\Sigma}_{\widetilde{\pi}\left(\boldsymbol{Z}_i^T\beta^0\right)} = \left(\pi_s(\boldsymbol{Z}_i^T\beta^0)\left(\delta_{st} - \pi_t\left(\boldsymbol{Z}_i^T\beta^0\right)\right)\right)_{s,t=1,\ldots,J}, \qquad i = 1, \ldots, N$$

and $\boldsymbol{V}_{\lambda,i}\left(\beta^0\right) = \lim_{n\to\infty} \boldsymbol{V}_{n,i}\left(\beta^0\right)$ with $\boldsymbol{V}_{n,i}$ given in (7.4).

Therefore, the residuals $\boldsymbol{r}_i^{\phi_2}$ can be standardized

$$\boldsymbol{r}_{i,S}^{\phi_2} = \boldsymbol{M}^{ii}\left(\widehat{\beta}_{\phi_2}\right)^{-1/2} \boldsymbol{r}_i^{\phi_2}.$$

However, the calculation of the square root of the matrix $\boldsymbol{\Sigma}_i^{-1}\left(\widehat{\beta}\right)$ can be avoided redefining $\boldsymbol{r}_i$ as

$$\widetilde{\boldsymbol{r}}_i = n\left(\boldsymbol{x}_i\right)^{-1/2} \widetilde{\boldsymbol{\Sigma}}_i^{-1/2}\left(\widehat{\beta}\right)\left(\widetilde{\boldsymbol{y}}_i - n\left(\boldsymbol{x}_i\right)\widetilde{\pi}\left(\boldsymbol{Z}_i^T\widehat{\beta}\right)\right)$$

where $\widetilde{\boldsymbol{\Sigma}}_i\left(\beta\right) = \left(\sigma_{sr}\left(\beta\right)\right)_{s,r=1,\ldots,J}$ with $\sigma_{sr}\left(\beta\right)$ defined in (7.5). A generalized inverse of $\widetilde{\boldsymbol{\Sigma}}_i\left(\beta\right)$ is given by

$$\operatorname{Diag}\left(\widetilde{\pi}\left(\boldsymbol{Z}_i^T\beta\right)^{-1}\right).$$

Therefore, a generalization of $\widetilde{\boldsymbol{r}}_i$ is obtained substituting the MLE by the minimum ϕ_2-divergence estimator

$$\widetilde{\boldsymbol{r}}_i^{\phi_2} = \operatorname{Diag}\left(\left(n\left(\boldsymbol{x}_i\right)\widetilde{\pi}\left(\boldsymbol{Z}_i^T\widehat{\beta}_{\phi_2}\right)\right)^{-1/2}\right)\widetilde{\boldsymbol{e}}_i^{\phi_2}.$$

Finally, a further generalization of residuals follows.

Definition 2. *The (ϕ_1, ϕ_2)-divergence residuals with ϕ_1, $\phi_2 \in \Phi$ are defined as the J-dimensional vector $\widetilde{\boldsymbol{r}}_i^{\phi_1,\phi_2}$ with sth coordinate*

$$\sqrt{\frac{2n(\boldsymbol{x}_i)}{\phi_1''(1)}}\operatorname{sign}\left(y_{si} - n(\boldsymbol{x}_i)\pi_s\left(\boldsymbol{Z}_i^T\widehat{\beta}_{\phi_2}\right)\right)\left(\pi_s\left(\boldsymbol{Z}_i^T\widehat{\beta}_{\phi_2}\right)\phi_1\left(\frac{y_{si}}{n(\boldsymbol{x}_i)\pi_s\left(\boldsymbol{Z}_i^T\widehat{\beta}_{\phi_2}\right)}\right)\right)^{1/2}.$$

Denoting by

$$\widetilde{\boldsymbol{R}}^{\phi_2} = \left(\left(\widetilde{\boldsymbol{r}}_1^{\phi_2}\right)^T, ..., \left(\widetilde{\boldsymbol{r}}_N^{\phi_2}\right)^T\right)^T$$

and

$$\widetilde{\boldsymbol{R}}^{\phi_1,\phi_2} = \left(\left(\widetilde{\boldsymbol{r}}_1^{\phi_1,\phi_2}\right)^T, ..., \left(\widetilde{\boldsymbol{r}}_N^{\phi_1,\phi_2}\right)^T\right)^T,$$

for $\phi_1(x) = \frac{1}{2}(x-1)^2$ and $\phi_2(x) = x\log x - x + 1$ we obtain $\widetilde{\boldsymbol{R}} = (\widetilde{\boldsymbol{r}}_1, ..., \widetilde{\boldsymbol{r}}_N)$.

The asymptotic distribution of $\widetilde{\boldsymbol{R}}^{\phi_2}$ and $\widetilde{\boldsymbol{R}}^{\phi_1,\phi_2}$ is given below.

Theorem 1. *The asymptotic distribution of $\widetilde{\boldsymbol{R}}^{\phi_2}$ and $\widetilde{\boldsymbol{R}}^{\phi_1,\phi_2}$ is normal with mean vector $\boldsymbol{0}$ and covariance matrix*

$$\operatorname{Diag}\left(\boldsymbol{p}_\lambda(\beta^0)^{-1/2}\right)\boldsymbol{\Sigma}_{\boldsymbol{p}_\lambda(\beta^0)}\operatorname{Diag}\left(\boldsymbol{p}_\lambda(\beta^0)^{-1/2}\right) - \boldsymbol{J}(\beta^0)$$

with $\boldsymbol{p}_\lambda(\beta) = \lim_{n\to\infty}\boldsymbol{p}(\beta)$ and

$$\boldsymbol{J}(\beta^0) = \operatorname{Diag}\left(\left(\boldsymbol{C}_{\lambda,i}^0\right)_{i=1,...,N}^T\right)\boldsymbol{Z}^T\boldsymbol{I}_{F,\lambda}(\beta^0)^{-1}\boldsymbol{Z}\operatorname{Diag}\left(\left(\boldsymbol{C}_{\lambda,i}^0\right)_{i=1,...,N}\right)$$

Proof. The asymptotic distribution of

$$\widetilde{\boldsymbol{R}}^{\phi_2} = \left(\left(\widetilde{\boldsymbol{r}}_1^{\phi_2}\right)^T, ..., \left(\widetilde{\boldsymbol{r}}_N^{\phi_2}\right)^T\right)^T$$

coincides with the asymptotic distribution of

$$\operatorname{Diag}\left(\boldsymbol{p}(\beta^0)^{-1/2}\right)\sqrt{n}\left(\widehat{\boldsymbol{p}} - \boldsymbol{p}\left(\widehat{\beta}_{\phi_2}\right)\right)$$

that can be easily obtained. Therefore

$$\widetilde{\boldsymbol{R}}^{\phi_2} \underset{n\to\infty}{\overset{L}{\to}} \mathcal{N}\left(\boldsymbol{0}_{JN\times 1}, \boldsymbol{\Sigma}_5(\beta^0)\right),$$

where

$$\begin{aligned}\boldsymbol{\Sigma}_5(\beta^0) = \boldsymbol{I} &- \operatorname{Diag}\left(\boldsymbol{p}_\lambda(\beta^0)^{1/2}\right)\boldsymbol{X}(\beta^0)\operatorname{Diag}\left(\boldsymbol{p}_\lambda(\beta^0)^{-1/2}\right)\\ &- \operatorname{Diag}\left(\boldsymbol{p}_\lambda(\beta^0)^{1/2}\right)\boldsymbol{L}_\lambda(\beta^0)^T\operatorname{Diag}\left(\boldsymbol{p}_\lambda(\beta^0)^{-1/2}\right)\end{aligned}$$

being

$$\boldsymbol{X}(\beta^0) = \boldsymbol{X}_0\left(\boldsymbol{X}_0^T\operatorname{Diag}(\boldsymbol{p}_\lambda(\beta^0))\boldsymbol{X}_0\right)^{-1}\boldsymbol{X}_0^T\operatorname{Diag}(\boldsymbol{p}_\lambda(\beta^0))$$

with

$$\boldsymbol{X}_0 = \begin{pmatrix} \mathbf{1}_{\boldsymbol{J}} & \mathbf{0} & \cdots & \mathbf{0} \\ \mathbf{0} & \mathbf{1}_{\boldsymbol{J}} & \cdots & \mathbf{0} \\ . & . & \cdots & . \\ \mathbf{0} & \mathbf{0} & \cdots & \mathbf{1}_{\boldsymbol{J}} \end{pmatrix}_{JN \times N}, \tag{7.6}$$

being $\mathbf{1}_{\boldsymbol{J}}$ the unit vector $J \times 1$-dimensional and $\boldsymbol{L}_\lambda\left(\beta^0\right) = \lim_{n\to\infty} \boldsymbol{L}_n\left(\beta^0\right)$ with

$$\begin{aligned} \boldsymbol{L}_n\left(\beta^0\right) = \boldsymbol{S}_n\left(\beta^0\right) \boldsymbol{Z}^T \boldsymbol{I}_{F,n}\left(\beta^0\right)^{-1} \boldsymbol{Z} \operatorname{Diag}\Big(\left(\boldsymbol{C}_{n,i}^0\right)_{i=1,\ldots,N}\Big) \\ \times \operatorname{Diag}\Big(\boldsymbol{p}\left(\beta^0\right)^{-1/2}\Big). \end{aligned}$$

Taking into account

$$\begin{aligned} &\boldsymbol{I} - \operatorname{Diag}(\boldsymbol{p}_\lambda(\beta^0)^{1/2})\, \boldsymbol{X}\left(\beta^0\right) \operatorname{Diag}(\boldsymbol{p}_\lambda(\beta^0)^{-1/2}) \\ &= \operatorname{Diag}\Big(\boldsymbol{p}_\lambda\left(\beta^0\right)^{1/2}\Big) \left(\boldsymbol{I} - \boldsymbol{X}\left(\beta^0\right)\right) \operatorname{Diag}\Big(\boldsymbol{p}_\lambda\left(\beta^0\right)^{-1/2}\Big) \\ &= \operatorname{Diag}\Big(\boldsymbol{p}_\lambda\left(\beta^0\right)^{-1/2}\Big) \operatorname{Diag}\left(\boldsymbol{p}_\lambda\left(\beta^0\right)\right) \left(\boldsymbol{I} - \boldsymbol{X}\left(\beta^0\right)\right) \operatorname{Diag}\Big(\boldsymbol{p}_\lambda\left(\beta^0\right)^{-1/2}\Big) \\ &= \operatorname{Diag}\Big(\boldsymbol{p}_\lambda\left(\beta^0\right)^{-1/2}\Big)\, \boldsymbol{\Sigma}_{\boldsymbol{p}_\lambda(\beta^0)} \operatorname{Diag}\Big(\boldsymbol{p}_\lambda\left(\beta^0\right)^{-1/2}\Big) \end{aligned}$$

so

$$\boldsymbol{\Sigma}_5\left(\beta^0\right) = \operatorname{Diag}\Big(\boldsymbol{p}_\lambda\left(\beta^0\right)^{-1/2}\Big)\, \boldsymbol{\Sigma}_{\boldsymbol{p}_\lambda(\beta^0)} \operatorname{Diag}\Big(\boldsymbol{p}_\lambda\left(\beta^0\right)^{-1/2}\Big) - \boldsymbol{J}\left(\beta^0\right)$$

with $\boldsymbol{J}\left(\beta^0\right) = \operatorname{Diag}(\boldsymbol{p}_\lambda(\beta^0)^{1/2})\, \boldsymbol{L}_\lambda\left(\beta^0\right)^T \operatorname{Diag}\left(\boldsymbol{p}_\lambda(\beta^0)^{-1/2}\right)$ or equivalently by the definition of $\boldsymbol{L}_\lambda\left(\beta^0\right)$ and the relation

$$\left(\frac{\partial \boldsymbol{p}}{\partial \beta}\right)^T_{\beta=\beta^0} = \boldsymbol{Z} \operatorname{Diag}\Big(\left(\boldsymbol{C}_{n,i}^0\right)_{i=1,\ldots,N}\Big) \operatorname{Diag}\Big(\boldsymbol{p}\left(\beta^0\right)^{1/2}\Big)$$

we obtain the expression of $\boldsymbol{J}\left(\beta^0\right)$.

The result for $\widetilde{\boldsymbol{R}}^{\phi_1,\phi_2}$ holds from

$$\frac{2n\left(\boldsymbol{x}_i\right)}{\phi_1''\left(1\right)} \pi_s\left(\boldsymbol{Z}_i^T \widehat{\beta}_{\phi_2}\right) \phi_1\left(\frac{y_{si}}{n\left(\boldsymbol{x}_i\right) \pi_s\left(\boldsymbol{Z}_i^T \widehat{\beta}_{\phi_2}\right)}\right) = \frac{\left(y_{si} - n\left(\boldsymbol{x}_i\right) \pi_s\left(\boldsymbol{Z}_i^T \widehat{\beta}_{\phi_2}\right)\right)^2}{n\left(\boldsymbol{x}_i\right) \pi_s\left(\boldsymbol{Z}_i^T \widehat{\beta}_{\phi_2}\right)} + o_P\left(1\right).$$

7.5 Numerical example

As an illustration of the new diagnostic tools presented in the previous sections we consider data on the perspectives of students. Psychology students at the University of Regensburg, Germany were asked if they expected to find adequate employment after getting their degree. The response categories were ordered with respect to their expectation. The responses were 'don't expect adequate employment' (category 1), 'not sure' (category 2), and 'immediately after the degree' (category 3). The data are given in Fahrmeir and Tutz (2001). As these data do not have any leverage point we modify the last observation to be this kind of point.

We fit a cumulative logistic model to the data taking the link function $\mathbf{g} = \mathbf{h}^{-1} = (g_1, g_2)$ with

$$g_1\left(\pi_1\left(\boldsymbol{Z}_i^T\beta\right), \pi_2\left(\boldsymbol{Z}_i^T\beta\right)\right) = \log\left(\frac{\pi_1\left(\boldsymbol{Z}_i^T\beta\right)}{1-\pi_1\left(\boldsymbol{Z}_i^T\beta\right)}\right),$$

$$g_2\left(\pi_1\left(\boldsymbol{Z}_i^T\beta\right), \pi_2\left(\boldsymbol{Z}_i^T\beta_i\right)\right) = \log\left\{\log\left(\frac{\pi_1\left(\boldsymbol{Z}_i^T\beta\right)+\pi_2\left(\boldsymbol{Z}_i^T\beta\right)}{1-\pi_1\left(\boldsymbol{Z}_i^T\beta\right)-\pi_2\left(\boldsymbol{Z}_i^T\beta\right)}\right)\right.$$
$$\left. - \log\left(\frac{\pi_1\left(\boldsymbol{Z}_i^T\beta\right)}{1-\pi_1\left(\boldsymbol{Z}_i^T\beta\right)}\right)\right\},$$

$\beta^T = (\alpha_1, \alpha_2, \beta_1)$ and the design matrix

$$\boldsymbol{Z}_i^T = \begin{pmatrix} 1\; 0\; \boldsymbol{x}_i^T \\ 0\; 1\; \boldsymbol{x}_i^T \end{pmatrix}.$$

To fit the model we consider the minimum ϕ-divergence estimator based on the power-divergence family. The power-divergence family, introduced by Cressie and Read (1984), is obtained from (7.2) considering the family of functions

$$\phi_{(a)}(x) = (a(a+1))^{-1}\left(x^{a+1} - x - a(x-1)\right); \qquad a \neq 0, \qquad a \neq -1,$$

with

$$\phi_{(0)}(x) = \lim_{a\to 0}\phi_{(a)}(x) = x\log x - x + 1$$
$$\phi_{(-1)}(x) = \lim_{a\to -1}\phi_{(a)}(x) = -\log x + x - 1.$$

It is interesting to note that for $a = 0$, we get the MLE which is obtained using $D_{\phi_{(0)}}$, that is, the Kullback–Leibler divergence and $a = 1$, the minimum chi-squared estimator which is obtained using $D_{\phi_{(1)}}$, that is, the Kagan divergence (Kagan (1963)).

After fitting a cumulative logistic model, we check for the adequacy of fit. For displaying diagnostic tools, index plots are generally suggested. Figure 7.1 gives the trace of the diagonal submatrices of the generalized hat matrix, Trace($\boldsymbol{H}^{ii}(\widehat{\beta}_{\phi_{(a)}})$), that we have defined as an indicator of the leverage of the points for $a = 0$ (MLE, the classical way) and $a = 1$ (minimum chi-squared estimator), plotted against the index i. Both measures identify Observations 2 and 3 as those having the highest leverage values. However, Fahrmeir and Tutz (2001) pointed out that they are not leverage points since their leverage values are primarily caused by the relatively local sample sizes. Observation 13 (this is the observation modified by us) is also detected as a leverage point but only by the measure corresponding to $a = 1$ not for that corresponding to $a = 0$.

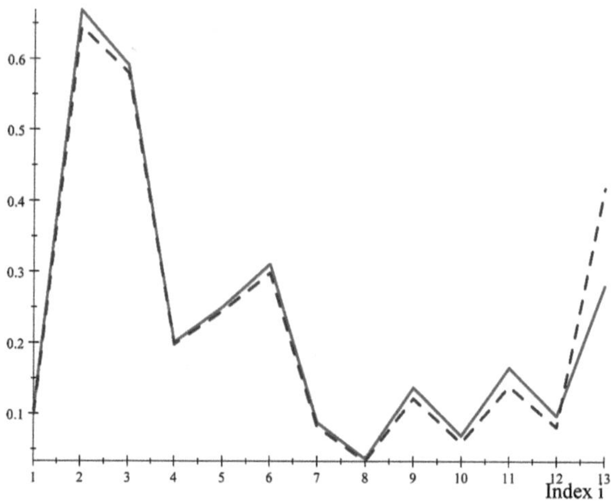

Figure 7.1. Index plot of $\mathrm{Trace}\left(\boldsymbol{H}^{ii}(\widehat{\beta}_{\phi_{(a)}})\right)$ as a function of a. Shown are ($a=0$, *solid line*), ($a=1$, *dashed line*)

After this, to study if Observation 13 has an adverse effect on the fit we calculate, for example, the standardized residuals based on $\phi_{(a)}$-divergences also for $a=0$ and $a=1$. From the index plot of Figure 7.2, it can be seen that this observation with high leverage is only fitted poorly by the model for $a=0$ not for $a=1$. This means that the cumulative logistic fit is sensitive to the estimation method. This can be seen clearly in a χ^2-probability plot drawn in Figure 7.3. The global fit of the model for $a=1$ is better (it approximates more to the straight line) than for $a=0$ and the influence of Observation 13 diminishes for $a=1$. Therefore, we prefer $a=1$ because no observation is fitted poorly or highly influential. That is to say, using the minimum $\phi_{(1)}$-divergence estimator the influential points over the fit are smoothed.

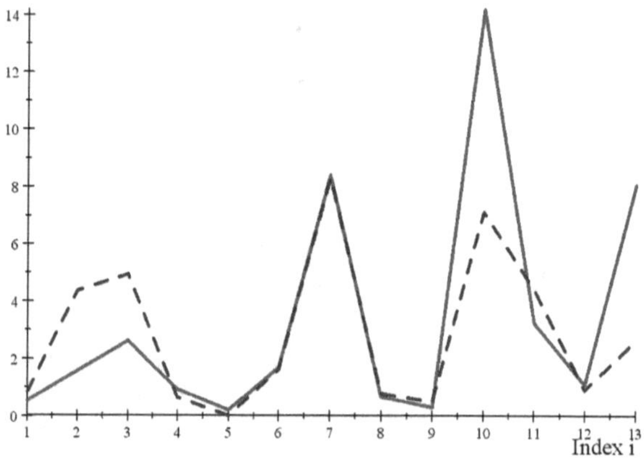

Figure 7.2. Index plot of $\left(\boldsymbol{r}_{i,S}^{\phi_{(a)}}\right)^T\left(\boldsymbol{r}_{i,S}^{\phi_{(a)}}\right)$ as a function of a. Shown are ($a=0$, *solid line*), ($a=1$, *dashed line*)

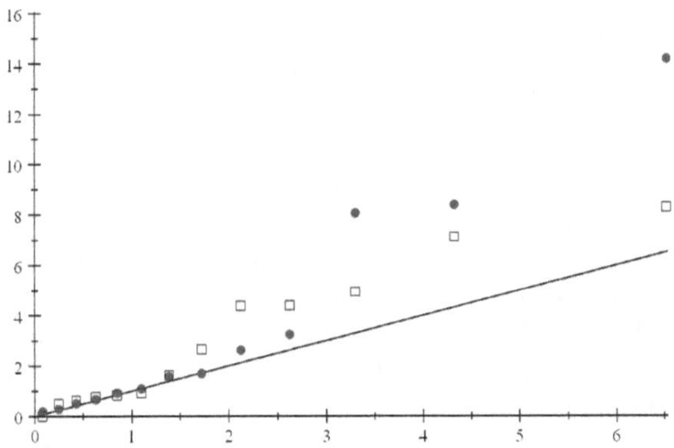

Figure 7.3. $\chi^2(2)$-probability plot of $\left(\boldsymbol{r}_{i,S}^{\phi(a)}\right)^T\left(\boldsymbol{r}_{i,S}^{\phi(a)}\right)$. Shown are ($a=0$, points), ($a=1$, *squares*)

7.6 Conclusion

In this chapter, we extend some diagnostic tools using the ϕ-divergence measure for generalized linear models with ordinal multinomial data. This has been done by Bedrick and Tsai (1993) only for the power-divergence parameter family that is a particular case of the ϕ-divergence measure and for binomial response models. The new tools introduced were diagnostics for detecting leverage points. In addition, residuals for detecting outliers were obtained. Their effectiveness was shown with real data but a simulation study would be necessary to make conclusions about choosing the best member of the family of diagnostic tools for the best fit or about measuring the sensitivity of the diagnostics. Finally, note that to implement the tools proposed the software requirement is a program for obtaining the minimum ϕ-divergence estimator which is similar to that for obtaining the MLE taking into account the expression (7.1).

Acknowledgements. This work was partially supported by Grants MTM2006-00892 and UCM-BSCH-2008-910707.

References

Ali, S.M. and Silvey, S.D. (1966). A general class of coefficients of divergence of one distribution from another. *Journal of the Royal Statistical Society* ***B***, **26**, 131–142.

Bedrick, E.J. and Tsai, C. (1993). Diagnostics for binomial response models using power divergence statistics. *Computational Statistics and Data Analysis*, **15**, 381–392.

Cook, R.D. (1986). Assessment of local influence (with discussion). *Journal of the Royal Statistical Society* ***B***, **48**, 133–169.

Cook, R.D. and Weisberg, S. (1982). *Residuals and Influence in Regression.* Chapman & Hall, New York.

Cramér, H. (1946). *Mathematical Methods of Statistics.* Princeton University Press, Princeton, NJ.

Cressie, N.A.C. and Read, T. (1984). Multinomial goodness-of-fit tests. *Journal of the Royal Statistical Society **B***, **46**, 440–464.

Fahrmeir, L. and Tutz, G. (2001). *Multivariate Statistical Modelling Based on Generalized Linear Models.* Springer-Verlag, New York.

Green, P.J. (1984). Iteratively reweighted least squares for maximum likelihood estimation, and some robust and resistant alternatives (with discussion). *Journal of the Royal Statistical Society **B***, **46**, 149–192.

Harville, D.A. (1997). *Matrix Algebra from Statistician's Perspective.* Springer-Verlag, New York.

Jorgensen, B. (1983). Maximum likehood estimation and large sample inference for generalized linear and non linear regresion models. *Biometrika*, **70**, 19–28.

Kagan, M. (1963). On the theory of Fisher's amount of information. *Sov. Math. Dokl.*, **4**, 991–993.

Lesaffre, E. and Albert, A. (1989). Multiple-group logistic regression diagnostics. *Applied Statistics,* **38**, 425–440.

Liu, I. and Agresti, A. (2005). The analysis of ordered categorical data: An overview and a survey of recent developments. *Test,* **14(1)**, 1–73.

McCullagh, P. (1980). Regression models for ordinal data (with discussion). *Journal of the Royal Statistical Society **B***, **42**, 109–142.

McCullagh, P. and Nelder, J.A. (1989). *Generalized Linear Models (2nd ed.).* Chapman & Hall, New York.

Pardo, L. (2006). *Statistical Inference Based on Divergence Measures.* Chapman & Hall, New York.

Pardo, M.C. (2008). Testing equality restrictions in generalized linear models. Metrika, DOI: 10.1007/500184-009-0275-y.

Pregibon, D. (1981). Logistic regression diagnostics. *Annals of Statistics,* **9**, 705–724.

Rao, C.R. (1973). *Linear Statistical Inference and ITS Applications.* John Wiley, New York.

Thomas, W. and Cook, R.D. (1990). Assesing influence on predictions from generalized linear models. *Technometrics*, **32**, 59–65.

Vajda, I. (1989). *Theory of Statistical Inference and Information.* Kluwer Academic, Dordrecht.

Williams, D.A. (1987). Generalized linear model diagnostics using the deviance and single case deletions. *Applied Statistics*, **36(2)**, 181–191.

8

On a Minimization Problem Involving Divergences and Its Applications

Athanasios P. Sachlas and Takis Papaioannou

Department of Statistics and Insurance Science, University of Piraeus, Greece

Abstract: In this chapter, motivated by the seminal paper of Brockett, "Information theoretic approach to actuarial science: A unification and extension of relevant theory and applications," *Transactions of the Society of Actuaries*, Vol. 43, 73–135 (1991), we review minimization of the Kullback–Leibler divergence $D^{KL}(\mathbf{u}, \mathbf{v})$ between observed (raw) death probabilities or mortality rates, $\mathbf{u}$, and the same entities, $\mathbf{v}$, to be graduated (or smoothed) subject to a set of reasonable constraints such as monotonicity, bounded smoothness, etc. Noting that the quantities $\mathbf{u}$ and $\mathbf{v}$, involved in the above minimization problem based on the Kullback–Leibler divergence, are nonprobability vectors, we study the properties of divergence and statistical information theory for $D^{KL}(\mathbf{p}, \mathbf{q})$, where $\mathbf{p}$ and $\mathbf{q}$ are nonprobability vectors. We do the same for the Cressie and Read power divergence between nonprobability vectors, solve the problem of graduation of mortality rates via Lagrangian duality theory, discuss the ramifications of constraints, tests of goodness-of-fit, and compare with other graduation methods, predominantly the Whittaker and Henderson method. At the end we provide numerical illustrations and comparisons.

Keywords and phrases: Kullback–Leibler divergence, Cressie–Read divergence, divergence with nonprobability vectors, graduation of mortality rates

8.1 Introduction

Measures of divergence appear everywhere in mathematics, probability, and statistics. They express the "distance" between two functions or vectors. There are several measures of divergence or diversity in the literature. The most well known are those of Kullback–Leibler, Csiszar (otherwise called ϕ-divergence), and Cressie and Read (otherwise known as power divergence). A less well-known divergence is Jensen's difference.

A measure of divergence is not a metric as it does not satisfy the antisymmetric property and for that reason it is sometimes called a measure of directed divergence.

A bivariate function $D(f, g)$ of two functions or vectors f, g is a measure of divergence if $D(f, g) \geq 0$ with equality if and only if $f = g$ (c.f. Basu et al., 1998). This is the minimal requirement for a measure $D(f, g)$ to be a "kind" of distance between

C.H. Skiadas (ed.), *Advances in Data Analysis*, Statistics for Industry and Technology, DOI 10.1007/978-0-8176-4799-5_8,

f and g. Pardo (2006) mentions that a coefficient with the property of increasing as the two distributions involved move "further from each other" is called a *divergence measure* between two probability distributions. For other requirements see Read and Cressie (1988) and Mathai and Rathie (1975).

There are situations where measures of divergence involve nonprobability distributions, either pdfs or probability vectors. One such situation is the graduation of mortality rates which is discussed in this chapter and is a minimization problem involving divergences.

Minimization of a metric or a distance or a divergence has a dominant position in statistics. It appears almost everywhere and this justifies the strong connections among statistics and mathematical programming and operations research. Some examples are least squares, L^1, L^2, L^∞ optimization, minimum discrimination information, AIC, minimum divergence estimation, Hellinger distance estimation, and Bayes prior estimation.

All these problems involve divergences with probability distributions either pdfs or finite/infinite probability vectors. The usual setup is as follows: g_0 is a known or empirical (i.e., data-based) distribution and we seek f to minimize the directed divergence $D(f, g_0)$ of f from g_0 subject to some necessary constraints on f because otherwise the solution is $f = g_0$. There are situations where the setup is reversed: f is known, data-based, or estimated; i.e., $f = f_0$, g is unknown, and the objective is to minimize $D(f_0, g)$ with g satisfying some constraints.

What happens if f and g are not probability distributions? In this chapter we address this problem. In Section 8.2 we discuss principles, ideas, and techniques of divergence minimization. In Section 8.3 we present the properties of divergences without probability vectors. In Section 8.4 we present the problem of graduating mortality rates via divergences. In Section 8.5 we present a numerical illustration involving primarily Jensen's difference while in Section 8.6 we give concluding remarks.

8.2 Minimization of divergences

The minimization of divergences follows two patterns. One may be called parametric and the other nonparametric.

For the first we have $f(x)$ known and $g(x,\theta) \equiv g_\theta$ known but depending on an unknown parameter θ, scalar or vector valued. Then we seek to estimate θ by minimizing, without or with constraints,

$$D(f, g_\theta).$$

This leads to minimum divergence or distance estimation. For example, if $D(f, g_\theta)$ is the Kullback–Leibler divergence and f is taken to be $\hat{f}$ (i.e., empirical or data based) we have to minimize over θ,

$$\int \hat{f}(x) \ln \left[\hat{f}(x)/g(x,\theta)\right] \, d\mu,$$

where μ is an appropriate measure. In this way we obtain minimum Kullback–Leibler divergence estimators which coincide (a.s.) with the maximum likelihood estimators for discrete (e.g., multinomial) models. If the divergence is the Kolmogorov distance or the L^∞ metric, then again we try to minimize over θ,

$$\sup |F_n(x) - F(x,\theta)|,$$

where F is the cdf of X and F_n the corresponding empirical cdf. A classical example of minimum distance estimation is the least squares estimation in linear models which has been handled in the literature algebraically and geometrically. Chi-square, Hellinger distance, Csiszar's ϕ divergence, and Cressie and Read's power divergences have been used for minimum distance or divergence estimation of θ.

Here one observes the use of "divergence of the data from the model." There is no particular reason to try to minimize $D(\hat{f}, g_\theta)$ over θ. We could minimize $D(g_\theta, \hat{f})$ and this leads to different estimators of θ. These estimators have not been investigated in the literature. For references on this topic see Read and Cressie (1988), Vos (1992), Cutler and Cordero-Brana (1996), and Pardo (2006).

For the nonparametric problem we have minimization of $D(f,g)$ with constraints on f for the purpose of determining f which is as close as possible to known g. If $g = \hat{g}$ (i.e., data based), then we try to estimate the stochastic model that is closest to the data. A classical example is the minimum discrimination distribution that is obtained by minimizing the Kullback–Leibler divergence

$$\int f \ln(f/g)\, dx,$$

with constraints on f, and leads to the minimum discrimination information.

Kullback's solution is obtained via Lagrangian and calculus of variation while in the discrete case we may employ mathematical programming results. The optimal solution $f^\star$ is exponential.

In model selection, through various model fittings, we seek to find the model $\hat{g}$, which is as close as possible to the true hypothesized distribution f of the data. Thus we try to minimize

$$D(f, \hat{g}) \text{ over } \hat{g}.$$

A classical example is the Akaike information criterion, AIC (Akaike, 1973).

Another important paradigm of variational use of divergences in statistics is through the concept of *information or divergence statistics*: we have the usual divergence statistics $D(\hat{f}, g)$ with g known or $D(f, \hat{g})$ with f known or $D(\hat{f}, \hat{g})$. But we also have minimum discrimination information statistics (c.f. Kullback, 1959, pp. 82, 85) which are derived for various classes of probability models. For many problems the two approaches coincide. There is a large literature on the topic and divergences provide the tool to develop test statistics, confidence intervals, to compare and select models, and in general develop an applied statistical inference. For an important recent reference on the topic see Pardo (2006).

8.3 Properties of divergences without probability vectors

In this section we list some of the properties of the Kullback–Leibler and the Cressie–Read divergences without probability vectors. Details can be found in Sachlas and Papaioannou (2009).

The *Kullback–Leibler directed divergence* between two $n \times 1$ nonprobability vectors $\mathbf{p}$ and $\mathbf{q}$, is defined by

$$D^{KL}(\mathbf{p},\mathbf{q}) = \sum_{i=1}^{n} p_i \ln \frac{p_i}{q_i}$$

where $\mathbf{p} = (p_1,\ldots,p_n)^T > \mathbf{0}$, $\mathbf{q} = (q_1,\ldots,q_n)^T > \mathbf{0}$ with $\sum_{i=1}^n p_i \neq 1$ and $\sum_{i=1}^n q_i \neq 1$.

Proposition 1. *(The nonnegativity property) For the Kullback–Leibler directed divergence we have*

$$D^{KL}(\mathbf{p},\mathbf{q}) \geq 0, \tag{8.1}$$

if one of the following conditions holds,

$$(i) \sum_{i=1}^{n} p_i \geq \sum_{i=1}^{n} q_i, \qquad (ii) \sum_{i=1}^{n} p_i < \sum_{i=1}^{n} q_i \quad and \quad \ln k > -D^{KL}(\mathbf{p}^*,\mathbf{q}^*),$$

where $k = \sum pi / \sum qi$ *and* $\mathbf{p}^*$, $\mathbf{q}^*$ *are probability vectors whose elements are the normalized elements of* $\mathbf{p}$ *and* $\mathbf{q}$*; i.e.,* $p_i^* = p_i/\sum_{i=1}^n p_i$ *and* $q_i^* = q_i/\sum_{i=1}^n q_i$, $i = 1,\ldots,n$. *Equality in* (8.1) *holds if* $\mathbf{p} = \mathbf{q}$ *or* $\ln k = -D^{KL}(\mathbf{p}^*,\mathbf{q}^*)$. *Moreover if* $\sum_{i=1}^n p_i = \sum_{i=1}^n q_i$ *then* $D^{KL}(\mathbf{p},\mathbf{q}) \geq 0$ *with equality if and only if* $\mathbf{p} = \mathbf{q}$.

The minimal requirement for using $D^{KL}(\mathbf{p},\mathbf{q})$ as a measure of divergence is $\sum_{i=1}^n p_i = \sum_{i=1}^n q_i$. The Kullback–Leibler directed divergence between two bivariate nonprobability functions p_1, p_2 is

$$D^{KL}_{X,Y}(p_1,p_2) = \sum_x \sum_y p_1(x,y) \ln \frac{p_1(x,y)}{p_2(x,y)}.$$

Conditional divergence can now be defined in an analogous way as in the case of probability vectors (see Kullback, 1959).

Proposition 2. *(Strong additivity) Let* p_1, p_2 *be two bivariate nonprobability functions associated with two discrete variables* X, Y *in* R^2, *with* $\sum_x \sum_y p_i(x,y) \neq 1$. *Then*

$$D^{KL}_{X,Y}(p_1,p_2) = D^{KL}_{X}(f_1,f_2) + D^{KL}_{Y|X}(h_1,h_2) = D^{KL}_{Y}(g_1,g_2) + D^{KL}_{X|Y}(r_1,r_2),$$

where $f_i = f_i(x)$, $h_i = h_i(y|x)$, $g_i = g_i(y)$, $r_i = r_i(x|y)$, $i = 1,2$ *are the corresponding marginal and conditional nonprobability functions.*

Proposition 3. *(Weak additivity) If* $h_i(y|x) = g_i(y)$ *and thus* $p_i(x,y) = f_i(x)g_i(y)$, $i = 1,2$, *we have that the random variables* X^*, Y^*, *produced by normalizing* X, Y, *are independent, and it holds that*

$$D^{KL}_{X,Y}(p_1,p_2) = D^{KL}_{X}(f_1,f_2) + D^{KL}_{Y}(g_1,g_2) - \xi \ln \eta,$$

where $\xi = \sum_y g_1(y) = \sum_x f_1(x)$ *and* $\eta = \sum_y g_1(y)/\sum_y g_2(y) = \sum_x f_1(x)/\sum_x f_2(x)$.

It is easy to see that weak additivity holds if $\sum_x f_1(x) = \sum_x f_2(x)$ or $\sum_y g_1(y) = \sum_y g_2(y)$.

Proposition 4. *(Maximal information and sufficiency) Let $Y = T(X)$ be a measurable transformation of X, then*

$$D_X^{KL}(p_1, p_2) \geq D_Y^{KL}(g_1, g_2),$$

with equality if and only if Y is "sufficient," where $p_i = p_i(x)$, $g_i = g_i(y)$, $i = 1, 2$.

A basic property of measures of information and divergence is the limiting property. This property means that the series of random variables converges when $n \to \infty$ if and only if $I_{X_n} \to I_X$, where I denotes the information measure. Under some conditions (Kullback, 1959) the limiting property holds for the Kullback–Leibler divergence. For Csiszar's measure of divergence (ϕ-divergence) see Zografos et al. (1989).

The next proposition investigates whether the limiting property holds in the case of the Kullback–Leibler divergence with nonprobability vectors.

Proposition 5. *(The limiting property) Let $\{\mathbf{p}_n\}$ be a bounded from above sequence of nonprobability vectors. Then $\mathbf{p}_n \to \mathbf{p}$ if and only if $D^{KL}(\mathbf{p}_n, \mathbf{p}) \to 0$; i.e., the limiting property holds for the Kullback–Leibler divergence with nonprobability vectors.*

As expected the Kullback–Leibler directed divergence $D^{KL}(\mathbf{p}, \mathbf{q})$ with nonprobability vectors $\mathbf{p}$, $\mathbf{q}$ does not in general share the properties of the Kullback–Leibler directed divergence with probability vectors $\mathbf{p}^*$, $\mathbf{q}^*$. Under certain conditions, some of them are satisfied. More precisely $D^{KL}(\mathbf{p}, \mathbf{q})$, is nonnegative, additive, invariant under sufficient transformations, and greater than $D^{KL}(\mathbf{p}^*, \mathbf{q}^*)$. It also satisfies the property of maximal information and the limiting one. So, $D^{KL}(\mathbf{p}, \mathbf{q})$, in general terms, can be regarded as a measure of divergence and therefore can be used whenever we do not have probability vectors, provided that $\sum_i p_i = \sum_i q_i$.

The *Cressie–Read power divergence* of order r for nonprobability vectors is given by

$$D^{CR}(\mathbf{p}, \mathbf{q}) = \frac{1}{r(r+1)} \sum_i p_i \left[\left(\frac{p_i}{q_i} \right)^r - 1 \right], \qquad r \in R$$

where $\mathbf{p} = (p_1, \ldots, p_n)^T > \mathbf{0}$ and $\mathbf{q} = (q_1, \ldots, q_n)^T > \mathbf{0}$ with $\sum_i p_i \neq 1$ and $\sum_i q_i \neq 1$.

In the sequel we assume that $r \neq 0$ and $r \neq -1$.

Proposition 6. *(The nonnegativity property) Let*

$$k = \sum_i p_i / \sum_i q_i \quad \text{and} \quad m = \frac{1 - k^r}{k^r} \frac{1}{r(r+1)}.$$

Then for the Cressie–Read power divergence with nonprobability vectors we have

$$D^{CR}(\mathbf{p}, \mathbf{q}) \geq 0,$$

if one of the following conditions holds.

(i) $\sum_i p_i = \sum_i q_i$;
(ii) $\sum_i p_i > \sum_i q_i$ *and* $r \notin (-1, 0)$.
(iii) $\sum_i p_i > \sum_i q_i$ *and* $m < D^{CR}(\mathbf{p}^*, \mathbf{q}^*)$;
(iv) $\sum_i p_i < \sum_i q_i$ *and* $r \in (-1, 0)$.
(v) $\sum_i p_i < \sum_i q_i$ *and* $m < D^{CR}(\mathbf{p}^*, \mathbf{q}^*)$.

As for the equality part we have the following.

(a) If $\sum_i p_i = \sum_i q_i$ *equality holds if* $\mathbf{p} = \mathbf{q}$.

(b) If $\sum_i p_i > \sum_i q_i$ *or* $\sum_i p_i < \sum_i q_i$ *equality holds if* $m = D^{CR}(\mathbf{p}^*, \mathbf{q}^*)$.

Finally if $\sum_i p_i = \sum_i q_i$ *then* $D^{CR}(\mathbf{p}, \mathbf{q}) \geq 0$ *with equality if and only if* $\mathbf{p} = \mathbf{q}$.

Here again we have that the minimal requirement for using $D^{CR}(\mathbf{p}, \mathbf{q})$ as a measure of divergence is $\sum_i p_i = \sum_i q_i$ regardless of the value of r.

Proposition 7. $D^{CR}(\mathbf{p}, \mathbf{q}) \geq D^{CR}(\mathbf{p}^*, \mathbf{q}^*)$ *when one of the following conditions holds. (i)* $\sum_i p_i = \sum_i q_i$, *(ii)* $\sum_i p_i > \sum_i q_i$ *and* $r \notin (-1, 0)$, *(iii)* $\sum_i p_i < \sum_i q_i$ *and* $r \in (-1, 0)$. *Equality holds if* $m = D^{CR}(\mathbf{p}^*, \mathbf{q}^*)$ *independently of the value of* r, *where* m *is as in Proposition 6.*

Bivariate and conditional Cressie–Read divergence are defined in a similar way as before. Strong additivity is not satisfied for the power divergence with probability vectors as one can easily see with a numerical example involving two trinomial distributions. A further numerical investigation reveals that when $r > 0$ the subadditivity property holds, while when $r < 0$ the superadditivity property holds. Equality holds only when $r = 0$, which is the case of the Kullback–Leibler divergence.

No convenient expression was obtained in the case of nonprobability vectors. For weak additivity we have the following proposition.

Proposition 8. *(Weak additivity) If* $h_i(y|x) = g_i(y)$ *and thus* $p_i(x, y) = f_i(x)g_i(y)$, $i = 1, 2$, *we have that the random variables* X^*, Y^*, *which are the "standardized" values of* X, Y, *are independent, then*

(a) $D^{CR}_{X,Y}(p_1, p_2) = D^{CR}_X(f_1, f_2) + D^{CR}_Y(g_1, g_2) + p_{1\cdot\cdot}(1 - \eta^r)\, 1/(r(r + 1))$
$+ p_{1\cdot\cdot}\eta^r r(r + 1) D^{CR}_{X^*}(f_1^*, f_2^*)\, D^{CR}_{Y^*}(g_1^*, g_2^*)$, *where* $p_{i\cdot\cdot} = \sum_x \sum_y p_i(x, y)$, $i = 1, 2$,

(b) $D^{CR}_{X,Y}(p_1, p_2) = D^{CR}_X(f_1, f_2) + D^{CR}_Y(g_1, g_2)$ *if* $\eta = 1$ *and if one of the marginal pairs* $(f_1^*, f_2^*), (g_1^*, g_2^*)$ *is identical where* $\eta = p_{1\cdot\cdot}/p_{2\cdot\cdot}$.

Proposition 9. *(Maximal information and sufficiency) Let* $Y = T(X)$ *be a measurable transformation of* X*; then*

$$D^{CR}_X(p_1, p_2) \geq D^{CR}_Y(g_1, g_2),$$

when $b > 1$, *where* $b = \left(\sum_x p_1(x) / \sum_x p_2(x)\right)^r$, *with equality if and only if* Y *is "sufficient," where* $p_i = p_i(x)$, $g_i = g_i(y)$, $i = 1, 2$.

Zografos et al. (1989) proved that, under some conditions, the limiting property holds for Csiszar's measure of divergence (ϕ-divergence) defined as

$$D^C(f_1, f_2) = \int f_2(x) \phi\left(\frac{f_1(x)}{f_2(x)}\right) dx,$$

where ϕ is a real-valued convex function satisfying certain conditions. Cressie and Read divergence can be obtained from Csiszar's measure by taking $\phi(x) = [r(r + 1)]^{-1}(x^{r+1} - x)$ (Pardo, 2006). So the limiting property holds for the Cressie and Read divergence as well. The next proposition states that the limiting property holds in the case where we do not have probability vectors.

Proposition 10. *(The limiting property) It holds that* $\mathbf{p}_n \to \mathbf{p}$ *iff* $D^{CR}(\mathbf{p}_n, \mathbf{p}) \to 0$ *with* $r \neq 0, -1$*; i.e., the limiting property holds for the Cressie–Read divergence with nonprobability vectors.*

Summarizing, the power-directed divergence $D^{CR}(\mathbf{p}, \mathbf{q})$, under some conditions is nonnegative, additive, greater than $D^{CR}(\mathbf{p}^*, \mathbf{q}^*)$, and invariant under sufficient transformations. It also shares the property of maximal information and the basic limiting property. So, we can regard $D^{CR}(\mathbf{p}, \mathbf{q})$ as a measure of divergence, provided that $\sum_i p_i = \sum_i q_i$.

In mathematics and statistics there exist many divergences (see, e.g., Read and Cressie, 1988; Liese and Vajda, 1987; Mathai and Rathie, 1975). One of them, which has a special appeal since it originates from Shannon's entropy (a well-known index of diversity) and the convexity property is *Jensen's difference* as it was called by Burbea and Rao (1982). The Jensen difference between two nonprobability vectors is given by

$$J(\mathbf{p}, \mathbf{q}) \equiv H\left(\tfrac{1}{2}(\mathbf{p} + \mathbf{q})\right) - \tfrac{1}{2}\left[H(\mathbf{p}) + H(\mathbf{q})\right],$$

where $H(\mathbf{p}) = -\sum_i p_i \ln p_i$ is the Shannon entropy between the nonprobability vectors $\mathbf{p} = (p_1, \ldots, p_n)^T$ and $\mathbf{q} = (q_1, \ldots, q_n)^T$.

Proposition 11. *Let* $\sum_i p_i = \sum_i q_i$*. Then* $J(\mathbf{p}, \mathbf{q}) \geq 0$ *if and only if* $\mathbf{p} = \mathbf{q}$*, where* $\mathbf{p}$ *and* $\mathbf{q}$ *are nonprobability vectors.*

The above proposition means that $\mathbf{p} = \mathbf{q}$ is again the minimal requirement for $J(\mathbf{p}, \mathbf{q})$ to be a divergence measure. Further properties of $J(\mathbf{p}, \mathbf{q})$ are under investigation.

8.4 Graduating mortality rates via divergences

In this section we describe how measures of divergence can be used in order to smooth raw mortality rates. We first start with some basic notions of actuarial graduation while in the sequel we provide a numerical illustration.

8.4.1 Divergence-theoretic actuarial graduation

In order to describe the actual but unknown mortality pattern of a population, the actuary calculates from raw data crude mortality rates, death probabilities, or forces of mortality, which usually form an irregular series. Because of this, it is common to revise the initial estimates with the aim of producing smoother estimates, with a procedure called *graduation*. There are several methods of graduation classified into parametric curve fitting and nonparametric smoothing methods. For more details on the topic the interested reader is referred to Benjamin and Pollard (1980), Haberman (1998), London (1985), Wang (1998), and references therein.

A method of graduation using information-theoretic ideas was first introduced by Brockett and Zhang (1986). More specifically, Zhang and Brockett (1987) tried to construct a smooth series of n death probabilities $\{v_x\}$, $x = 1, 2, \ldots, n$ which is as close as possible to the observed series $\{u_x\}$ and in addition they assumed that the true but unknown underlying mortality pattern is (i) smooth, (ii) increasing with age x (i.e., monotone), and (iii) more steeply increasing in higher ages (i.e., convex). They also assumed that (iv) the total number of deaths in the graduated data equals the total number of deaths in the observed data, and (v) the total age of death in the graduated data equals the total age of death in the observed data. By total age of death we mean the sum of the product of the number of deaths at every age by the corresponding age. The last two constraints imply that the average age of death is required to be the same for the observed and graduated mortality data.

Mathematically the five constraints are written as follows: (i) $\sum_x (\Delta^3 v_x)^2 \leq M$, where M is a predetermined positive constant and $\Delta^3 v_x = -v_x + 3v_{x+1} - 3v_{x+2} + v_{x+3}$; (ii) $\Delta v_x \geq 0$, where $\Delta v_x = v_{x+1} - v_x$; (iii) $\Delta^2 v_x \geq 0$, where $\Delta^2 v_x = v_x - 2v_{x+1} + v_{x+2}$; (iv) $\sum\limits_x l_x v_x = \sum\limits_x l_x u_x$, where l_x is the number of people at risk in the age x; and (v) $\sum\limits_x x l_x v_x = \sum\limits_x x l_x u_x$. In matrix notation the constraints can be written as: (i) $(\mathbf{Av})^T (\mathbf{Av}) = \mathbf{v}^T \mathbf{A}^T \mathbf{Av} \leq M$, where $\mathbf{A}$ is an $(n-3) \times n$ matrix with rows of the form $(0 -1\ 3\ -3\ 1\ 0 \ldots 0)$; (ii) $\mathbf{Bv} \geq \mathbf{0}$, where $\mathbf{B}$ is an $(n-1) \times n$ matrix with rows of the form $(0 -1\ 1\ 0\ 0 \ldots 0)$; (iii) $\mathbf{Cv} \geq \mathbf{0}$, where $\mathbf{C}$ is an $(n-2) \times n$ matrix with rows of the form $(0\ 1\ -2\ 1\ 0 \ldots 0)$; (iv) $\mathbf{d}^T\mathbf{v} = \mathbf{d}^T\mathbf{u}$, where $\mathbf{d} = (l_x, l_{x+1}, \ldots, l_{x+n-1})^T$; and (v) $\mathbf{e}^T\mathbf{v} = \mathbf{e}^T\mathbf{u}$, where $\mathbf{e} = (xl_x, (x+1)l_{x+1}, \ldots, (x+n-1)l_{x+n-1})^T$, respectively. For more details see Zhang and Brockett (1987). It is easy to see that the constraints (i)–(v) may be written in the form of $g_i(\mathbf{v}) = \frac{1}{2}\mathbf{v}^T\mathbf{D}_i\mathbf{v} + \mathbf{b}_i^T\mathbf{v} + c_i \leq 0$, $i = 1, 2, \ldots, m$, where, for each i, $\mathbf{D}_i$, $\mathbf{b}_i$, c_i are a positive semidefinite matrix and constants, respectively, easily written down from (i)–(v) and in this case we have $m = 2(n+1)$ constraints, where n is the number of ungraduated values.

In order to obtain the graduated values, Brockett (1991) minimizes the Kullback–Leibler divergence between the crude death probabilities $\mathbf{u} = (u_1, u_2, \ldots, u_n)^T$ and the new death probabilities $\mathbf{v} = (v_1, v_2, \ldots, v_n)^T$,

$$D^{KL}(\mathbf{v}, \mathbf{u}) = \sum_x v_x \ln \frac{v_x}{u_x},$$

subject to the constraints (i)–(v).

However, the mortality rates (death probabilities) $\mathbf{u}$ and $\mathbf{v}$ are not probability vectors since we have $\sum_{x=1}^n u_x > 1$ and $\sum_{x=1}^n v_x > 1$. Brockett (1991, p. 104) states that "$D^{KL}(\mathbf{v}, \mathbf{u}) = \sum_{x=1}^n v_x \ln(v_x/u_x)$ is still a measure of fit even in the nonprobability situation because the mortality rates are nonnegative and because of the assumed constraints." In view of the discussion and results in Section 8.3 the appropriate constraint to be used here is (vi)

$$\sum_{x=1}^n v_x = \sum_{x=1}^n u_x$$

and not conditions (iv) and (v). It is easy to see via counterexamples that conditions (iv) and (v) do not imply (vi). It may be necessary, however, to use them on actuarial grounds. Constraint (vi) can be written in notation form and thus in the form of $g_i(\mathbf{v})$ as $\mathbf{e}^T\mathbf{v} = \mathbf{e}^T\mathbf{u}$, where $\mathbf{e} = (1, 1, \ldots, 1)^T$.

A new and unifying way to obtain the graduated values v_x is to minimize the Cressie–Read divergence

$$D^{CR}(\mathbf{v},\mathbf{u}) = \frac{1}{r(r+1)} \sum_x v_x \left[\left(\frac{v_x}{u_x} \right)^r - 1 \right]$$

for given r subject to constraints (i)–(v) and/or (vi); i.e., $\mathbf{v} \geq \mathbf{0}$ and $g_i(\mathbf{v}) = \frac{1}{2}\mathbf{v}^T\mathbf{D}_i\mathbf{v} + \mathbf{b}_i^T\mathbf{v} + c_i \leq 0$, $i = 1, 2, \ldots, m+1$, where $m = 2(n+1)$. The minimization is done for various values of the parameter r and in this way we can interpret the resulting series of the graduated values, as the series that satisfies the constraints and is least distinguishable in the sense of the Cressie–Read directed divergence from the series of the crude values $\{u_x\}$.

Another method to obtain the graduated values is to minimize the Jensen's difference

$$J(\mathbf{u},\mathbf{v}) = -\sum_{i=1}^n \frac{1}{2}(u_i + v_i)\ln\left(\frac{1}{2}(u_i+v_i)\right) + \frac{1}{2}\left[\sum_{i=1}^n v_i \ln v_i + \sum_{i=1}^n u_i \ln u_i\right]$$

between the crude and the graduated mortality rates $\mathbf{u}$ and $\mathbf{v}$ under the constraints (i)–(v) and/or (vi).

8.4.2 Lagrangian duality results for the power divergence

It is easily seen that the minimization of either the Kullback–Leibler measure or the power divergence is a difficult task as a lot of constraints are involved on it. For this reason, Zhang and Brockett (1987) derived duality results for the quadratically constrained problem by using an approximation technique. More specifically, they first converted the problem of minimizing the Kullback–Leibler divergence into a sequence of nonlinear programs with linear constraints and then by taking a limit they were led to a dual problem. Teboulle (1989) produced the same dual problem by a simple application of Lagrangian duality (Boyd and Vandenberghe, 2006).

The quadratically constrained Cressie–Read graduation problem was defined before. We restate it as primal problem: Find $\mathbf{v} \in R^n$ which solves the primal problem

$$(P) \qquad \min \frac{1}{r(r+1)} \sum_{j=1}^n v_j \left[\left(\frac{v_j}{u_j} \right)^r - 1 \right]$$

subject to

$$g_i(\mathbf{v}) = \frac{1}{2}\mathbf{v}^T\mathbf{D}_i\mathbf{v} + \mathbf{b}_i^T\mathbf{v} + c_i \leq 0, \qquad i = 1, 2, \ldots, m, \qquad \mathbf{v} \geq \mathbf{0}.$$

One may solve (P) using constrained optimization methods or revert to its Lagrangian dual problem given below which does not involve constraints.

The *Lagrangian dual problem* of the $D^{CR}(\mathbf{v},\mathbf{u})$ minimization under constraints $g_i(\mathbf{v})$ is given by

$$(D) \qquad \sup_{\lambda \in R_+^m, \mathbf{y}_i \in R^{n_i}} \left\{ \sum_{j=1}^n u_j \left(\frac{1}{r+1} - r \sum_{i=1}^m (\lambda_i \mathbf{b}_i^T + \mathbf{y}_i^T \mathbf{A}_i)_j \right)^{1/r} \right.$$

$$\left. \cdot \left(r(r+1) \sum_{i=1}^m (\lambda_i \mathbf{b}_i^T + \mathbf{y}_i^T \mathbf{A}_i)_j - 1 \right) - \frac{1}{2} \sum_{i=1}^m \frac{\|\mathbf{y}_i\|^2}{\lambda_i} + \lambda^T \mathbf{c} \right\}.$$

The justification and solution of the problem (D) is given in the theorem below.

Theorem 1. *(a) If (P) is feasible then* inf (P) *is attained and* min(P) = sup(D). *Moreover, if there exists a* $\mathbf{v} \in R^n$ *satisfying* $\mathbf{v} > 0$, $g_i(\mathbf{v}) < 0$, $i = 1, \ldots, m$, *then* sup (D) *is attained and* min (P) = max (D).

(b) If $\mathbf{v}^*$ *solves the primal problem (P) and* $\mathbf{y}_i^* \in R^{n_i}$, $\lambda^* \in R_+^m$ *solve the dual problem (D), then*

$$v_j^* = u_j \left[\frac{1}{r+1} - r \sum_{i=1}^{m} (\lambda_i^* \mathbf{b}_i^T + \mathbf{y}_i^{*T} \mathbf{A}_i)_j \right]^{1/r}, \qquad j = 1, 2, \ldots, n.$$

It is obvious that if we choose $r = 0$, we perform graduation through the Kullback–Leibler directed divergence that Zhang and Brockett (1987) described. In the following section we provide a numerical illustration.

8.5 Numerical investigation

As mentioned above, the problem of graduation is to find the best fitting values v_x that satisfy the mathematical and actuarial constraints (i)–(v) and are the least distinguishable from the initial estimates u_x. In this section we illustrate the constrained minimization of Cressie–Read divergence, which can be easily solved by using any of the readily available nonlinear programming codes.

For the illustration, we use a dataset of death probabilities from London (1985, p. 20). It consists of 15 death probabilities belonging to ages 70–84 (computed from a total of 2073 observations). The raw data along with the graduations made by London and Brockett are presented in Table 8.1a. We note that London performed his graduation by graphic means and then revised his results by a linear transformation of the graduated values. Brockett computed the graduated values via the minimization of the Kullback–Leibler divergence subject to constraints (i)–(v). He also graduated the raw data without the constraint of convexity (iii). In the same table we present, for comparison reasons, the results obtained via the Whittaker–Henderson method of graduation. The Whittaker–Henderson method (London, 1985) is a well-known and frequently used method of graduation, where a function $F + hS$ is minimized without constraints. The value of the smoothness measure S and the goodness-of-fit measures (F, deviance, log-likeilihood and χ^2) are given in Table 8.1b.

We first graduated the crude values via the minimization of the Jensen difference. The minimization was conducted subject to constraints (i)–(v), proposed by Brockett (1991), the additional constraint (vi) that Sachlas and Papaioannou (2009) proposed, and finally subject to constraints (i)–(iii) and (vi). The relevant results are presented along with the raw data in Table 8.2a. The results are equivalent to those presented by London and Brockett. The value of the smoothness measure and the goodness-of-fit measures are given in Table 8.2b.

Finally we conducted graduation via the minimization of the Cressie–Read power divergence subject to constraints (i)–(v) and using the dual results of Section 8.4.2. The results for the values of $r = 0.2, 2/3, 1$, and 1.5 are given in Table 8.3a. For comparison reasons we give the values of smoothness and goodness of fit measures in Table 8.3b.

Table 8.1. Graduations by London, Brockett, and Whittaker–Henderson

(a) Graduated values

x	u_x	v_x (London)	v_x (Brockett w/o convexity)	v_x (Brockett)	v_x (W–H)
70	0.044	0.065	0.071	0.068	0.049
71	0.084	0.068	0.072	0.068	0.068
72	0.071	0.072	0.073	0.072	0.070
73	0.076	0.076	0.074	0.075	0.065
74	0.040	0.080	0.076	0.079	0.065
75	0.104	0.085	0.077	0.083	0.079
76	0.160	0.090	0.083	0.088	0.089
77	0.058	0.095	0.098	0.093	0.088
78	0.110	0.103	0.111	0.103	0.093
79	0.093	0.114	0.123	0.118	0.105
80	0.139	0.130	0.135	0.135	0.127
81	0.154	0.153	0.152	0.156	0.153
82	0.183	0.185	0.174	0.179	0.181
83	0.206	0.213	0.206	0.207	0.210
84	0.239	0.240	0.249	0.244	0.239

(b) Smoothness and goodness-of-fit values

	London	Brockett w/o convexity	Brockett	W–H
S	0.0002	0.00023	0.000099	0.00094
F	17.53	21.67	18.49	15.72
Deviance	17.02	20.41	17.77	14.35
log-likelihood	-713.43	-715.13	-713.81	-712.10
χ^2	17.49	21.63	18.46	15.74

Concluding the numerical illustration, we can say that all the above mentioned graduation methods give almost the same results. From Tables 8.1b, 8.2b, and 8.3b we see that all the graduations are equivalent in terms of smoothness. The graduations are also equivalent as far as goodness-of-fit is concerned. The overall winner is the graduation through the minimization of the Jensen difference subject to constraints (i)–(v).

For a related numerical investigation see Sachlas and Papaioannou (2009). Graduation methods are usually compared in the literature by applying them to a specific small or large dataset and employing or using several bestness criteria (Debon et al., 2005, 2006).

8.6 Conclusions and comments

The Kullback–Leibler $D^{KL}(\mathbf{p}, \mathbf{q})$ and the Cressie–Read power divergence $D^{CR}(\mathbf{p}, \mathbf{q})$ involving nonprobability vectors share some of the properties of Kullback–Leibler and Cressie–Read power divergence, with probability vectors. Under some conditions, they are nonnegative, additive, and invariant under sufficient transformations. The property

Table 8.2. Several graduations through Jensen difference

(a) Graduated values

x	u_x	v_x (5 constraints)	v_x (6 constraints)	v_x (4 constraints)
70	0.044	0.062	0.054	0.059
71	0.084	0.066	0.061	0.064
72	0.071	0.071	0.068	0.069
73	0.076	0.075	0.075	0.073
74	0.040	0.080	0.082	0.078
75	0.104	0.086	0.089	0.085
76	0.160	0.093	0.097	0.092
77	0.058	0.099	0.104	0.098
78	0.110	0.106	0.112	0.105
79	0.093	0.113	0.119	0.112
80	0.139	0.131	0.138	0.132
81	0.154	0.156	0.159	0.157
82	0.183	0.182	0.180	0.184
83	0.206	0.209	0.201	0.212
84	0.239	0.238	0.222	0.242

(b) Smoothness and goodness-of-fit values

	5 constraints	6 constraints	4 constraints
S	0.000199	0.0002	0.0002
F	16.62	16.70	16.93
Deviance	16.40	16.89	16.48
log-likelihood	−713.12	−713.37	−713.16
χ^2	16.59	16.68	16.93

of maximal information and the limiting property are satisfied as well. Thus, we may regard $D^{KL}(\mathbf{p}, \mathbf{q})$ and $D^{CR}(\mathbf{p}, \mathbf{q})$ as measures of information. A minimal requirement for $D^{KL}(\mathbf{p}, \mathbf{q})$ and $D^{CR}(\mathbf{p}, \mathbf{q})$ to be measures of divergence is $\sum_i p_i = \sum_i q_i$. The Jensen difference between nonprobability vectors can also be regarded as a measure of divergence provided that $\sum_i p_i = \sum_i q_i$.

As an application of the previous results we explored the use of the Jensen difference and the general Cressie–Read power divergences in order to obtain graduated values. The minimization of the Jensen difference and the power divergence for various values of r, with constraints (i)–(v) and/or (vi), gave equivalent results, in terms of smoothness, to those of other methods of graduation such as the widely used Whittaker–Henderson method. For our numerical investigation, the overall winner is the graduation through the minimization of the Jensen difference subject to constraints (i)–(v).

The similarity of results between the methods of Whittaker–Henderson and power divergence or Jensen's difference minimization under the said constraints allows us to claim that the two graduation methods are nearly equivalent. This is supported not only by the numerical investigation but also from the fact that in Whittaker–Henderson we minimize a form of the Lagrangian function $F + hS$, while in power divergence we minimize F subject to, among others, a constraint on S which in turn leads to a similar Lagrangian.

Table 8.3. Several graduations through power divergence

(a) Graduated values

x	u_x	v_x $(r = 0.2)$	v_x $(r = 2/3)$	v_x $(r = 1)$	v_x $(r = 1.5)$
70	0.044	0.058	0.062	0.065	0.067
71	0.084	0.063	0.066	0.067	0.069
72	0.071	0.068	0.070	0.070	0.071
73	0.076	0.073	0.073	0.073	0.073
74	0.040	0.078	0.077	0.076	0.074
75	0.104	0.086	0.084	0.082	0.081
76	0.160	0.094	0.091	0.089	0.088
77	0.058	0.101	0.098	0.096	0.094
78	0.110	0.109	0.107	0.106	0.105
79	0.093	0.118	0.118	0.117	0.118
80	0.139	0.138	0.139	0.139	0.140
81	0.154	0.160	0.161	0.161	0.162
82	0.183	0.181	0.183	0.184	0.185
83	0.206	0.202	0.206	0.206	0.208
84	0.239	0.224	0.228	0.229	0.231

(b) Smoothness and goodness-of-fit values

	$r = 0.2$	$r = 2/3$	$r = 1$	$r = 1.5$
S	0.00019	0.00019	0.00017	0.00019
F	16.99	17.29	17.90	18.35
Deviance	16.79	16.84	17.17	17.42
log-likelihood	−713.32	−713.34	−713.51	−713.63
χ^2	16.97	17.27	17.87	18.31

References

Akaike, H. (1973). Information theory and an extension of the maximum likelihood principle. In B.N. Petrov and F. Csaki (Eds.), *2nd International Symposium on Information Theory*, 267–281, Akademiai Kiado, Budapest.

Basu, A., Harris, I.R., Hjort, N.L., Jones, M.C. (1998). Robust and efficient estimation by minimising a density power divergence, *Biometrika*, 85, 3, 549–559.

Benjamin, P., Pollard, J.H. (1980). *The Analysis of Mortality and Other Actuarial Statistics*, Heinemann, London.

Boyd, S., Vandenberghe, L. (2006). *Convex Optimization*, Cambridge University Press, Cambridge, UK.

Brockett, P.L. (1991). Information theoretic approach to actuarial science: A unification and extension of relevant theory and applications, *Transactions of the Society of Actuaries*, 43, 73–114.

Brockett, P.L., Zhang, J. (1986). Information theoretical mortality graduation, *Scandinavian Actuarial Journal*, 3–4, 131–140.

Burbea, J., Rao, C.R. (1982). On the convexity of some divergence measures based on entropy functions, *IEEE Transactions on Information Theory*, 28, 3, 489–495.

Cressie, N.A.C., Read, T.R.C. (1984). Multinomial goodness-of-fit tests, *Journal of Royal Statistical Society, B*, 46, 3, 440–464.

Cutler, A., Cordero-Brana, O.I. (1996). Minimum Hellinger distance estimation for finite mixture models, *Journal of American Statistical Association*, 91, 1716–1723.

Debon, A., Montes, F., Sala, R. (2005). A comparison of parametric models for mortality graduation. Application to mortality data for the Valencia region (Spain), *SORT*, 29, 2, 269–288.

Debon, A., Montes, F., Sala, R. (2006). A comparison of nonparametric methods in the graduation of mortality: Application to data from the Valencia region (Spain), *International Statistical Review*, 74, 2, 215–233.

Haberman, S. (1998). Actuarial Methods, *Encyclopedia of Biostatistics*, 1, (P. Armitage and T. Colton (Eds.), 37–49, John Wiley & Sons, New York.

Kullback, S. (1959). *Information Theory and Statistics*, John Wiley & Sons, New York.

Liese F., Vajda, I. (1987). *Convex Statistical Distances*, B.G. Teubner, Leipzig.

London, D. (1985). *Graduation: The Revision of Estimates*, ACTEX, Winsted, Connecticut.

Mathai, A.M., Rathie, P.N. (1975). *Basic Concepts in Information Theory and Statistics*, Wiley, New Delhi.

Miller, M.D. (1949). *Elements of Graduation*, Actuarial Society of America, New York.

Pardo, L. (2006). *Statistical Inference Based on Divergence Measures*, Chapman & Hall/CRC, Boca Raton, Florida.

Read, T.R.C., Cressie, N.A.C. (1988). *Goodness-of-Fit Statistics for Discrete Multivariate Data*, Springer-Verlag, New York.

Sachlas, A.P., Papaioannou, T. (2009). Divergences without probability vectors and their applications, *Applied Stochastic Models in Business and Industry* (to appear).

Teboulle, M. (1989). A simple duality proof for quadratically constrained entropy functionals and extension to convex constraints, *SIAM Journal of Applied Mathematics*, 49, 6, 1845–1850.

Vos, P.D. (1992). Minimum f-divergence estimators and quasi-likelihood functions, *Annals of the Institute of Statistical Mathematics*, 44, pp. 261–279.

Wang, J.L. (1998). Smoothing Hazard Rates, *Encyclopedia of Biostatistics*, 5, P. Armitage and T. Colton (Eds.), 4140–4150, John Wiley & Sons, New York.

Zhang, J., Brockett, P.L. (1987). Quadratically constrained information theoretic analysis, *SIAM Journal of Applied Mathematics*, 47, 4, 871–885.

Zografos, K., Ferentinos, K., Papaioannou, T. (1989). Limiting properties of some measures of information, *Annals of the Institute of Statistical Mathematics, B*, 41, 3, 451–460.

Part III

Asymptotic Behaviour of Stochastic Processes and Random Fields

9

Remarks on Stochastic Models Under Consideration

Ekaterina V. Bulinskaya

Department of Mathematics and Mechanics, Moscow State University, Russia

Abstract: The results obtained and the methods used in the chapters included in Part III are discussed with emphasis on possible applications.

Keywords and phrases: Stochastic models, input-output systems, branching process in random environment, asymptotics of time compression in queueing systems, cost approach, asymptotically optimal policies, dependent random fields, central limit theorem, approximation accuracy, branching random walks, conditional limit theorems

9.1 Introduction

Part III is devoted to analysis of models arising in various domains comprising some biological systems, queueing networks, insurance, interacting particles systems, and others. All these diverse models are described by stochastic processes or random fields.

Two chapters treat the branching processes. Namely, Afanasyev studies the asymptotic behaviour of the conditional branching process in a random environment (BPRE). The novel feature of his chapter is the assumption that the population size exceeds a high level whereas the usual hypothesis is the process survival. New types of invariance principles are established for the critical BPRE. In the chapter by Yarovaya the combinations of random walks on a lattice $\mathbb{Z}^d$ and a branching process with a single source (a point where births and deaths can occur) are considered. Limit behaviour of critical and subcritical branching random walks is studied in this setting. An important problem considered is the effect of lattice dimension on reaching the critical regime. The above-mentioned research direction admits various applications in biological and physical as well as queueing models.

Two other chapters deal with the so-called input-output systems. Afanas'eva assumes the input flow to be an integer-valued stochastic process with nondecreasing trajectories (in particular, a doubly stochastic Poisson or a regenerative process), the output being a renewal one. As usual in queueing theory, the main object of interest is the system state $q(t)$, i.e., the number of customers present at time t. The asymptotically Gaussian behaviour of $q(t)$ is proved under appropriate time compression. Bulinskaya studies the problem of stochastic control to optimize the system performance in the framework of the cost approach. The principal advantage of this approach proposed

C.H. Skiadas (ed.), *Advances in Data Analysis*, Statistics for Industry and Technology, DOI 10.1007/978-0-8176-4799-5_9,

by the author for using in actuarial sciences is the possibility to study more realistic insurance models taking into account the investment activity and dividend payment (thus going beyond the traditional ruin problems). Asymptotically optimal controls are also introduced and their explicit form is established along with sensitivity analysis.

The remaining two chapters are concerned with the central limit theorem (CLT) for dependent random fields comprising positively or negatively associated ones. Bulinski proves the CLT for a two-scale procedure when a dependent random field is defined on subsets of $\mathbb{R}^d$ growing in the Van Hove sense and simultaneously the grids of observations become increasingly dense. To this end a generalisation of the dependence condition for random fields on a lattice $\mathbb{Z}^d$ is proposed. The statistical versions of the CLT obtained under the random normalisation (different from studentisation) are important for applications to dependent data analysis. Shashkin studies a (BL, θ)-dependent random field on a graph. Such a model has a number of interesting applications discussed in Section 9.3 below. Moreover, the convergence rate in the CLT is estimated (the Berry–Esseen type theorem).

Thus, each chapter emphasizes a different aspect of limit behaviour; consequently, various methods for the study of asymptotics are demonstrated. The results obtained are not only important from a theoretical viewpoint but they are useful for applications as well.

9.2 Results and methods

Now we dwell on the contributions of the authors of this part in alphabetical order. The list of references accompanying these introductory remarks contains only the sources supplementing those mentioned by the authors.

In Afanasyev (Chapter 10) the popular model of BPRE is considered. Its basic distinction from the classic Galton–Watson model consists in discarding the homogeneity requirement that the reproduction laws are identical for all generations. Moreover, in the model of BPRE it is supposed that these laws are generated by a random mechanism with given probabilistic characteristics.

Let ξ_n be the size of the nth generation of BPRE. Originally the process $\{\xi_n\}$ was investigated under the condition that the population survives for a long time; i.e., $\xi_n > 0$, $n \to \infty$. The asymptotics of the nonextinction probability was obtained and different limit theorems were proved including invariance principles which are similar to the Donsker–Prohkorov one for a random walk. For a critical BPRE (the most difficult to study from a mathematical point of view) one can refer in this regard to Afanasyev et al. (2005).

The present chapter studies the population assuming that its size becomes arbitrarily large. The author concentrates on new invariance principles for a critical BPRE $\{\xi_n\}$ and for the hitting time $T(x) = \min\{n : \xi_n \geq x\}$ entailing the conditional limit theorems proved in Afanasyev (1999, 2006) for occupation and hitting times. The processes are considered in two timelines: absolute and relative (in which the process lifetime is taken as a unit). Weak convergence technique in $D[0, \infty)$ and the invariance principle previously obtained by the author (under condition that the population survives for a long time) are used to establish the new ones. The distributions of the limit processes are expressed in terms of the laws for certain functionals in Brownian excursion (see, e.g., Borodin and Salminen, 2002 for definitions). Moreover, the author gives the explicit form of the finite-dimensional distributions for one of the limit processes.

The chapter by Afanas'eva (Chapter 11) is devoted to analysis of a multichannel queueing system in asymptotics of time compression (i.e., instead of a function $X(t)$ where $t \geq 0$ one uses $X_T(t) = X(tT)$ as $T \to \infty$). Under some natural assumptions weak convergence of one-dimensional distributions of the normalized customer's number (in an infinite-channel system) to the Gaussian law is established. Normalising coefficients are expressed in terms of input flow moments, such as mean and covariance function, and the service time distribution. As an interesting illustration, systems with input given by a doubly stochastic Poisson process (DSPP) and a regenerative one are considered. It is also proved that the loss system with n_T servers behaves in the same way as an infinite-channel one if n_T grows fast enough, as $T \to \infty$. The author applies the characteristic functions technique and properties of the regenerative, doubly stochastic, and Markov-modulated processes.

We mention in passing that DSPP is also called a Cox or conditional Poisson process. For basic definitions one can use, e.g., Thorisson (2000), Serfozo (1999), Daley and Vere-Jones (1988), and Brémaud (1981).

Chapter 12 by Bulinskaya develops a cost approach in actuarial sciences introduced by the author in 2003. It is well known that up to now the reliability approach (i.e., the study of the ruin probability) prevailed because the insurance companies were created to share and transfer risks. However, during the last decade the investments and dividend payments attracted the attention of many researchers (see, e.g., Taksar and Hunderup, 2007; Gerber et al., 2007;Bulinskaya et al., 2007;Dickson and Waters, 2004, and references therein).

The optimal policies of an insurance company minimizing its expected losses during a fixed planning horizon are found under the realistic assumptions of boundedness of assets and loan amounts. The Bellman method is used for this purpose. The optimal policies obtained are of threshold nature and depend on cost parameters and planning horizons. For users' convenience the asymptotically optimal policies are proposed as well. Their main advantage is stationarity (critical levels are the same for each step). To prove these results the properties of renewal processes are employed. It is worth mentioning that the sensitivity analysis with respect to fluctuations of the cost parameters and perturbation of the underlying process distribution is also carried out. Here the differential importance measure, Sobol decomposition and probability metrics (see, e.g., Zolotarev, 1997; Rachev, 1991) are involved. The so-called empirical asymptotically optimal policies based on observations are indicated as a future research direction.

The study of dependent random fields became a very important branch of modern probability theory. Starting from the pioneering papers by Harris, Lehmann, Esery, Proschan, Walkup, Fortuin, Kastelein, and Ginibre the theory of positively or negatively dependent random systems and their modifications was developed. The main sources of interest here are mathematical statistics, reliability theory, statistical physics, and percolation (see the quite recent book by Bulinski and Shashkin, 2007 and references therein). In the chapter by Bulinski (Chapter 13) the new CLTs for random fields are established comprising the classical Newman theorem for associated fields. Namely, the author uses the (BL, θ)-dependence, i.e., the specified inequalities for covariances of the test (bounded Lipschitz) functions in observations. For random fields on a lattice $\mathbb{Z}^d$ this approach was introduced and employed in Bulinski and Suquet (2001). Now the generalisation of this dependence condition is proposed to study random fields defined on subsets of $\mathbb{R}^d$. The author uses a two-scale procedure mentioned in the Introduction. An important achievement of this research is the new version of the CLT for stationary random fields with random normalization. The main tools here are characteristic functions, covariance inequalities, and truncation of initial random variables.

The chapter by Shashkin (Chapter 14) contains interesting results for random fields defined on a graph. Likewise for the random fields defined on an Euclidean lattice, the first results in this domain describe the phase transition and cluster existence properties for interacting particle models. In probability theory there are a number of deep results, having various statistical applications, concerned with the limit behaviour of sums of random variables.

As in practice all the sets of observations are finite, one has not only to prove the CLT but also to establish an estimate of accuracy of the Gaussian approximation, so a Berry–Esseen type theorem is provided. To prove the main result for a (BL, θ)-dependent random field on a graph the Stein–Tikhomirov technique is applied.

Numerous traditional applications of branching processes (see, e.g., Athreya and Ney, 1972) in various areas of natural science have demonstrated the necessity of developing more realistic stochastic models in which population evolution depends on the structure of a medium (see, e.g., Molchanov, 1994). Chapter 15 by Yarovaya is concerned with one such model taking into account that the particles multiply obeying a branching process and simultaneously move in space according to a specified random walk. The important problem considered there is the study of particle population evolution. As known (see, e.g., Gartner and Molchanov, 1990), the inhomogeneity of a medium plays an essential role in the formation of abnormal properties of transport processes. Consequently, interest in branching random walks under the assumption that the birth and the death of particles occur at a single lattice point (i.e., a source) has increased, see, e.g., Vatutin and Xiong (2007).

Behaviour of branching random walks depends on reproduction intensity at the source. In this connection, the definition of criticality for a branching random walk is introduced. Special attention in this chapter is paid to critical and subcritical processes. The asymptotic behaviour of survival probability is obtained and used to prove conditional limit theorems for the population size (under assumption of the process survival). The results differ for low ($d = 1, 2$) and high ($d \geq 3$) dimensions. The Laplace transforms and subsequent application of Tauberian theorems are heavily exploited in proofs.

9.3 Applications

We start with applications in domains mainly related to economics.

Queueing and insurance models, as well as those arising in finance, dams, inventories, reliability, or evolution of biological populations belong to the class of input-output models. Thus the results obtained for one research domain are of value for the others. Sometimes it is enough to give another interpretation to underlying processes; see, e.g., Bulinskaya (2007a). For decision making it is necessary to compare various strategies (policies) to control the system (see, e.g., Bulinskaya, 2007b) and therefore it is desirable to choose that which is rather simple and not too sensitive to fluctuations of various parameters (see also Pappenberger et al., 2008). In particular, due to climate changes, the problem of floods and dam construction becomes quite important. Using the loss functions one can formalise the problem and establish the specified critical levels.

Evidently investigation of queueing systems with complicated input flows is motivated by applications. Thus the number of calls to flying control officers from aircraft in the airport area are well described by a Markov modulated process. Another example is a Poisson process with periodic intensity having a random amplitude (see, e.g., Afanas'eva, 1984). One can also recall the models for road traffic in cities. Consideration of multiphase service leads to complex input processes even if the input to the first phase is Poissonian. Note that the greater part of results in queueing theory is proved for the case of Poisson input flow. It cannot be explained only by simplicity of such models. In fact, due to well-known theorems concerning the sums of large numbers of independent flows, many real situations are well approximated by systems with Poisson inputs (see, e.g., Grigelionis, 1963). On the other hand, if input flow is complicated then, with rare exception, one cannot obtain explicit formulae for a system's operational characteristics (mean waiting time, loss probability, average queue length). Hence, in the study of such systems one can indicate the following two directions. The first one deals with estimation of various important characteristics of the model (see, e.g., Rolski, 1986). The second one comprises the investigation of extreme regimes, namely, heavy and light traffic. We mention only pioneering works by Kingman, Rise, Prokhorov, Viskov, and Borovkov in the heavy-traffic case, and Kovalenko, Assmussen, Daley, and Rolski in the light-traffic one, referring for details to Gnedenko and Kovalenko (2005). The critical and subcritical queues are considered in Peköz and Blanchet (2006), whereas the martingale approach is treated in Pang et al. (2007).

Due to inequalities $q^0(t) \le q(t) \le q^\infty(t)$, for any t, with $q(t)$, $q^0(t)$, and $q^\infty(t)$ denoting the customer's number in an infinite-channel system, a loss system, and a system with unbounded waiting time (possessing n channels), respectively, investigation of infinite-channel systems provides the upper or lower bound for the n-channel systems (with loss or unbounded waiting time). These facts are also useful for applications.

BPRE is applied to model evolution of biological populations. In particular, biologists and mathematicians use it nowadays to solve the problem of the nearest mutual ancestor of humans (see, e.g., O'Connel, 1995). Note that the study of conditional processes assuming that the population size becomes large is connected with new applied areas of this research. So, for example, a challenging problem is to reveal the asymptotic properties of a Galton–Watson branching process given that its total progeny is equal to n, $n \to \infty$. This is connected directly with the theory of random trees (see, e.g., Vatutin and Dyakonova, 2002; Geiger and Kauffmann, 2004). It seems that this scheme could be useful in some stochastic models related to medicine (concerning immunology). As has been shown recently various types of branching random walks are used as approximations of catalytic superprocesses (see, e.g., Topchii et al., 2003); their theory forms a domain of active development (see Fleishman and Le Gall, 1995; Greven et al., 1999). It is worth mentioning that the concept of "strong centers" is used for the interpretation of the intermittency phenomenon in the theory of random media. This research may also attract attention due to possible applications in the survival analysis of cell populations.

The mathematical models in radiobiology and oncology are of great importance. This is a vast domain for stochastic and statistical analysis of very complicated problems. Recall that there are various stochastic models describing the tissue (or organ) response under irradiation. One uses, e.g., the single-hit, multiple-hit, or LQ-models for probability that a cell after irradiation of a certain dose will be alive. Due to Withers et al. (1988) the idea of independent functional subunits (FSUs) was introduced for biological modeling. The approach based on this idea and involving the binomial distribution was developed during the last two decades; see, e.g., Stavrev et al. (2001) and references therein. Namely, the tissue (or organ) consists of N functional subunits that behave "statistically independently" under irradiation (it is also possible to consider the irradiation of part of the tissue or organ). One assumes that there exists "the functional reserve" M (or critical volume), i.e., the number of structural elements that must be damaged to cause a failure in the structure of interest.

In Bulinski and Khrennikov (2005) a generalization of the critical volume model was proposed. The idea of the dependent (in particular, independent) FSUs was expressed in terms of dependent (mixing) random fields defined on a lattice $\mathbb{Z}^d$. In Bulinski (Chapter 13) the next important step is made. Namely, the author considers the dependence structures based on positive or negative association (comprising independent random variables) and dwells on general (BL, θ)-dependence. The impact of the irradiation on the FSU can be described now not only in terms of the indicator functions to take into account the intermediate cases between the killed and alive FSUs. Moreover, the general case of growing in the van Hove sense domains in $\mathbb{R}^d$ with increasing density of grids of observations is investigated. The results obtained show that the approximation of random sums describing the collective effects of the summand behaviour should involve the possible dependence structure of FSUs. Hence, instead of the traditional normal approximation for independent summands other explicit formulae are proposed. Thus for strictly stationary random fields one can construct the approximate confidence intervals for unknown mean value. The problem of nonindividual, say, population response and the problems of nonuniform irradiation are interesting and important for further development.

Dependent random systems indexed by points of a regular graph are an important instrument in the study of natural or social phenomena; see, e.g., Newman et al. (2002). They include such classical notions as the Ising model and percolation on the finite-dimensional Euclidean lattice, which are useful as models for ferromagnetic materials and liquid distribution in a porous medium. Thus it is interesting to obtain theorems concerning the behaviour of such systems.

In recent years much attention has been drawn to Ising models, contact process, and similar dependent systems indexed by graphs which cannot be embedded in an Euclidean lattice. Besides pure mathematical interest, this is also enforced by applications in new models coming into consideration (see, e.g., the overview in Schonmann, 2001). For example, in physics and chemistry the structure of some newly analysed molecular formation is so complicated that it should be considered as a nonamenable transitive graph. In theoretical radiotherapy it is important to study the formation of the vascular system in a tumour (which can also be represented as an interacting system on a graph), since its dimension is the crucial characteristic of the proliferation and destruction of the tumour under irradiation (Sabo et al., 2001). Finally, in economic applications a regular graph may serve as a good model of producer–salesman–customer relations. This construction is known as a Petri net which is a graph made of actors and transactions; see, e.g., Artigues and Roubellat (2001). In the mathematical version of all these models it is natural to impose some dependence conditions. It is known that association of random variables arises naturally if the random variables involved are positively dependent, as in the classical Ising model. Associated random systems exhibit behaviour similar to that of independent ones if their covariance function decreases rapidly at infinity. The latter condition is usually easily verifiable. Moreover, if that is the case then an associated system has a property called (BL, θ)-dependence, which appears also in negatively associated systems, as well as in some other natural situations. For example, such a simple process as autoregression (possibly a nonlinear one but Lipshitz) often satisfies this definition; see the book by Bulinski and Shashkin (2007). The same is true for its multiparametric extensions. So, if autoregression is used to describe the graph-indexed system, then its main characteristics can be estimated via the central limit theorem for (BL, θ)-dependent systems whereas the Berry–Esseen type theorem allows us to estimate the accuracy of approximation.

Consequently, the results obtained are useful for practice.

Acknowledgements. It is a pleasure to express my gratitude to Professor C. H. Skiadas for help in organizing the special session "Asymptotic Behaviour of Stochastic Processes and Random Fields" at the ASMDA-2007 conference and to all the participants of the present Part III for valuable contributions and friendly discussions of the problems under consideration.

References

Artigues, Ch., Roubellat, F. (2001). Petri net model and a general method for on and off-line multi-resource shop floor scheduling with setup times. *Int. J. Prod. Econ.*, 74:63–75.

Athreya, K., Ney P. (1972). *Branching Processes*, Springer Verlag, New York.

Borodin, A.N. and Salminen, P. (2002). *Handbook on Brownian Motion–Facts and Formulae*, 2nd ed., Birkhäser, Basel-Boston-Berlin.

Brémaud, P. (1981). *Point Processes and Queues: Martingale Dynamics*, Springer-Verlag, New York.

Bulinskaya, E., Karapetyan, N., Yartseva, D. (2007). On dividends payment by insurance company. *Transactions of XXVI International Seminar on Stability of Stochastic Models*, Nahariya, Israel, 1:44–51.

Daley, D. and Vere-Jones, D. (1988). *An Introduction to the Theory of Point Processes.* Springer, New York.

Geiger, J., Kauffmann, L. (2004). The shape of large Galton-Watson trees with possibly infinite variance. *Random Struct. Algorithms*, 25:311–335.

Gerber, H.U., Shiu, E.S.W., Smith, N. (2007). Methods for estimating the optimal dividend barrier and the probability of ruin. *Insu.: Math. Econ.*, 42:243–254.

Gnedenko, B.V., Kovalenko, I.N. (2005). *Introduction to Queueing Theory* Komkniga, Moscow (in Russian).

Grigelionis, B. (1963). On the convergence of sums of random step processes to a Poisson process. *Teor. Veroyatn. Primen.*, 8:189–194.

Newman, M.E.J., Watts, D.J., Strogatz, S.H. (2002). Random graph models of social networks. *Proc. Nat. Acad. Sci.*, 99:2566–2572.

O'Connel, N. (1995). The genealogy of branching processes and the age of our most recent common ancestor. *Adv. Appl. Prob.*, 27:418–442.

Pang, G., Talreja, R., Whitt, W. (2007). Martingale proofs of many-server heavy-traffic limits for Markovian queues. *Probab. Surv.*, 4:193–267.

Pappenberger, F., Beven, K.J., Ratto, M., Matgen, P. (2008). Multi-method global sensitivity analysis of flood inundation models. *Adv. Water Resour.*, 31:1–14.

Peköz, E.A., Blanchet, J. (2006). Heavy traffic limits via Brownian embedding. *Probab. Eng. Inf. Sci.*, 20:595–598.

Rachev, S.T. (1991). *Probability Metrics and the Stability of Stochastic Models.* Wiley, Chichester.

Sabo, E., Boltenko, A., Sova, Y., Stein, A., Kleinhaus, S., Resnick, M.B. (2001). Microscopic analysis and significance of vascular architectural complexity in renal cell carcinoma. *Clin. Cancer Res.*, 7:533–537.

Schonmann, R. (2001). Multiplicity of phase transitions and mean-field criticality on highly non-amenable graphs. *Commun. Math. Phys.*, 219:271–322.

Serfozo, R. (1999). *Introduction to Stochastic Networks.* Springer, New York.

Taksar, M., Hunderup, C.L. (2007). The influence of bankruptcy value on optimal risk control for diffusion models with proportional reinsurance. *Ins.: Math. Econ.*, 40:11–21.

Thorisson, H. (2000). *Coupling, Stationarity and Regeneration.* Probability and its Applications. Springer, New York.

Vatutin, V.A., Dyakonova, E.E. (2002). Reduced branching processes in random environment. In *Mathematics and Computer Science II: Algorithms, Trees, Combinatorics and Probabilities.* Ed. B. Chauvin et al., Birkhäuser, Basel, pp. 455–467.

Withers, H.R., Taylor, J.H., Maciejewski, B. (1988). Treatment volume and tissue tolerance. *Int. J. Radiat. Oncol. Biol. Phys.*, 14:751–759.

Zolotarev, V.M. (1997). *Modern Theory of Summation of Independent Random Variables.* VSP, Utrecht.

10

New Invariance Principles for Critical Branching Process in Random Environment

Valeriy I. Afanasyev

Department of Discrete Mathematics, Steklov Institute, Moscow, Russia

Abstract: The invariance principles are proved for a branching process in a random environment provided that the size of its population reaches a high level.

Keywords and phrases: Galton–Watson branching process, branching process in a random environment, conditional invariance principles and limit theorems, Brownian meander, Brownian excursion

10.1 Introduction

In a classical Galton–Watson branching process particles of any generation split independently of the others and the history of the process according to the same reproduction law. Suppose now that particles of the nth generation, $n \in \mathbb{N}_0 := \{0, 1, \ldots\}$, have their own reproduction law $\varPi_n = \left\{p_0^{(n)}, p_1^{(n)}, \ldots\right\}$. It means that

$$\mathbf{P}\left(\text{a particle of } n\text{th generation splits into } k \text{ particles}\right) = p_k^{(n)}$$

for any $k \in \mathbb{N}_0$. The set $\{\varPi_0, \varPi_1, \ldots\}$ is called a *varying environment* in contrast to an *invariable environment* for a Galton–Watson branching process.

This model of the branching process in a varying environment has a shortcoming. It is difficult to keep the information about the countable set of reproduction laws. Therefore, it is convenient to suppose that these laws are created by means of some random mechanism and if these laws are fixed, the particles split as stated above.

Suppose henceforth that the random sequences $\varPi_0$, $\varPi_1, \ldots$ are independent and identically distributed. The set $\{\varPi_0, \varPi_1, \ldots\}$ is called a *random environment.*

To formalize the above description of the process, introduce the generating functions of $\varPi_n$:

$$\varphi_n(s) = \sum_{k=0}^{\infty} p_k^{(n)} s^k, \qquad s \in [-1, 1], \qquad n \in \mathbb{N}_0 \ .$$

C.H. Skiadas (ed.), *Advances in Data Analysis*, Statistics for Industry and Technology, DOI 10.1007/978-0-8176-4799-5_10,

In the case of a branching process in a varying environment the generating function of the number of particles in the nth generation equals $\varphi_0\left(\varphi_1\left(\varphi_2\left(\ldots\varphi_{n-1}(s)\ldots\right)\right)\right)$.

A *branching process in a random environment* (BPRE) is defined by the relation

$$\mathbf{E}\left(s^{\xi_n} \mid \Pi_0, \Pi_1, \ldots, \Pi_{n-1}\right) = \varphi_0\left(\varphi_1\left(\varphi_2\left(\ldots\varphi_{n-1}(s)\ldots\right)\right)\right)$$

for $s \in [-1,1]$, $n \in \mathbb{N}$, where ξ_n is the number of particles in the nth generation, $\xi_0 = 1$. Hence, if the random environment is fixed, we get a branching process in a varying environment.

At present BPRE is a popular model to study the evolution of biological populations. This model reflects many important features of biological populations that do not occur in the classical Galton–Watson branching process.

Consider the random variables

$$X_n = \ln \varphi'_{n-1}(1), \qquad \eta_n = \frac{\varphi''_{n-1}(1)}{\left(\varphi'_{n-1}(1)\right)^2}, \qquad n \in \mathbb{N}.$$

Suppose the process $\{\xi_n\}$ to be critical, i.e., $\mathbf{E}X_1 = 0$, and satisfy the following conditions:

$$0 < \mathbf{E}X_1^2 := \sigma^2 < +\infty, \qquad \mathbf{E}\ln^q\left(\eta_1 \vee 1\right) < +\infty, \tag{10.1}$$

for some $q > 2$.

It is proved in Afanasyev et al. (2005), for a critical BPRE $\{\xi_n\}$ satisfying the conditions (10.1), that, as $n \to \infty$,

$$\mathbf{P}\left(\xi_n > 0\right) \sim \frac{c_1}{\sqrt{n}}, \tag{10.2}$$

where c_1 is a positive constant and

$$\left\{\frac{\ln \xi_{\lfloor nt \rfloor}}{\sigma\sqrt{n}}, \ t \in [0,1] \;\middle|\; \xi_n > 0\right\} \xrightarrow{D} \left\{W^+(t), \ t \in [0,1]\right\},$$

the random process $\{W^+(t)\}$ is a *Brownian meander*; the sign $\xrightarrow{D}$ means the convergence in distribution in the space $D[0,1]$ with the Borel σ-algebra in the Skorokhod topology.

As a consequence of these results we can get (see Afanasyev, 2006) the following two invariance principles, as $n \to \infty$, for a critical BPRE $\{\xi_n\}$ satisfying the conditions (10.1).

(1) *In the absolute timeline*:

$$\left\{\frac{\ln\left(\xi_{\lfloor nt \rfloor} \vee 1\right)}{\sigma\sqrt{n}}, \ t \in [0,+\infty) \;\middle|\; \xi_n > 0\right\} \xrightarrow{D} \left\{\frac{W_0^+\left(\alpha^2 t\right)}{\alpha}, \ t \in [0,1]\right\}; \tag{10.3}$$

(2) *In the relative timeline* (depending on $T := \min\{n : \xi_n = 0\}$, the lifetime of the process $\{\xi_n\}$):

$$\left\{\frac{\ln\left(\xi_{\lfloor tT \rfloor} \vee 1\right)}{\sigma\sqrt{n}}, \ t \in [0,+\infty) \;\middle|\; \xi_n > 0\right\} \xrightarrow{D} \left\{\frac{W_0^+(t)}{\alpha}, \ t \in [0,1]\right\}, \tag{10.4}$$

where the random process $\{W_0^+(t)\}$ is a *Brownian excursion* ($W_0^+(t) = 0$ for $t > 1$); α is a random variable uniformly distributed in $(0,1)$ and independent of $\{W_0^+(t)\}$.

Sign $\xrightarrow{D}$ in relations (10.3), (10.4) means the convergence in distribution in the space $D[0,+\infty)$ with the Borel σ-algebra in the Skorokhod topology.

10.2 Main results

Let $T(x)$ be the hitting time of the semi-axis $[x,+\infty)$, $x \in [0,+\infty)$, by the process $\{\xi_n\}$; i.e.,

$$T(x) = \min\{n : \xi_n \geq x\}.$$

In some probability theory problems a necessity can arise to consider the process $\{\xi_n\}$ conditioned by $T(x) < +\infty$ (it means that the size of some BPRE generation is not less than x). In Afanasyev (1999, 2006) for a critical BPRE $\{\xi_n\}$ satisfying the conditions (10.1), it is shown that, as $x \to +\infty$,

$$\mathbf{P}(T(x) < +\infty) \sim \frac{c_2}{\ln x}, \tag{10.5}$$

where $c_2 = c_1\sigma\sqrt{\pi/2}$, and the limit theorems are proved for the random variables $T/\ln^2 x$, $T(x)/\ln^2 x$, $T(x)/T$, $\mu(x)/\ln^2 x$, $\nu(x)/\ln^2 x$, $\mu(x)/T$, $\nu(x)/T$ (here $\mu(x)$, $\nu(x)$ are the capacities of the sets $\{i : \xi_i > x\}$ and $\{i : \xi_i \leq x\}$, respectively) conditioned by $T(x) < +\infty$.

The aim of the chapter is to obtain the invariance principles for the critical BPRE conditioned by $T(x) < +\infty$ entailing all these theorems.

Introduce the functionals for a Brownian excursion $\{W_0^+(t)\}$:

(1) *The maximum*

$$h_0^+ = \sup_{t\in[0,1]} W_0^+(t).$$

(2) *The hitting time of the semi-axis* $[x,+\infty)$,

$$\tau_0^+(x) = \inf\{t : W_0^+(t) \geq x\}, \qquad x \in [0,+\infty).$$

(3) *The local time at the level* t,

$$l_0^+(t) = \lim_{\varepsilon\downarrow 0}\frac{1}{\varepsilon}\int_0^1 I_{[t,t+\varepsilon]}\left(W_0^+(s)\right)ds, \qquad t \in [0,+\infty),$$

where $I_{[t,t+\varepsilon]}(\cdot)$ is the indicator function of the set $[t,t+\varepsilon]$.

Theorem 1. *Let* $\{\xi_n\}$ *be a critical BPRE satisfying the conditions* (10.1). *Then, as* $x \to +\infty$,

$$\left\{\frac{\ln(\xi_{\lfloor tx^2\sigma^{-2}\rfloor} \vee 1)}{x},\ t \in [0,+\infty) \,\middle|\, T(\exp x) < +\infty\right\} \xrightarrow{D} \{Y(t),\ t \in [0,+\infty)\},$$

where $\{Y(t)\}$ *is a random process with continuous trajectories and distribution given by*

$$\boldsymbol{P}(\{Y(t)\} \in A) = \sqrt{\frac{2}{\pi}} \int_0^{+\infty} \boldsymbol{P}\left(\left\{\frac{W_0^+(tv^2)}{v}\right\} \in A,\ h_0^+ > v\right) dv$$

for any Borel set A *of the space* $D[0,+\infty)$.

Theorem 2. *Let* $\{\xi_n\}$ *be the critical BPRE satisfying the conditions* (10.1). *Then, as* $x \to +\infty$,

$$\left\{\frac{\ln(\xi_{\lfloor tT \rfloor} \vee 1)}{x},\ t \in [0,+\infty) \,\middle|\, T(\exp x) < +\infty\right\} \xrightarrow{D} \left\{\widetilde{Y}(t),\ t \in [0,+\infty)\right\},$$

where $\left\{\widetilde{Y}(t)\right\}$ *is a random process with continuous trajectories and distribution given by*

$$\boldsymbol{P}\left(\left\{\widetilde{Y}(t)\right\} \in A\right) = \sqrt{\frac{2}{\pi}} \int_0^{+\infty} \boldsymbol{P}\left(\left\{\frac{W_0^+(t)}{v}\right\} \in A,\ h_0^+ > v\right) dv$$

for any Borel set A *of the space* $D[0,+\infty)$.

The absolute timeline is considered in Theorem 1, while Theorem 2 deals with the relative timeline. It is interesting to mention that the invariance principle for the critical Galton–Watson branching process has a similar form. Let Z_n be the number of particles in the nth generation of this process. It turned out that (see Afanasyev, 2005) if $\mathbf{Var} Z_1 := 4\lambda \in (0,+\infty)$, then, as $n \to \infty$,

$$\left\{\frac{Z_{\lfloor nt \rfloor}}{\lambda n},\ t \in [0,+\infty) \,\middle|\, Z_n > 0\right\} \xrightarrow{D} \left\{Y^+(t),\ t \in [0,+\infty)\right\},$$

where $\{Y^+(t)\}$ is a random process with continuous trajectories and distribution given by

$$\mathbf{P}\left(\{Y^+(t)\} \in A\right) = \sqrt{\frac{2}{\pi}} \int_0^{+\infty} \mathbf{P}\left(\left\{\frac{l_0^+(tv)}{v}\right\} \in A,\ h_0^+ > v\right) dv$$

for any Borel set A of the space $D[0,+\infty)$.

If we depict the graph of the process $\{\xi_n\}$ in the co-ordinate system yOz (n on the axis Oy, and ξ_n on the axis Oz), then the condition $\xi_n > 0$ means that we are interested in trajectories of this process elongated along the axis Oy, whereas in relations (10.3) and (10.4) their form along the axis Oz is considered. The condition $T(x) < +\infty$ means that we are interested in trajectories elongated along the axis Oz. If we are interested in their form along the axis Oy, then we need the next two invariance principles (the first one is for the absolute timeline and the second one is for the relative timeline).

Theorem 3. *Let* $\{\xi_n\}$ *be a critical BPRE satisfying the conditions* (10.1). *Then, as* $x \to +\infty$,

$$\left\{\frac{\sigma^2 T(ux)}{\ln^2 x},\ u \in [0,+\infty) \,\middle|\, T(x) < +\infty\right\} \xrightarrow{D} \left\{H(u),\ u \in [0,+\infty)\right\},$$

where $\{H(u)\}$ *is a random process with trajectories from the space* $D[0,+\infty)$ *and distribution given by*

$$\boldsymbol{P}(\{H(u)\} \in A) = \sqrt{\frac{2}{\pi}} \int_0^{+\infty} \boldsymbol{P}\left(\left\{\frac{\tau_0^+(uv)}{v^2}\right\} \in A,\ h_0^+ > v\right) dv$$

for any Borel set A *of the space* $D[0,+\infty)$.

Theorem 4. *Let* $\{\xi_n\}$ *be a critical BPRE satisfying the conditions* (10.1). *Then, as* $x \to +\infty$,

$$\left\{\frac{T(ux)}{T},\ u \in [0,+\infty)\ \middle|\ T(x) < +\infty\right\} \xrightarrow{D} \left\{\widetilde{H}(u),\ u \in [0,+\infty)\right\},$$

where $\{\widetilde{H}(u)\}$ *is a random process with trajectories from the space* $D[0,+\infty)$ *and distribution given by*

$$\boldsymbol{P}\left(\left\{\widetilde{H}(u)\right\} \in A\right) = \sqrt{\frac{2}{\pi}} \int_0^{+\infty} \boldsymbol{P}(\{\tau_0^+(uv)\} \in A,\ h_0^+ > v)\, dv$$

for any Borel set A *of the space* $D[0,+\infty)$.

10.3 Proof of Theorem 1

Introduce for $x \in (1,+\infty)$ the random process Y_x:

$$Y_x(t) = \frac{\ln(\xi_{\lfloor t\sigma^{-2}\ln^2 x\rfloor} \vee 1)}{\ln x}, \qquad t \in [0,+\infty),$$

and for $v \in (0,\infty)$ the random process Z_v:

$$Z_v(t) = \frac{W_0^+(v^2 t)}{v}, \qquad t \in [0,\infty).$$

It is required to show that, for arbitrary Borel set A from $D[0,+\infty)$ satisfying the condition $\mathbf{P}(Y \in \partial A) = 0$ (∂A is the boundary of the set A, $Y = \{Y(t)\}$), the following equality is valid...

$$\lim_{x\to+\infty} \mathbf{P}(Y_x \in A \mid T(x) < +\infty) = \mathbf{P}(Y \in A).$$

By the definition of the process Y this relation can be written as follows,

$$\lim_{x\to+\infty} \mathbf{P}(Y_x \in A \mid T(x) < +\infty) = \sqrt{\frac{2}{\pi}} \int_0^{+\infty} \mathbf{P}(Z_v \in A,\ h_0^+ > v)\, dv. \tag{10.6}$$

Note that for arbitrary $\varepsilon > 0$

$$\mathbf{P}\left(Y_x \in A\ ,\ T(x) < +\infty\right) = P_1\left(x, \varepsilon\right) + P_2\left(x, \varepsilon\right), \tag{10.7}$$

where

$$P_1\left(x, \varepsilon\right) = \mathbf{P}\left(Y_x \in A\ ,\ T(x) < +\infty,\ T > \varepsilon\sigma^{-2}\ln^2 x\right),$$
$$P_2\left(x, \varepsilon\right) = \mathbf{P}\left(Y_x \in A\ ,\ T(x) < +\infty,\ T \leq \varepsilon\sigma^{-2}\ln^2 x\right).$$

The essential role further is played by the following statement which is proved in Afanasyev (2006).

Lemma 1. *If the conditions (10.1) hold, then*

$$\lim_{\varepsilon \downarrow 0} \limsup_{x \to +\infty} \boldsymbol{P}\left(T \leq \varepsilon \ln^2 x \mid T(x) < +\infty\right) = 0.$$

Since

$$P_2\left(x, \varepsilon\right) \leq \mathbf{P}\left(T(x) < +\infty,\ T \leq \varepsilon\sigma^{-2}\ln^2 x\right),$$

it follows from Lemma 1 that

$$\lim_{\varepsilon \downarrow 0} \limsup_{x \to +\infty} \frac{P_2\left(x, \varepsilon\right)}{P\left(T(x) < +\infty\right)} = 0. \tag{10.8}$$

Obviously,

$$\{T(x) < +\infty\} = \left\{\sup_{t \in [0,1]} \xi_{\lfloor tT \rfloor} > x\right\} = \left\{\sup_{t \in [0,1]} \ln\left(\xi_{\lfloor tT \rfloor} \vee 1\right) > \ln x\right\},$$

therefore we conclude, letting $n = \varepsilon\sigma^{-2}\ln^2 x$, that

$$P_1\left(x, \varepsilon\right)$$
$$= \mathbf{P}\left(\left\{\sqrt{\varepsilon}\frac{\ln\left(\xi_{\lfloor nt/\varepsilon \rfloor} \vee 1\right)}{\sigma\sqrt{n}},\ t \geq 0\right\} \in A\ ,\ \sup_{t \in [0,1]} \left(\xi_{\lfloor tT \rfloor} \vee 1\right) > \frac{\sigma\sqrt{n}}{\sqrt{\varepsilon}},\ T > n\right). \tag{10.9}$$

It follows from (10.2) and (10.5) that if $n = \varepsilon\sigma^{-2}\ln^2 x$, then

$$\lim_{x \to +\infty} \frac{\mathbf{P}\left(T > n\right)}{\mathbf{P}\left(T(x) < +\infty\right)} = \frac{\sigma c_1}{\sqrt{\varepsilon} c_2} = \sqrt{\frac{2}{\varepsilon\pi}}. \tag{10.10}$$

Applying to the right-hand side of relation (10.9) the invariance principle (10.3) and using (10.10), we obtain that

$$\lim_{x \to +\infty} \frac{P_1\left(x, \varepsilon\right)}{\mathbf{P}\left(T(x) < +\infty\right)}$$
$$= \sqrt{\frac{2}{\varepsilon\pi}} \mathbf{P}\left(\left\{\sqrt{\varepsilon}\frac{W_0^+\left(\alpha^2 t/\varepsilon\right)}{\alpha},\ t \geq 0\right\} \in A\ ,\ \sup_{t \in [0,1]} \frac{W_0^+\left(t\right)}{\alpha} > \frac{1}{\sqrt{\varepsilon}}\right). \tag{10.11}$$

Transform the probability in the right-hand side of relation (10.11), taking into account that the random variable α is uniformly distributed in $(0,1)$ and independent of $\{W_0^+(t)\}$:

$$\mathbf{P}\left(\left\{\sqrt{\varepsilon}\frac{W_0^+(\alpha^2 t/\varepsilon)}{\alpha},\ t\ge 0\right\}\in A\ ,\ \sup_{t\in[0,1]}\frac{W_0^+(t)}{\alpha}>\frac{1}{\sqrt{\varepsilon}}\right)$$

$$=\int_0^1\mathbf{P}\left(\left\{\sqrt{\varepsilon}\frac{W_0^+(u^2 t/\varepsilon)}{u},\ t\ge 0\right\}\in A\ ,\ \sup_{t\in[0,1]}\frac{W_0^+(t)}{u}>\frac{1}{\sqrt{\varepsilon}}\right)du$$

$$=\sqrt{\varepsilon}\int_0^{1/\sqrt{\varepsilon}}\mathbf{P}\left(\left\{\frac{W_0^+(v^2 t)}{v},\ t\ge 0\right\}\in A\ ,\ \sup_{t\in[0,1]}W_0^+(t)>v\right)dv.$$

Thus we have, recalling the definitions of the random process Z_v and the random variable h_0^+, that

$$\lim_{x\to+\infty}\frac{P_1(x,\varepsilon)}{\mathbf{P}(T(x)<+\infty)}=\sqrt{\frac{2}{\pi}}\int_0^{1/\sqrt{\varepsilon}}\mathbf{P}\left(Z_v\in A\ ,\ h_0^+>v\right)dv.$$

Passing to limit as $\varepsilon\downarrow 0$, we obtain that

$$\lim_{\varepsilon\downarrow 0}\lim_{x\to+\infty}\frac{P_1(x,\varepsilon)}{\mathbf{P}(T(x)<+\infty)}=\sqrt{\frac{2}{\pi}}\int_0^{+\infty}\mathbf{P}\left(Z_v\in A\ ,\ h_0^+>v\right)dv. \tag{10.12}$$

The validity of relation (10.6) follows from (10.7), (10.8), and (10.12). However in these reasonings there is a defect as the key relation (10.11) is valid if

$$\mathbf{P}\left(\left\{\sqrt{\varepsilon}W_0^+(\alpha^2 t/\varepsilon)/\alpha,\ t\ge 0\right\}\in\partial A\right)=0,$$

but we know only that $\mathbf{P}(Y\in\partial A)=0$.

To overcome this difficulty, first we consider the cylinder subsets of $D[0,+\infty)$:

$$A=\{y: y(t_i)\le a_i,\ i=1,\dots,m\},$$

where $m\in\mathbb{N}$, $t_1,t_2,\dots,t_m\in[0,+\infty)$ and $a_1,a_2,\dots,a_m\in\mathbb{R}$. For these subsets relations (10.11) and (hence) (10.6) are valid because of the absolute continuity of the finite-dimensional distributions of a Brownian excursion.

Now we consider the following subsets of $D[0,+\infty)$,

$$A=\{y: w_y(\delta;a,b)\ge\varepsilon\},$$

where $0\le a<b<+\infty$, ε,δ are positive numbers, $w_y(\delta;a,b)$ is the standard modulus of continuity of a function $y(t)$, $t\in[0,+\infty)$; i.e.,

$$w_y(\delta;a,b)=\sup_{t,s:\ |t-s|\le\delta}|y(t)-y(s)|\qquad(t,s\in[a,b]).$$

It is clear that the mapping $y\to w_y(\delta;a,b)$, $y\in D[0,+\infty)$, is continuous if $y(t)$ is a continuous function on $[0,+\infty)$. Whence, taking into account that trajectories of a

Brownian excursion are continuous a.s., we are convinced of the validity of relation (10.11) for all sets A under consideration and for all positive ε with the exception of no more than a countable set. But it means the validity of relation (10.6), which in this case has the following form,

$$\lim_{x\to+\infty} \mathbf{P}\left(w_{Y_x}(\delta;a,b) \geq \varepsilon \mid T(x) < +\infty\right)$$

$$= \sqrt{\frac{2}{\pi}} \int_0^{+\infty} \mathbf{P}\left(w_{Z_v}(\delta;a,b) \geq \varepsilon\ ,\ h_0^+ > v\right) dv. \tag{10.13}$$

It is well known (see Durrett and Iglehart (1977)) that

$$\int_0^{+\infty} \mathbf{P}\left(h_0^+ > v\right) dv = \mathbf{E}h_0^+ = \sqrt{\frac{\pi}{2}}.$$

Since trajectories of the process Z_v are continuous a.s.,

$$\lim_{\delta\to 0} \mathbf{P}\left(w_{Z_v}(\delta;a,b) \geq \varepsilon\ \right) = 0.$$

Hence by the dominated convergence theorem we obtain from (10.13) that

$$\lim_{\delta\to 0}\lim_{x\to+\infty} \mathbf{P}\left(w_{Y_x}(\delta;a,b) \geq \varepsilon \mid T(x) < +\infty\right) = 0. \tag{10.14}$$

It is well known (see Billingsley, 1968), that the convergence of the finite-dimensional distributions of the process Y_x and relation (10.14) mean the validity of Theorem 1 (including the statement about the continuity of trajectories of the process Y).

10.4 Finite-dimensional distributions

Note that the distributions of the limit processes in Theorems 1 and 2 are determined by means of the joint distribution of a Brownian excursion $\left\{W_0^+(t)\right\}$ (with some scale changes) and its maximum h_0^+. We can consider the process $\left\{\tau_0^+(x)\,,\ x \in [0,+\infty)\right\}$ as (in some sense) *the inverse process* for a Brownian excursion. The distributions of the limit processes in Theorems 3 and 4 are determined by means of the joint distribution of the inverse process for a Brownian excursion and h_0^+.

However, there are other examples of the limit processes description. For instance, it is possible to determine the finite-dimensional distributions or to express these processes in terms of known random processes. We give here the finite-dimensional distributions of the limit process in Theorem 1.

We introduce some notations:

(1) for $t > 0$, $x > 0$, $y > 0$

$$f(t,x) = \sqrt{\frac{2}{\pi t^3}}\, x \exp\left(-\frac{x^2}{2t}\right),$$

$$g(t,x,y) = \frac{1}{\sqrt{2\pi t}}\left[\exp\left(-\frac{(y-x)^2}{2t}\right) - \exp\left(-\frac{(y+x)^2}{2t}\right)\right];$$

(2) for $t > 0, y \in (0,1]$

$$p_1(0,0;t,y) = \sum_{k=-\infty}^{+\infty} f(t, 2k+y),$$

for $0 < s < t, x \in (0,1], y \in (0,1]$

$$p_1(s,x;t,y) = \sum_{k=-\infty}^{+\infty} g(t-s, x, 2k+y);$$

(3) for $t > 0$

$$p_2(0,0;t,y) = \begin{cases} \sum\limits_{k\in\mathbb{Z}\setminus\{0\}} f(t, 2k-y), & y \in (0,1], \\ f(t,y), & y > 1, \end{cases}$$

for $0 < s < t, x \in (0,1]$

$$p_2(s,x;t,y) = \begin{cases} \sum\limits_{k\in\mathbb{Z}\setminus\{0\}} g(t-s, -x, 2k+y), & y \in (0,1], \\ g(t-s,x,y), & y > 1. \end{cases}$$

Then for arbitrary numbers $m \in \mathbb{N}$, $t_1, t_2, \ldots, t_m \in [0, +\infty)$ $(t_1 < t_2 < \cdots < t_m)$ and $a_1, a_2, \ldots, a_m \in [0,1)$

$$\begin{aligned} &\mathbf{P}(Y(t_i) > a_i,\ i = 1,\ldots,m) \\ =&\sum_{j=0}^{m} \int\limits_{G_j(a_1,\ldots,a_m)}\!\!\cdots\!\int \prod_{i=0}^{j-1} p_1(t_i,x_i;t_{i+1},x_{i+1})\, p_2(t_j,x_j;t_{j+1},x_{j+1}) \\ &\times \prod_{i=j+1}^{m-1} g(t_{i+1}-t_i, x_i, x_{i+1})\, dx_1,\ldots,dx_m, \end{aligned} \tag{10.15}$$

where $t_0 = 0$, $x_0 = 0$, $t_{m+1} = +\infty$, $x_{m+1} = +\infty$, $p_2(t_m,x_m;t_{m+1},x_{m+1}) = x_m$, and for $j = 0,1,\ldots,m$,

$$G_j(a_1,\ldots,a_m) = \{(x_1,\ldots,x_m) : a_i < x_i \le 1,\ i = 1,\ldots,j;\ a_i < x_i,\ i = j+1,\ldots,m\}.$$

We note that as usual a product over an empty set of indices is equal to 1. In relation (10.15) for the finite-dimensional distributions of the process Y there are the $m+1$ summands each of which is similar to the finite-dimensional distributions of a Markov process and $p_1(s,x;t,y)$, $p_2(s,x;t,y)$, $g(t,x,y)$ play the role of the transition densities.

But if even one of the numbers $a_1, \ldots, a_m$ is more than or equal to 1 then

$$\begin{aligned} &\mathbf{P}\left(Y\left(t_i\right) > a_i,\ i = 1, \ldots, m\right) \\ &= \int_{G_0(a_1,\ldots,a_m)} \cdots \int f\left(t_1, x_1\right) \prod_{i=1}^{m-1} g\left(t_{i+1} - t_i, x_i, x_{i+1}\right) dx_1, \ldots, dx_m. \end{aligned} \tag{10.16}$$

As the state 0 is absorbing for the process Y it is reasonable to consider for positive numbers $a_1, \ldots, a_m$ the probability $\mathbf{P}\left(Y\left(t_i\right) > a_i,\ i = 1, \ldots, m-1;\ Y\left(t_m\right) = 0\right)$.

It is obvious that this probability is equal to

$$\mathbf{P}\left(Y\left(t_i\right) > a_i,\ i = 1, \ldots, m-1\right) - \mathbf{P}\left(Y\left(t_i\right) > a_i,\ i = 1, \ldots, m-1;\ Y\left(t_m\right) > 0\right),$$

and the last two probabilities can be found from (10.15) and (10.16).

10.5 Conclusion

Finally, we indicate how to find the finite-dimensional distributions of the process Y. The main role is played here by the following relation,

$$\begin{aligned} &\mathbf{P}\left(Y\left(t_i\right) > a_i,\ i = 1, \ldots, m\right) \\ &= \sqrt{\frac{2}{\pi t_m}} \mathbf{P}\left(W^+\left(t_i/t_m\right) > a_i/\sqrt{t_m},\ i = 1, \ldots, m;\ h^+ > 1/\sqrt{t_m}\right), \end{aligned}$$

where $a_1, \ldots, a_m$ are nonnegative numbers, $h^+ = \sup_{t \in \mathbb{R}} W^+(t)$. In addition we use here the formulas for the joint distribution of minimum, maximum, and location at the last moment of a Brownian motion on time interval $[0,1]$ (see Billingsley, 1968) and the joint distribution of maximum and location at the last moment of a Brownian meander on a time interval $[0,1]$ (see Durrett and Iglehart, 1977), as well as the following result. On time interval $[0,t], t \in (0,1]$ (see Durrett and Iglehart, 1977), as well as the following result.

Lemma 2. *For any $t_1, t_2 \in (0,1]$ $(t_1 < t_2)$, positive x_1, x_2 and Borel set A from the space $D\left[t_1, t_2\right]$,*

$$\begin{aligned} &\boldsymbol{P}\left(W^+ \in A \mid W^+\left(t_1\right) = x_1,\ W^+\left(t_2\right) = x_2\ \right) \\ =&\boldsymbol{P}\left(W \in A \,\middle|\, W\left(t_1\right) = x_1,\ W\left(t_2\right) = x_2,\ \inf_{t \in [t_1, t_2]} W\left(t\right) \geq 0\ \right). \end{aligned}$$

Acknowledgements. This work was supported by the RFBR grant 08-01-00078.

References

Afanasyev, V. (1999). On the passage time of a fixed level by a critical branching process in a random environment. *Discrete Mathematics and Applications*, 9:627–643.

Afanasyev, V. (2005). On the conditional invariance principle for a critical Galton-Watson branching process. *Discrete Mathematics and Applications*, 15:17–32.

Afanasyev, V. (2006). Arcsine law for branching processes in random environment and Galton-Watson processes. *Teoriya Veroyatnostey i ee Primeneniya*, 51:449–464.

Afanasyev, V., Geiger, J., Kersting, G., and Vatutin, V. (2005). Criticality for branching processes in random environment. *Annals of Probability*, 33:645–673.

Billingsley, P. (1968). *Convergence of probability measures.* John Wiley & Sons, New York–London–Sydney–Toronto.

Durrett, R. and Iglehart, D. (1977). Brownian meander and Brownian excursion. *Annals of Probability*, 5:130–135.

11

Gaussian Approximation for Multichannel Queueing Systems

Larisa G. Afanas'eva

Department of Mathematics and Mechanics, Moscow State University, Russia

Abstract: For queueing systems with a large number of channels we prove convergence of one-dimensional distributions of the customer's number in the system to the Gaussian ones. Convergence conditions are given in terms of moments of the input flow, which is of a rather general character. We find normalising coefficients considering regenerative and doubly stochastic Poisson flows.

Keywords and phrases: Gaussian approximation, doubly stochastic Poisson process, regenerative process

11.1 Introduction

Asymptotic analysis of queueing systems under the assumption that the number of customers in the system grows attracted the attention of many researchers; see, e.g., Afanas'eva (1984), Afanasieva and Bashtova (2004), Asmussen (1991), and references therein. The A. A. Borovkov monograph (Borovkov, 1984) develops the theory of limit behaviour of the processes describing the operation of such systems under the most general restrictions on an input flow, service times, and a structure of such a system. An arrival process, after a certain normalisation, is required to converge in one or another sense to a stochastic process $\xi(t)$. Then the normalised number of customers $q(t)$ in the system converges to a stochastic process $\zeta(t)$ related to $\xi(t)$. It is established that if the number n of servers in the system grows rather fast then asymptotic properties of $q(t)$ for loss systems or those with unbounded waiting time are the same as for infinite-channel systems. Applying these theorems to systems with input flows usual in queueing theory (such as doubly stochastic Poisson, Markov-modulated, regenerative, semi-Markov, and others) causes the following problems. Firstly, one has to prove convergence of an input flow to a stochastic process $\xi(t)$, secondly, to find normalising coefficients, and finally, to express normalising coefficients for the process $q(t)$ in terms of input flow and service characteristics. This chapter provides sufficient conditions for convergence of one-dimensional distributions of the process $q(t)$, normalised in a special way, to the Gaussian ones in terms of moments of the input flow (mathematical expectation, correlation function, and the third mixed moment); normalising coefficients are

C.H. Skiadas (ed.), *Advances in Data Analysis*, Statistics for Industry and Technology, DOI 10.1007/978-0-8176-4799-5_11,

also obtained with the help of these characteristics. We study an infinite-channel system. However, it follows from Borovkov (1984) that similar statements are also true for systems with n channels, if n grows fast enough. The approach proposed can be utilised as well for proving convergence of finite-dimensional distributions. Difficulties that arise are of a technical character. As an illustration, systems with a doubly stochastic Poisson flow and a regenerative input process are considered. In the former case normalising coefficients are expressed by means of the average intensity and correlation function of a stochastic intensity, while the moments of the cumulative process specifying the regenerative one are involved in the latter case.

11.2 Model description

Let the arrival process $\{X(t), t \in (-\infty, +\infty)\}$ be an integer-valued stochastic process having nondecreasing trajectories and $X(0) = 0$. If $s < t$ the increment $X(s,t) = X(t) - X(s)$ determines the number of customers arriving at the system on time interval (s,t). An increasing sequence $\{t_j\}_{j=-\infty}^{+\infty}$ where $t_0 \le 0$, $t_1 > 0$ defines the jump times of the process $X(t)$. Assume that the system contains an infinite number of channels. Service times are independent random variables not depending on the input flow with a common distribution function $B(x)$. We suppose that the service time is not a constant and has a finite mean $b = \int_0^\infty \bar{B}(x)\,dx < \infty$, where $\bar{B}(x) = 1 - B(x)$. The asymptotic behaviour of process $q(t)$ that equals the number of customers in the system at time t under the assumption that the system operation started in the infinitely remote past is studied. The asymptotic case of time compression is considered when the input flow is defined by the relation

$$X_T(t) = X(Tt) \quad \text{and} \quad T \to \infty.$$

Let us assume $\mathsf{E}|X(t)|^3 < \infty$ and introduce functions $H(t) = \mathsf{E}X(t)$, $R(t,s) = \mathsf{E}\hat{X}(t)\hat{X}(s)$, $G(t,y,u) = \mathsf{E}\hat{X}(t)\hat{X}(y)\hat{X}(u)$, where $\hat{X}(t) = X(t) - H(t)$. We suppose that

$$\int_0^\infty \int_0^\infty |R(s,t)|\, dB(t)\, dB(s) < \infty. \tag{11.1}$$

This condition provides the existence of all integrals that are introduced below in the mean square sense. They exist a.s. due to the process $X(t)$ properties. Let $q_T(t)$ be the process $q(t)$ for the system with the input flow $X_T(t)$. The following theorem establishes that, for every fixed t, the r.v. $q_T(t)$ is asymptotically Gaussian, as $T \to \infty$.

11.3 The basic theorem

Theorem 1. *Let the following conditions hold.*

(1) $\sqrt{|t|}\,((H(t)/t) - \lambda) \to 0$, *as* $|t| \to \infty$, *for a certain* $\lambda \in (0, \infty)$,
(2) There exists $\lim_{T\to\infty} T^{-1}R(Tt, Ts) = g(t,s)$ *with* $g(t,s)$ *satisfying* (11.1),
(3) $\lim_{T\to\infty} T^{-3/2} G(Tt, Tu, Tv) = 0$.

Then for any t the distribution of

$$\hat{q}_T(t) = \frac{q_T(t) - \lambda bT}{\sqrt{T}\tilde{\sigma}(t)}$$

weakly converges to the Gaussian one with parameters (0, 1) *and*

$$\tilde{\sigma}^2(t) = \lambda b - \lambda \int_0^\infty [\bar{B}(y)]^2 dy + u(t),$$

where

$$u(t) = g(t,t) - 2\int_0^t g(t, t-y)\, dB(y) + 2\int_0^\infty \int_0^y g(t-y, t-s)\, dB(s)\, dB(y). \tag{11.2}$$

Proof. We consider the generating function

$$P(z,t) = \mathsf{E}z^{q(t)} = \mathsf{E}\prod_{t_j \le t} (1 + (z-1)\bar{B}(t-t_j)) = \mathsf{E}f(z), \qquad |z| < 1. \tag{11.3}$$

Let us introduce the following integrals

$$J_k^T(t) = \int_{-\infty}^t (\bar{B}(t-y))^k dX_T(y) = X(tT) - k\int_0^\infty X(T(t-y))(\bar{B}(y))^{k-1}\, dB(y),\ k = 1,2,3. \tag{11.4}$$

First we prove an auxiliary proposition.

Lemma 1. *Let the following conditions be fulfilled, as* $T \to \infty$.

1. $\sqrt{T}\left(T^{-1}\mathsf{E}J_1^T(t) - \lambda b\right) \to 0$.
2. $T^{-1}\mathsf{Var}J_1^T(t) \to \sigma_J^2(t) > 0$.
3. $T^{-1}\mathsf{E}J_2^T(t) \to a_2$.
4. (a) $T^{-3/2}\left(\mathsf{E}\left(J_1^T(t) - \mathsf{E}J_1^T(t)\right)^3\right) \to 0$.
   (b) $T^{-3/2}\mathrm{cov}\left(J_1^T(t), J_2^T(t)\right) \to 0$.

Then for any t the distribution of

$$\hat{q}_T(t) = \frac{q_T(t) - \lambda bT}{\sqrt{T}\sigma_q(t)}$$

weakly converges to the Gaussian one with parameters (0, 1) *and*

$$\sigma_q^2(t) = \lambda b + \sigma_J^2(t) - a_2. \tag{11.5}$$

Proof. Since t is fixed, sometimes we omit it, writing, for example, J_k^T instead of $J_k^T(t)$. From the relation (11.3) we find

$$P_T\left(e^{is/\sqrt{T}},t\right) = \mathsf{E}e^{isq_T/\sqrt{T}} = 1-\left(1-e^{is/\sqrt{T}}\right)\mathsf{E}J_1^T$$
$$+\frac{(1-e^{is/\sqrt{T}})^2}{2}\left(\mathsf{E}\left(J_1^T\right)^2-\mathsf{E}J_2^T\right)-\frac{(1-e^{is/\sqrt{T}})^3}{3!}\mathsf{E}f_T'''\left(\theta e^{is/\sqrt{T}}\right),$$
$$|\theta|\le 1$$

where

$$f_T'''(z) = \sum_{t_j,t_k,t_m\le Tt} P_j^T P_k^T P_m^T \mathbf{1}\{A_{j,k,m}\}\prod_{t_s\le Tt}\left(1+(z-1)P_s^T\right)\mathbf{1}\{s\ne j,k,m\}$$

and $P_j^T = \bar{B}(t-t_j/T)$. Here $A_{j,k,m}$ means that all the indices j,k,m are different.
With the help of rather complicated calculations one can obtain the estimate

$$\mathsf{E}f_T'''(\theta e^{is/\sqrt{T}}) = \left(\mathsf{E}\left(J_1^T\right)^3-\mathsf{E}J_1^TJ_2^T+2\mathsf{E}J_3^T\right)(1+\varepsilon_1(\theta s/\sqrt{T})), \tag{11.6}$$

where $\varepsilon_1(s)\to 0$, as $s\to 0$.
Now we write the expansion of

$$\varphi_T(s) = \ln \mathsf{E}e^{is((q_T-\lambda bT)/\sqrt{T})}$$

as follows

$$-is\lambda b\sqrt{T}+is\frac{H_T(t)}{\sqrt{T}}-\frac{s^2}{2T}\left(\mathsf{Var}J_1^T-\mathsf{E}J_2^T+\mathsf{E}J_1^T\right)-\frac{s^3}{T^{3/2}}g_T(\theta s/\sqrt{T}). \tag{11.7}$$

The estimation of $g_T(\theta s/\sqrt{T})$ is based on relation (11.6) and inequalities

$$|1-e^{-\beta}+\beta|\le\frac{|\beta|^2}{2},\qquad |1-e^{-\beta}+\beta-\frac{\beta^2}{2}|\le\frac{|\beta|^3}{6}\qquad (Re\beta\ge 0).$$

We give here the final result

$$|g_T(s/\sqrt{T})|\le C\left(|\mathsf{E}\left(J_1^T-\mathsf{E}J_1^T\right)^3|+3|\mathsf{cov}\left(J_1^T,J_2^T\right)+2\mathsf{E}J_3^T|\right)(1+u_T(s/\sqrt{T})), \tag{11.8}$$

where $u_T(s/\sqrt{T})\to 0$, as $T\to\infty$. From (11.7), (11.8), and conditions of Lemma 1 we get

$$\varphi_T(s)\to-\frac{s^2}{2\sigma_q^2(t)}\qquad (T\to\infty),$$

where $\sigma_q^2(t)$ is given by (11.5). ■

We complete the proof of the theorem by the following lemmas.

Lemma 2. *Under condition* (1) *of Theorem* 1,

$$\lim_{T\to\infty}\sqrt{T}\left(T^{-1}\mathsf{E}J_1^T-\lambda b\right)=0; \tag{11.9}$$

$$\lim_{T\to\infty}T^{-1}\mathsf{E}J_2^T=\lambda\int_0^\infty\left[\bar{B}(y)\right]^2dy. \tag{11.10}$$

Proof. The proof immediately follows from the inequality

$$\sqrt{T}\left|\frac{\mathsf{E}J_1^T}{T}-\lambda b\right|\le\int_0^\infty y\sqrt{T}\left|T^{-1}H(-yT)-\lambda\right|dB(y)$$

and relation (11.4) for J_2^T. ■

Lemma 3. *Under condition* (2) *of Theorem* 1

$$\lim_{T\to\infty}T^{-1}\mathsf{Var}J_1^T(t)=u(t),$$

where $u(t)$ is given by relation (11.2)

Proof. From (11.4) and condition (2) we get, as $T\to\infty$,

$$T^{-1}\mathsf{Var}J_1^T=T^{-1}[R(Tt,Tt)-2\int_0^\infty R(Tt,T(t-y))\,dB(y)$$

$$+2\int_0^\infty\int_0^y R(T(t-s),T(t-y))\,dB(s)\,dB(y)]\to u(t).$$

■

Lemma 4. *Under conditions* (1)–(3) *of Theorem* 1

$$T^{-3/2}g_T(s/\sqrt{T})\to 0\qquad(T\to\infty).$$

Proof. In view of (11.9), (11.10), and (11.8) it is sufficient to verify that

$$\lim_{T\to\infty}T^{-3/2}\mathsf{E}\left(J_1^T-\mathsf{E}J_1^T\right)^3=0. \tag{11.11}$$

Since

$$\mathsf{E}\left(J_1^T-\mathsf{E}J_1^T\right)^3=G(Tt,Tt,Tt)+3\int_0^\infty G(Tt,Tt,T(t-y))\,dB(y)$$

$$+6\int_0^\infty\int_0^y G(Tt,T(t-y),T(t-s))\,dB(s)\,dB(y)$$

$$+6\int_0^\infty\int_0^y\int_0^u G(T(t-y),T(t-v),T(t-u))\,dB(u)\,dB(v)\,dB(y)$$

relation (11.11) follows from condition (3) of the theorem. ■

Combination of these lemmas proves the theorem.

Corollary 1. *If conditions of Theorem* 1 *hold and*

$$g(t,s)=\left(\sigma_X^2\min(|t|,|s|)+C\right)\mathbf{1}\{ts\ge 0\} \tag{11.12}$$

then $\hat{q}_T(t)$ has the asymptotically Gaussian distribution with

$$\tilde{\sigma}_q^2=\lambda b+\left(\sigma_X^2-\lambda\right)\int_0^\infty(\bar{B}(y))^2\,dy. \tag{11.13}$$

Proof. It is sufficient to calculate $u(t)$ using (11.2) and (11.12). ■

11.4 A limit theorem for a regenerative arrival process

We begin by introducing notation for a regenerative input flow.

We say that input flow $\{X(t),\, t \in (-\infty, +\infty)\}$ is regenerative if there exists a nondecreasing sequence of r.v.s $\{\theta_j\}_{j=-\infty}^{+\infty}$ $(\theta_0 = 0)$ such that

$$\{\theta_{j+1} - \theta_j, X(\theta_j + t) - X(\theta_j),\; t \in [0, \theta_{j+1} - \theta_j)\}_{j=-\infty}^{+\infty},$$

is a sequence of i.i.d. random elements. As usually, we introduce the following r.v.s.

$$\tau_j = \theta_{j+1} - \theta_j, \qquad \varkappa_j(t) = X(\theta_j + t) - X(\theta_j), \qquad \xi_j = \varkappa_j(\tau_j)$$

and set

$$a = \mathsf{E}\xi_i, \qquad \mu = \mathsf{E}\tau_i, \qquad \sigma_\xi^2 = \mathsf{Var}\ \xi_i, \qquad \sigma_\tau^2 = \mathsf{Var}\ \tau_i, \qquad r_{\xi,\tau} = \mathsf{cov}(\xi_i, \tau_i).$$

This class of processes is rather broad. In particular, it includes recurrent, semi-Markov, Markov-modulated, Markov arrival processes, and others.

To apply some renewal theory results we assume that the distribution of τ_i has an absolutely continuous component.

Theorem 2. *Let $\{X(t),\, t \in (-\infty, +\infty)\}$ be a regenerative flow and $\mathsf{E}\xi_j^4 < \infty$, $\mathsf{E}\tau_j^4 < \infty$. Then the distribution of $\hat{q}_T(t)$ weakly converges to the Gaussian one with parameters $(0, 1)$ and σ_q^2 is given by (11.13) where*

$$\lambda = \frac{a}{\mu}, \qquad \sigma_X^2 = \frac{\sigma_\xi^2}{\mu} + \frac{a^2\sigma_\tau^2}{\mu^3} - \frac{2ar_{\xi,\tau}}{\mu^2}. \tag{11.14}$$

Proof. It is sufficient to verify conditions (1)–(3) of Theorem 1 for $\{X(t),\, t \geq 0\}$. Consider stochastic processes

$$N(t) = \max\{j \geq 0 \colon \theta_j \leq t\}, \qquad \gamma_t = t - \theta_{N(t)}.$$

Process $X_T(t)$ can be represented as

$$X_T(t) = Z(Tt) + \varkappa_{N(Tt)}(\gamma_{Tt}), \tag{11.15}$$

where $Z(u) = \sum_{j=0}^{N(u)} \xi_j$. It is well known (see, e.g., Cox (1963)) that r.v.s γ_t, $\varkappa_{N(t)}(\gamma_t)$, and $\tau_{N(t)}$ have a limit distribution as $t \to \infty$. If $F(x,y) = P(\xi_i \leq x, \tau_i \leq y)$ then

$$\lim_{t\to\infty} P(\xi_{N(t)} \leq x) = \mu^{-1} \int_0^\infty y F(x, dy) = F(x). \tag{11.16}$$

Taking into account that $\varkappa_{N(t)}(\gamma_t) \leq \xi_{N(t)}$ we get from relation (11.15)

$$\mathsf{E}Z(t) \leq H(t) = \mathsf{E}X(t) \leq \mathsf{E}Z(t) + \mathsf{E}\xi_{N(t)}.$$

Now condition (1) of Theorem 1 follows from asymptotic results for $\mathsf{E}Z(t)$ (see, e.g., Cox, 1963; Smith, 1955) and (11.16). For the correlation function we find

$$R(Tt,Ts) = \mathsf{cov}(Z(Ts),Z(Tt)) + \mathsf{cov}\left(Z(Tt),\varkappa_{N(Ts)}(\gamma_{Ts})\right) + \mathsf{cov}\left(Z(Ts),\varkappa_{N(Tt)}(\gamma_{Tt})\right) + \mathsf{cov}\left(\varkappa_{N(Ts)}(\gamma_{Ts}),\varkappa_{N(Tt)}(\gamma_{Tt})\right) = g_1+g_2+g_3+g_4.$$

Since $\mathsf{cov}(Z(Ts),Z(Tt)-Z(Ts)) \le C$ $(t \ge s)$ one can obtain that

$$\lim_{T\to\infty} T^{-1}\mathsf{cov}(Z(Ts),Z(Tt)) = \sigma_X^2 s \qquad (t \ge s \ge 0),$$

where σ_X^2 is defined by (11.14) (see, e.g., Cox, 1958; Smith, 1958), C is a constant. Besides,

$$|g_2| = \left|\mathsf{cov}\left(\xi_{N(Ts)},\varkappa_{N(Ts)}(\gamma_{Ts})\right)\right| \le C_2 < \infty$$

and similarly $|g_3| \le C_3 < \infty$, $|g_4| \le C_4 < \infty$. It means that condition (2) holds and $g(t,s)$ is given by (11.12).

Concerning condition (3) we note that

$$G(y,u,v) = G_z(y,u,v) + d(y,u,v) + d(u,y,v) + d(v,y,u) + \mathsf{E}\hat{\eta}_u\hat{\eta}_v\hat{\eta}_y, \tag{11.17}$$

where $G_z(y,u,v)$ is a function $G(y,u,v)$ for process $Z(t)$ and

$$\hat{\eta}_y = \varkappa_{N(y)}(\gamma_y) - \mathsf{E}\varkappa_{N(y)}(\gamma_y), \qquad \hat{Z}(y) = Z(y) - \mathsf{E}Z(y),$$
$$d(y,u,v) = \mathsf{E}[\hat{\eta}_y\hat{Z}(u)\hat{Z}(v) + \hat{\eta}_y\hat{\eta}_v\hat{Z}(v)].$$

We have

$$|\mathsf{E}\hat{\eta}_y\hat{Z}(u)\hat{Z}(v)| \le \sqrt{\mathsf{E}\hat{\eta}_y^2}\,\sqrt[4]{\mathsf{E}(\hat{Z}(u))^4\mathsf{E}(\hat{Z}(v))^4} \le C_5\sqrt{uv} + \tilde{C}_6$$

and condition (3) follows from the asymptotic behaviour of moments of $\hat{Z}(t)$, as $t \to \infty$; see, e.g., Cox (1958, 1963), and Smith (1958).

The general case when the coordinate origin is not a point of regeneration can be reduced to the previous one. ■

11.5 Doubly stochastic poisson process (DSPP)

To begin with we recall the definition of a DSPP. Let $\{\Lambda(t),\, t \in (-\infty,+\infty)\}$ be a stochastic process with nondecreasing trajectories and values in $\mathbb{R}$. All trajectories also have limits on the right, are left continuous, and $\Lambda(0) = 0$. The random time substitution with the help of $\Lambda(t)$ leads to a DSPP (Grandell, 1976); i.e.,

$$X(t) = A(\Lambda(t))$$

where $\{A(t),\, t \in (-\infty,+\infty)\}$ is a standard Poisson process not depending on $\Lambda(t)$. Assuming $\mathsf{E}|\Lambda(t)|^3 < \infty$ we introduce functions

$$H_\Lambda(t) = \mathsf{E}\Lambda(t), \qquad R_\Lambda(t,s) = \mathsf{E}\hat{\Lambda}(t)\hat{\Lambda}(s), \qquad G_\Lambda(t,y,u) = \mathsf{E}\hat{\Lambda}(t)\hat{\Lambda}(y)\hat{\Lambda}(u).$$

Theorem 3. *Let $\{X(t),\, t \in (-\infty, +\infty)\}$ be a DSPP with integrated intensity function $\Lambda(t)$. If for $H_\Lambda(t)$, $R_\Lambda(t,s)$, and $G_\Lambda(t,y,u)$ conditions (1)–(3) of Theorem 1 hold, then distribution of $\hat{q}_T(t)$ weakly converges to a Gaussian one, for any t. The normalising constant is given by the relation*

$$\tilde{\sigma}^2(t) = \lambda b + u_\Lambda(t) \int\limits_0^\infty \left[\bar{B}(y)\right]^2 dy,$$

where $\lambda = \lim_{t\to\infty}(H_\Lambda(t)/t)$ and $u_\Lambda(t)$ is a function $u(t)$ for the process $\Lambda(t)$.

Proof. Theorem 3 follows from Theorem 1 as there are explicit relations among functions $H_\Lambda(t)$, $R_\Lambda(t,s)$, $G_\Lambda(t,y,u)$ for $\Lambda(t)$ and the corresponding functions for the process $X(t)$. For example, $H_\Lambda(t) = H(t)$ and

$$R(t,s) = R_\Lambda(t,s) + H(\min(|s|,|t|))\mathbf{1}\{ts \geq 0\}.$$

The expression for $G(t,y,u)$ being more complicated is not given here. ∎

Furthermore we assume that

$$\Lambda(t) = \int_0^t \lambda(y)\, dy,$$

where $\lambda(y)$ is a nonnegative locally integrable stationary stochastic process in a wide sense with mean λ; moreover, $P(\lambda(t) \leq \lambda^*, t \in (-\infty, +\infty)) = 1,\ \lambda^* < \infty$.

We write $\lambda = \mathsf{E}\lambda(y)$, $\hat{\lambda}(y) = \lambda(y) - \lambda$ and introduce the functions

$$r(y) = \mathsf{E}\hat{\lambda}(0)\hat{\lambda}(y), \qquad G_\lambda(u,v) = \mathsf{E}\hat{\lambda}(0)\hat{\lambda}(u)\hat{\lambda}(v).$$

Corollary 2. *If, for $T \to \infty$,*

$$|r(tT)| \leq C|tT|^{-\alpha}, \quad \textit{where} \quad \alpha > 2,\ 0 < C < \infty; \tag{11.18}$$

$$T^{-3/2} G_\lambda(Ts, Tt) \to 0 \qquad (T \to \infty), \qquad ts \neq 0, \tag{11.19}$$

then $\hat{q}_T(t)$ is asymptotically normal and

$$\sigma_q^2 = \lambda b + 2\int\limits_0^\infty (\bar{B}(y))^2\, dy \int\limits_0^\infty r(u)\, du. \tag{11.20}$$

Proof. We use Lemma 1 to prove this proposition. Since $H_\lambda(t) = \lambda t$ and condition 4. (a) follows from (11.19), it is sufficient to verify condition 2. To this end we write

$$\begin{aligned} T^{-1}\mathsf{Var}\, J_T(t) &= 2T \int\limits_{-\infty}^{t} \int\limits_{-\infty}^{y} \bar{B}(t-u)\bar{B}(t-y)\, r(T(y-u))\, du\, dy \\ &= 2T \int\limits_{-\infty}^{t} \bar{B}(t-y) \int\limits_{y-T^{-\beta}}^{y} \bar{B}(t-u)\, r(T(y-u))\, du\, dy \\ &\quad + 2T \int\limits_{-\infty}^{t} \bar{B}(t-y) \int\limits_{-\infty}^{y-T^{-\beta}} \bar{B}(t-u)\, r(T(y-u))\, du\, dy = I_1^\beta + I_2^\beta. \end{aligned}$$

For $\beta \in (0,1)$, by virtue of condition (11.18),

$$I_2^\beta \to 0 \quad (T \to \infty)$$

and for I_1^β we get

$$I_1^\beta = 2T \int_{-\infty}^{t} \bar{B}(t-y)\bar{B}(t-y-\theta T^{-\beta}) \int_{y-T^{-\beta}}^{y} r(T(y-u))\, du\, dy,$$

where $|\theta| < 1$. Changing variables $v = T(y-u)$ and passing to the limit as $T \to \infty$ proves (11.20). ■

For a Markov-modulated process (see, e.g., Asmussen, 1991), the intensity has the form

$$\lambda(t) = \sum_{k=1}^{\infty} \lambda_k \mathbf{1}\{U(t) = k\}, \tag{11.21}$$

where $U(t)$ is a homogeneous Markov chain with a finite or a countable set of states and $\{\lambda_k, k = 0, 1, \dots\}$ is a collection of nonnegative numbers such that $\lambda_k \le \lambda^*$ for any k. Let $P_{ij}(t) = P(U(t) = j | U(0) = i)$ for $t \ge 0$.

Corollary 3. *Let $\lambda(t)$ be specified by (11.21) where $U(t)$ is a stationary ergodic Markov chain and*

$$\pi_j = P(U(t) = j), \qquad j = 0, 1, \dots.$$

If

$$|P_{ij}(t) - \pi_j| \le \frac{C\pi_j}{t^\alpha}, \qquad \alpha > 2,\ 0 < C < \infty, \tag{11.22}$$

then $\hat{q}_T(t)$ is asymptotically normal and

$$\sigma_q^2 = \lambda b + 2\sum_{k=0}^{\infty}\sum_{j=0}^{\infty} \lambda_k \lambda_j \pi_k \int_0^{\infty} (P_{kj}(y) - \pi_j)\, dy \int_0^{\infty} \left(\bar{B}(u)\right)^2 du, \tag{11.23}$$

$$\lambda = \sum_{k=0}^{\infty} \lambda_k \pi_k.$$

Proof. We have to verify (11.18) and (11.19). In view of the relation

$$r(y) = \sum_{k=0}^{\infty}\sum_{j=0}^{\infty} \lambda_k \lambda_j \pi_k (P_{kj}(y) - \pi_j)$$

(11.18) follows from (11.22). Since

$$G_\lambda(s,t) = \sum_{k=0}^{\infty}\sum_{j=0}^{\infty}\sum_{m=0}^{\infty} (\lambda_k - \lambda)(\lambda_j - \lambda)(\lambda_m - \lambda)\pi_k P_{kj}(s) P_{jm}(t)$$

from inequality (11.22) one can easily obtain the estimate

$$|G_\lambda(s,t)| \le 8CT^{3-\alpha}(\lambda^*)^3 (t^{-\alpha} + s^{-\alpha} + (Tst)^{-\alpha})$$

for $s > 0$, $t > 0$, and $T \to \infty$.

A similar estimate is valid for negative s and t and for the case $st < 0$. It means that (11.19) is fulfilled. ■

Another example of applying Corollary 2 is given for a stationary regenerative process $\lambda(t)$. Let $\{\theta_i\}_{i=-\infty}^{+\infty}$ be the moments of regeneration for $\lambda(t)$, $\theta_0 \le 0 < \theta_1$, and $\tau_i = \theta_{i+1} - \theta_i$, $\lambda_i(t) = \lambda(\theta_i + t)$ for $t \in (0, \tau_i)$. We suppose τ_i $(i \neq 0)$ to have a common d.f. $F(x)$ and $\tau = \mathsf{E}\tau_1 < \infty$. Consider stochastic processes

$$N(t) = \max\{j : \theta_j \le t\}, \qquad t \in (-\infty, +\infty),$$

and

$$\gamma_t = t - \theta_{N(t)}, \qquad \chi_t = \theta_{N(t)+1} - t.$$

If the distribution of (γ_0, χ_0) satisfies the following condition

$$P(\gamma_0 > x, \chi_0 > y) = \tau^{-1} \int_{x+y}^{\infty} \bar{F}(u)du \tag{11.24}$$

then the process $N(t)$ has stationary increments. Besides, the distributions of $\tau_{N(t)}$ and of (γ_t, χ_t) do not depend on t and

$$P(\tau_{N(t)} > x) = P(\tau_0 > x) = \tau^{-1}\left[x\bar{F}(x) + \int_x^{\infty} \bar{F}(y)dy\right]. \tag{11.25}$$

It follows from (11.24) and (11.25) that the conditional distribution of γ_t for a fixed $\tau_{N(t)}$ is uniform on $(0, \tau_{N(t)})$.

Corollary 4. *Let $\lambda(t)$ be a regenerative process satisfying* (11.24). *Then $\lambda(t)$ is a stationary process. If also $\mathsf{E}\tau_1^3 < \infty$ then conditions* (11.18) *and* (11.19) *are fulfilled.*

Proof. Due to stationarity of $(\tau_{N(t)}, \gamma_t, \chi_t)$ we have $\mathsf{E}\lambda(t)$ equal to

$$\mathsf{E}\lambda_{N(t)}(\gamma_t) = \mathsf{E}\left(\mathsf{E}(\lambda_{N(t)}(\gamma_t) \mid \tau_{N(t)})\right) = \frac{1}{\tau}\int_0^{\infty}\int_0^{x} \mathsf{E}(\lambda_0(u) \mid \tau_0 = x)\, du\, dF(x) = \lambda. \tag{11.26}$$

Since $\lambda(u)$ and $\lambda(v)$ are independent if v and u fall into different regeneration periods, it is easy to see that for $v > u$

$$\begin{aligned} r(u,v) &= \mathsf{E}\left(\lambda_{N(u)}(\gamma_u) - \lambda\right)\left(\lambda_{N(u)}(\gamma_u + v - u) - \lambda\right)\mathbf{1}(v - u < \chi_u) \\ &= \frac{1}{\tau}\int_{v-u}^{\infty}\int_0^{x-(v-u)} \mathsf{E}\left((\lambda_0(y) - \lambda)(\lambda_0(y + v - u) - \lambda) \mid \tau_0 = x\right) dy\, dF(x) = r(v-u). \end{aligned} \tag{11.27}$$

It means that $\lambda(t)$ is a stationary process in a wide sense. As far as $\lambda(t)$ is bounded a.s. and $\mathsf{E}\tau_i^3 < \infty$, estimates (11.18) and (11.19) take place. ∎

If $\lambda_i(t)$ and τ_i are independent, relations (11.26) and (11.27) can be rewritten as

$$\lambda = \tau^{-1}\int_0^{\infty} \mathsf{E}\lambda_0(y)\bar{F}(y)\, dy, \qquad r(t) = \tau^{-1}\int_0^{\infty} \mathsf{E}(\lambda_0(x) - \lambda)(\lambda_0(x+t) - \lambda)\bar{F}(x+t)\, dx. \tag{11.28}$$

For illustration let $\{\xi_i\}_{i=-\infty}^{+\infty}$ be a sequence of i.i.d. nonnegative random variables not depending on $\{\theta_i\}_{i=-\infty}^{+\infty}$ and $a = \mathsf{E}\xi_i$, $a_2 = \mathsf{Var}\ \xi_i$. Assuming $\lambda(t) = \xi_i$ on the ith regeneration period, one can easily deduce from (11.28) that

$$\lambda = a, \qquad r(t) = a_2\tau^{-1}\int_t^\infty \bar{F}(x)\,dx$$

hence σ_q^2 is determined by the equality

$$\sigma_q^2 = \lambda b + \frac{a\,\mathsf{E}\tau_1^2}{\tau}\int_0^\infty \left(\bar{B}(y)\right)^2 dy.$$

Now consider a loss system and let n_T be the number of servers in it. If $q_T^0(t)$ is the number of customers in the system at time t, the following assertion is true.

Theorem 4. *Let conditions* (1)–(3) *of Theorem* 1 *hold and*

$$\frac{n_T - mT}{\sqrt{T}} \to \infty \qquad (T \to \infty) \tag{11.29}$$

for some $m \in (0, \infty)$. *Then*

$$\hat{q}_T^0(t) = \frac{q_T^0(t) - \lambda bT}{\sqrt{T}\tilde{\sigma}(t)}$$

is asymptotically normal with parameters $(0, 1)$.

The proof is almost evident as random variables $q_T(t)$ and $q_T^0(t)$ are asymptotically equivalent when (11.29) is true.

11.6 Conclusion

Considering queueing systems with a rather complicated input flow, in particular, a doubly stochastic Poisson process (DSPP), one cannot, with rare exceptions, obtain explicit expressions for their operation characteristics (average length of the queue, loss probability, etc.). The study of such systems develops in two directions. One of them is to estimate these characteristics (see Afanas'eva, 1984; Asmussen, 1991; Rolski, 1986, 1989), and the other one is to study the extreme cases: heavy and light traffic situations (see Afanasieva and Bashtova, 2004; Iglehart and Whitt, 1970). We have considered systems with an arbitrary input flow in asymptotics of time compression. The asymptotical behaviour of q_T for an infinite-channel queueing system has been studied. Under some natural conditions convergence of one-dimensional distributions of the process $q_T(t)$, normalised in a special way, to Gaussian ones was proved. Normalising coefficients were given in terms of the first and second moments of the input flow.

The results concerning the systems with an increasing number of channels n_T are based on Theorem 4 or on the following inequalities

$$q_T^0(t) \le q_T(t) \le q_T^\infty(t).$$

Here $q_T^0(t)$ and $q_T^\infty(t)$ denote the number of customers, at time t, in queueing systems with n_T channels and loss or unbounded waiting time, respectively.

For application of our results it is necessary to calculate the normalising coefficient σ_q^2. It is not so easy a problem even for DSPP. However, it is possible to obtain statistical estimates for σ_q^2 when we investigate a real situation.

Acknowledgements. This research was partially supported by the RFBR grant 07-01-00362.

References

Afanas'eva, L.G. (1984). On waiting-time process in periodic queues. *Lecture Notes in Mathematics. Stability Problems for Stochastic Models*, 1155, Springer, New York, 1–20.

Afanasieva, L.G., Bashtova, E.E. (2004). The queue with periodic doubly stochastic Poisson input. In *Transactions of XXIV International Seminar on Stability Problems for Stochastic Models, Jurmala, Latvia*, pp. 80–87.

Asmussen, S. (1991) Ladder heights and Markov-modulated $M|G|1$ queue. *Stochast. Proc. Their Appl.*, 37:313–326.

Borovkov, A. (1984). *Asymptotic Methods in Queueing Theory.* Wiley, Chichester.

Cox, D.R. (1958). Discussion of paper by W.L.Smith. *J. R. Statist. Soc. B.*, 20:286–287.

Cox, D.R. (1963). *Renewal Theory.* John Wiley and Sons, New York.

Grandell, J. (1976). Doubly stochastic Poisson processes. *Lecture Notes in Mathematics*, 529, Springer, New York, 1–276.

Iglehart, D.L., Whitt, W. (1970). Multiple channel queues in heavy traffic. *Adv. Appl. Probab.*, 2:150–177.

Rolski, T. (1986). Upper bounds for single server queues with doubly stochastic Poisson arrivals. *Math. Oper. Res.*, 11:442–450.

Rolski, T. (1989). Queues with nonstationary input. *Queue. Syst.*, 5:113–130.

Smith, W.L. (1955). Regenerative stochastic processes. *J. R. Statist. Soc. A.*, 242:6–31.

Smith, W.L. (1958). Renewal theory and its ramifications. *J. R. Statist. Soc. B.*, 20:243–302.

12

Stochastic Insurance Models, Their Optimality and Stability

Ekaterina V. Bulinskaya

Department of Mathematics and Mechanics, Moscow State University, Russia

Abstract: A class of discrete-time stochastic insurance models is investigated in the framework of cost approach, the aim being maximization of profit (or minimization of loss) during a finite or infinite time interval. Optimal and asymptotically optimal controls are established under the assumption that probability distributions of the claim process and premium inflow are known. Sensitivity analysis of the models with respect to cost parameter fluctuations and distribution perturbations is also provided.

Keywords and phrases: Stochastic insurance models, cost approach, optimal and asymptotically optimal policies

12.1 Introduction

Many centuries ago insurance companies were created for risk sharing and transferring. By paying a fixed money amount a risk averse policyholder obtains the guarantee of indemnification in case of an insured event (or risk) realization. According to legislation, fulfillment of liabilities to its clients is the *primary task* of any insurance company. Therefore by the beginning of the twentieth century the study of ruin probability was initiated by F. Lundberg and H. Cramér in the framework of collective risk theory. This subject was further developed by many researchers. Thus during the last century the reliability approach dominated in actuarial sciences.

Being a corporation, an insurance company has obligations to its shareholders as well. So, the *secondary* but very important task is to get profit and pay dividends. This problem has also attracted the attention of actuaries. The pioneering work of de Finetti (1957) was followed by many others, especially during the last decade. Investment policies and borrowing were treated as well; see, e.g., Schmidli (1994), Bulinskaya (2004, 2005). In particular, the company functioning after its "ruin" and reinstatement of solvability by shareholders, using their money to raise the company capital to some positive level, was treated, e.g., in Dickson and Waters (2004).

Investigation of insurance models in the framework of cost approach was initiated in Bulinskaya (2003). We concentrate below on a class of discrete-time insurance models

C.H. Skiadas (ed.), *Advances in Data Analysis*, Statistics for Industry and Technology, DOI 10.1007/978-0-8176-4799-5_12,

generalizing those introduced in Bulinskaya (2007a,b). Namely, we take into account that asset amounts to sell and money amounts to borrow cannot be infinite. It is reasonable to study discrete-time models since the financial balance is struck by the end of a calendar year and the duration of a reinsurance treaty is usually a year as well.

12.2 Model description

The aim of this research is to establish the optimal (and asymptotically optimal) policies of an insurance company minimizing its expected losses during a fixed planning horizon of n years, $n \le \infty$.

Suppose that $\{\xi_i\}_{i\ge 1}$ is a sequence of i.i.d. nonnegative r.v.s with a finite mean and a density $\varphi(s) > 0$ for s belonging to some finite or infinite interval $[\underline{\kappa}, \overline{\kappa}] \subset \mathsf{R}_+$. The corresponding distribution function $F(t) = \mathsf{P}(\xi_i \le t) = \int_0^t \varphi(s)\, ds$. Here ξ_i is the excess of claims over premiums in the year i.

Assume that by the end of a year the company can make one of the following decisions: I, sell some assets (immediately); II, borrow some money, the loan being available by the end of the next year; III, sell assets and borrow money.

Let x be the initial capital (if $x < 0$ its absolute value is the company debt) whereas c_1 is the loss incurred by selling the assets unit, c_2 is the interest rate while borrowing, r is the penalty for delay of a claim unit payment, and h is the inflation rate. For simplicity, we set the discount factor $\alpha = 1$. In contrast with the above-mentioned papers, here we take into account the following parameters: a_1, the assets amount available for sale, and a_2, the upper bound for a loan. Thus we have $z_i \le a_i$, $i = 1, 2$, where z_1 is the amount sold and z_2 is the amount borrowed.

12.3 Optimal control

Denote by $f_n(x)$ the minimal expected n-year costs. According to the Bellman optimality principle (see, e.g., Bellman, 1957), for $n \ge 1$,

$$f_n(x) = \min_{0 \le z_i \le a_i\ i=1,2} [c_1 z_1 + c_2 z_2 + L(x + z_1) + \mathsf{E} f_{n-1}(x + z_1 + z_2 - \xi_1)]$$

where $f_0(x) \equiv 0$, $L(v) = \mathsf{E}[h(v - \xi_1)^+ + r(\xi_1 - v)^+]$ and E stands for expectation.

Putting $v = x + z_1$, $u = v + z_2$, and

$$G_n(u, v) = (c_1 - c_2)v + c_2 u + L(v) + \int_0^\infty f_{n-1}(u - s)\varphi(s)\, ds$$

one gets

$$f_n(x) = -c_1 x + \min_{(u,v)\in D_x} G_n(u, v). \tag{12.1}$$

The minimum in (12.1) can be attained either inside or on the boundary of the optimization set $D_x = \{(u,v) : x \le v \le x + a_1, v \le u \le v + a_2\}$. Therefore to establish an optimal control, that is, u and v providing the minimum, we introduce the following notation,

$$K_n(v) = \frac{\partial G_n(u,v)}{\partial v} = c_1 - c_2 + L'(v) := K(v),$$
$$S_n(u) = \frac{\partial G_n(u,v)}{\partial u} = c_2 + \int_0^\infty f'_{n-1}(u-s)\varphi(s)\,ds.$$

Moreover, $T_n(v) = S_n(v)+K(v)$ and $B_n(v) = S_n(v+a_2)+K(v)$ represent $dG_n(v,v)/dv$ and $dG_n(v, v + a_2)/dv$, respectively, while

$$\mathcal{H}_a(u) = c_2 - c_1 + \int_0^{u-\overline{v}} K(u-s)\varphi(s)\,ds + \int_{u+a-\overline{v}}^\infty K(u+a-s)\varphi(s)\,ds.$$

We introduce $\overline{v}$, u_n, v_n, w_n, and $\widehat{u}_a$ as the roots of the following equations,

$$K(\overline{v}) = 0, \qquad S_n(u_n) = 0, \qquad T_n(v_n) = 0, \qquad B_n(w_n) = 0, \qquad \mathcal{H}_a(\widehat{u}_a) = 0, \quad (12.2)$$

provided the solutions exist for a given set of cost parameters. Otherwise, if $K(v) > 0$ for all v, set $\overline{v} = -\infty$, and if $K(v) < 0$ for all v, set $\overline{v} = +\infty$; the same agreement holds for $S_n(u)$, $T_n(v)$, $B_n(w)$, and u_n, v_n, w_n, $n \ge 1$. Also let $\overline{t}$ and $\widehat{t}$ be defined by $F(\overline{t}) = r/(r+h)$ and $F^{2*}(\widehat{t}) = r/(r+h)$, respectively.

Theorem 1. *For $a_i = \infty$, $i = 1, 2$, the optimal behaviour at the first step of an n-step process has the form:*

(I) If $c_2 < c_1 - r$, $(k-1)r < c_2 \le kr$, $k \ge 1$, then $u = v = x$ for $n \le k$ and $v = x$, $u = \max(u_n, x)$ for $n > k$. The sequence $\{u_n\}$ is bounded, increasing, and $\lim_{n\to\infty} u_n = \widehat{t}$.

(II) If $c_2 \ge c_1$, $(l-1)r < c_1 \le lr$, $l \ge 1$, then $u = v = x$ for $n < l$ and $u = v = \max(v_n, x)$ for $n \ge l$. The sequence $\{v_n\}$ is bounded, increasing, and $\lim_{n\to\infty} v_n = \overline{t}$.

(III) If $\max(c_1 - r, (m-2)c_1/(m-1)) \le c_2 \le (m-1)c_1/m$, $m \ge 2$, $(l-1)r < c_1 \le lr$, $l \ge 1$ (hence $m \ge l$ and $u_n \ge \overline{v}$ for $n \ge m$), then $u = v = x$ for $n < l$ and $v = \max(\overline{v}, x)$, $u = \max(u_n, x)$ for $n \ge m$. The sequence $\{u_n\}$ is bounded, increasing, and $\lim_{n\to\infty} u_n = \widehat{u}_\infty$.

If $l \le n < m$, then the optimal decision may be determined either by parameters $(u_n, \overline{v})$, or v_n, moreover, if v_{n_0} is optimal for some n_0, then v_n is also optimal for $l \le n < n_0$.

The proof can be found in Bulinskaya (2007a). A more thorough analysis, undertaken below, allows us to refine statement (III) of Theorem 1 and to obtain new results under restrictions $a_1 < \infty$ and/or $a_2 < \infty$.

First of all, it is useful to establish that all the functions under consideration are nondecreasing. Since $S_{n+1}(u) = S_n(u) + H_n(u)$, $T_{n+1}(v) = T_n(v) + H_n(v)$, $B_{n+1}(v) = B_n(v) + H_n(v + a_2)$, where $H_n(u) = (f'_n - f'_{n-1}) * F(u)$, to prove that the sequences $\{u_n\}$, $\{v_n\}$, $\{w_n\}$ are increasing we check that $H_n(u) < 0$ for $u = u_n, v_n, w_n$, $n \ge 1$.

For given r and h consider $\Gamma^- = \{(c_1, c_2) : c_2 < c_1 - r\}$, $\Gamma^+ = \{(c_1, c_2) : c_2 > c_1 + h\}$, $\Gamma = \{(c_1, c_2) : c_1 - r \le c_2 \le c_1 + h\}$, $\Gamma_n^- = \{(c_1, c_2) \in \Gamma : S_n(\overline{v}) < 0\}$,

$\Gamma_n^+ = \{(c_1,c_2) \in \Gamma : S_n(\overline{v}) > 0\}$ and $\Gamma_n^0 = \{(c_1,c_2) \in \Gamma : S_n(\overline{v}) = 0\}$. Put also $\Delta_l^k = \{(c_1,c_2) : (l-1)r < c_1 \le lr, (k-1)r < c_2 \le kr\}$, $k \ge 1$, $l \ge 1$, $\Delta_l = \{(c_1,c_2) : (l-1)r < c_1 \le lr\}$, $\Delta_{>l} = \{(c_1,c_2) : c_1 > lr\}$, Δ^k and $\Delta^{>k}$ are defined, for c_2, similarly. Clearly $\mathsf{R}_+^2 = \cup_{l\ge1} \cup_{k\ge1} \Delta_l^k = \Gamma^- \cup \Gamma \cup \Gamma^+$ and $\Gamma = \Gamma_n^- \cup \Gamma_n^+ \cup \Gamma_n^0$, for any $n \ge 1$.

As $K(\overline{v}) = 0$ is equivalent to $F(\overline{v}) = (r + c_2 - c_1)/(r+h)$, it follows immediately that $\overline{v}$, existing if and only if $(c_1,c_2) \in \Gamma$, increases from $\underline{\kappa}$ to $\overline{\kappa}$, as $c_2 - c_1$ increases from $-r$ to h. Obviously, $S_n(\overline{v}) = T_n(\overline{v})$ and $K(v) < 0$ for $v < \overline{v}$, whereas $K(v) > 0$ for $v > \overline{v}$. Moreover, it is possible to verify that $\overline{v} < v_n < u_n$ in Γ_n^-, $\overline{v} = v_n = u_n$ in Γ_n^0, $\overline{v} > v_n > u_n$ in Γ_n^+, similarly, $-\infty = \overline{v} < v_n < u_n$ in Γ^- and $+\infty = \overline{v} > v_n > u_n$ in Γ^+.

Now we formulate one of the new results.

Theorem 2. *Let $a_1 < \infty$, $a_2 = \infty$; then the optimal decision at the first step of the n-step process has the form:*

$$z_1^{(n)} = \min[a_1, (\min(v_n, \overline{v}) - x)^+], \qquad z_2^{(n)} = \left(u_n - x - z_1^{(n)}\right)^+.$$

The sequences $\{u_n\}$ and $\{v_n\}$, defined by (12.2), are nondecreasing. There exist $\lim_{n\to\infty} u_n$ equal to $\widehat{u}_{a_1}$ in Γ and $\widehat{t}$ in Γ^-, whereas $\lim_{n\to\infty} v_n \ge \overline{t}$ in Γ^+.

The proof is by induction on n. At first take $n = 1$. Since $S_1(u) = c_2 > 0$, $u_1 = -\infty$ and it is optimal to take $u = v$. On the other hand, $T_1(v) = c_1 - r + (r+h)F(v)$ therefore $v_1 = -\infty$ in $\Delta_{>1}$ and in Δ_1 there exists $v_1 \in [0, \overline{v}]$ such that $F(v_1) = (r - c_1)/(r+h)$. In the latter case the optimal decision is $u = v = x + a_1$ for $x < v_1 - a_1$, $u = v = v_1$, for $x \in [v_1 - a_1, v_1)$, and $u = v = x$ for $x \ge v_1$.

For further investigation we need only to know $f_1'(x)$ which is equal to $L'(x) = -r + (r+h)F(x)$ in $\Delta_{>1}$ whereas in Δ_1

$$f_1'(x) = -c_1 + \begin{cases} T_1(x + a_1), & x < v_1 - a_1, \\ 0, & x \in [v_1 - a_1, v_1), \\ T_1(x), & x \ge v_1. \end{cases} \tag{12.3}$$

It is obvious that $f_1'(x)$ is nondecreasing, hence the same is true of $S_2(u)$ and $T_2(v)$ taking values in $[c_2 - r, c_2 + h]$ and $[c_1 - 2r, c_1 + 2h]$, respectively. Hence, $u_2 = -\infty$ in $\Delta^{>1}$, $v_2 = -\infty$ in $\Delta_{>2}$, and, for $n = 2$, the optimal decision is $u = v = x$ if $(c_1,c_2) \in A_2 = \Delta_{>2} \cap \Delta^{>1}$.

Proceeding in the same way we establish that $u_n = v_n = -\infty$, $n \le k$, in $A_k = \Delta_{>k} \cap \Delta^{>k-1}$, $k > 2$, $u = v = x$ is optimal for all $n \le k$ and $f_n'(x) = -nr + (r+h)\sum_{l=1}^n F^{l*}(x)$. Moreover, in Δ_{k+1}^k there exist $u_{k+1} \ge \underline{\kappa}$ satisfying $\sum_{l=2}^{k+1} F^{l*}(u_{k+1}) = (kr - c_2)/(r+h)$ and $v_{k+1} \ge \underline{\kappa}$ satisfying $\sum_{l=1}^{k+1} F^{l*}(v_{k+1}) = ((k+1)r - c_1)/(r+h)$.

Since $\Gamma^- \subset \cup_{k\ge1} A_k$, it is not difficult to see that, for $n > k$,

$$f_n'(x) = -c_2 + L'(x) + \begin{cases} 0, & x < u_n, \\ S_n(x), & x \ge u_n, \end{cases}$$

if $(c_1,c_2) \in \Gamma^- \cap \Delta^k$. Verifying that $f_n'(x) - f_{n-1}'(x) < 0$ for $x < u_n$, one obtains $u_{n+1} > u_n$. Furthermore, for all u and $n > k$, $S_{n+1}(u)$ is given by

$$\int_0^\infty L'(u-s)\varphi(s)\,ds + \int_0^{u-u_n} S_n(u-s)\varphi(s\,ds) \ge -r + (r+h)F^{2*}(u),$$

therefore $u_n \le \widehat{t}$, for all n. Obviously, there exists $\lim_{n\to\infty} u_n$ and it is not difficult to show that it is equal to $\widehat{t}$.

Next, consider the set Γ. For each $k > 1$, it is divided into subsets Γ_k^- and Γ_k^+ by a curve $c_2 = g_k(c_1)$ defined implicitly by equality $S_k(\overline{v}) = 0$. The point $(kr, (k-1)r)$ on the boundary of Γ, corresponding to $\overline{v} = \underline{\kappa}$, belongs to $g_k(c_1)$, since $S_k(\underline{\kappa}) = T_k(\underline{\kappa}) = 0$, for such (c_1, c_2). Moreover, according to the rule of implicit function differentiation and the form of $S_k(\cdot)$ in Δ_k^{k-1},

$$g_k'(c_1) = \sum_{l=2}^{k} \varphi^{l*}(\overline{v}) \Big/ \sum_{l=1}^{k} \varphi^{l*}(\overline{v}), \tag{12.4}$$

whence it is obvious that $0 < g_k'(c_1) < 1$. The last result is valid for other values of c_1, although the expression of $g_k'(c_1)$ is more complicated than (12.4).

For $n > k$ and $(c_1, c_2) \in \Gamma_k^-$,

$$f_n'(x) = -c_1 + \begin{cases} K(x+a_1), & x < \overline{v} - a_1, \\ 0, & x \in [\overline{v} - a_1, \overline{v}), \\ K(x), & x \in [\overline{v}, u_n), \\ T_n(x), & x \geq u_n. \end{cases}$$

It is easy to verify that $\Gamma_n^- \subset \Gamma_{n+1}^-$ and

$$S_{n+1}(u) = \mathcal{H}_{a_1}(u) + \int_0^{u-u_n} S_n(u-s)\varphi(s)\,ds \geq \mathcal{H}_{a_1}(u), \tag{12.5}$$

entailing $u_n \leq \widehat{u}_{a_1}$, for all n.

Since $\mathcal{H}_\infty(u) \geq \mathcal{H}_{a_1}(u) \geq \mathcal{H}_0(u) = -r + (r+h)F^{2*}(u)$, for any u and $a_1 > 0$, one has $\widehat{u}_\infty < \widehat{u}_{a_1} < \widehat{u}_0 = \widehat{t}$. It is not difficult to establish that $\lim_{n\to\infty} u_n = \widehat{u}_{a_1}$.

On the other hand, if $(c_1, c_2) \in \Gamma_k^+$, then

$$f_k'(x) = -c_1 + \begin{cases} K(x+a_1), & x < u_k - a_1, \\ T_k(x+a_1), & x \in [u_k - a_1, v_k - a_1), \\ 0, & x \in [v_k - a_1, v_k), \\ T_k(x), & x \geq v_k. \end{cases}$$

It follows immediately that $u_{k+1} > u_k$, $v_{k+1} > v_k$, and $\Gamma_{k+1}^+ \subset \Gamma_k^+$. Thus, for any $(c_1, c_2) \in \Gamma$, there exists such $n_0(c_1, c_2)$ that (12.5) is valid for all $n \geq n_0$.

Finally, for Γ^+ it is optimal to take $v = x + a_1$, $u = u_n$ for $x < u_n - a_1$, $u = v = x + a_1$ for $x \in [u_n - a_1, v_n - a_1)$, $u = v = v_n$ for $x \in [v_n - a_1, v_n)$, and $u = v = x$ for $x \geq v_n$, if u_n and v_n are finite. ■ Below we study the influence of the other restriction.

Corollary 1. *If $a_1 \leq \infty$, $a_2 < \infty$, and $(c_1, c_2) \in \Gamma^+$, then*

$$z_1^{(n)} = \min[a_1, (v_n - x)^+], \qquad z_2^{(n)} = \left(u_n - x - z_1^{(n)}\right)^+.$$

Proof. Being similar to that of Theorem 2 it is omitted. ■

Theorem 3. *Suppose that $a_1 \leq \infty$, $a_2 < \infty$, and $(c_1, c_2) \in \Gamma^-$. Then the optimal decision at the first step of the n-step process is given by*

$$z_1^{(n)} = \min[a_1, (w_n - x)^+], \qquad z_2^{(n)} = \min\left[a_2, \left(u_n - x - z_1^{(n)}\right)^+\right],$$

where w_n and u_n are defined by (12.2). There exist $\lim_{n\to\infty} u_n = \widehat{t}$ and $\lim_{n\to\infty} w_n \leq \overline{t}$.

Proof. Begin by treating the case $a_1 = \infty$, $a_2 < \infty$. It follows easily from assumptions that $u_n > \max(v_n, w_n + a_2) \geq \min(v_n, w_n + a_2) > w_n$. Moreover, $v_1 = w_1 = -\infty$ and $f_1'(x) = L'(x)$. Then, if $(c_1, c_2) \in \Delta_2^1$, it is not difficult to verify that there exist finite u_n and w_n, $n \geq 2$. Hence it is optimal to take $v = w_n$, $u = w_n + a_2$, for $x < w_n$, $v = x$, $u = x + a_2$, for $x \in [w_n, u_n - a_2)$, $v = x$, $u = u_n$, for $x \in [u_n - a_2, u_n)$, and $v = x$, $u = x$, for $x \geq u_n$. Consequently, one gets

$$f_n'(x) = -c_1 + \begin{cases} 0, & x < w_n, \\ B_n(x), & x \in [w_n, u_n - a_2), \\ K(x), & x \in [u_n - a_2, u_n), \\ T_n(x), & x \geq u_n \end{cases} = -c_2 + L'(x) + \begin{cases} -K(x), \\ S_n(x + a_2), \\ 0, \\ S_n(x), \end{cases} \tag{12.6}$$

and $B_n(v) \geq L'(v)$. That means, $w_n \leq \bar{t}$ for all n and a_2. Using (12.6) one also obtains $\lim_{n\to\infty} u_n = \hat{t}$.

If $(c_1, c_2) \in \Delta_l^1$, $l > 2$, there exists $w_l > -\infty$, whereas $w_m = -\infty$, for $m < l$. Thus

$$f_n'(x) = -c_1 + \begin{cases} B_n(x), & x < u_n - a_2, \\ K(x), & x \in [u_n - a_2, u_n), \\ T_n(x), & x \geq u_n, \end{cases}$$

for $1 < n < l$, and $f_n'(x)$ has the form (12.6) for $n \geq l$.

The subsets Δ_l^k corresponding to $k \geq 2$ are treated in the same way giving also $z_1^{(n)} = (w_n - x)^+$, $z_2^{(n)} = \min\left[a_2, \left(u_n - x - z_1^{(n)}\right)^+\right]$.

Changes necessary under assumption $a_1 < \infty$ are almost obvious so the details are omitted. ■

Remark 1. If $c_2 > c_1$ the optimal behaviour for $a_1 = \infty$, $a_2 < \infty$ has the same form as that for $a_1 = a_2 = \infty$ given in Theorem 1.

12.4 Sensitivity analysis

We begin studying the impact of model parameters on the optimal decision by the motivating

Example 1. Assume $\underline{\kappa} = 0$, $\overline{\kappa} = d$, and $\varphi(s) = d^{-1}$, $s \in [\underline{\kappa}, \overline{\kappa}]$; that is, distribution of ξ_i is uniform. Obviously, $F(u) = u/d$, $u \in [0, d]$, and $\overline{v} = d(r + c_2 - c_1)/(r + h)$, while $F^{2*}(u) = u^2/2d^2$, $u \in [0, d]$, $F^{2*}(u) = 1 - (u - 2d)^2/2d^2$, $u \in [d, 2d]$. Suppose also that $a_1 < \infty$.

According to (12.3) the form of $g_2(c_1)$, given by the relation $S_2(\overline{v}) = 0$, depends on a_1 for $(c_1, c_2) \in \Delta_1^1$. Moreover, $c_2 - r + (r + h)F^{2*}(u) = S_2^{(0)}(u) \leq S_2^{(a_1)}(u) \leq S_2^{(\infty)}(u) = c_2 + \int_0^{u - v_1} L'(u - s)\varphi(s)\, ds$, whence it follows that the domain Γ_2^- decreases as a_1 increases.

On the other hand, the curve $g_2(c_1)$ is the same for all a_1 if $(c_1, c_2) \in \Delta_2^1$. It is determined by equation $S_2^{(0)}(\overline{v}) = 0$, which can be rewritten in the form $2(r + h)(r - c_2) = (r + c_2 - c_1)^2$, for $h \geq r$. Thus, $g_2^{(0)}(c_1)$ does not depend on d. It starts from the point $c_1 = 2r$, $c_2 = r$ and crosses the line $c_1 = r$ at $c_2 = -(2r + h) + \sqrt{5r^2 + 4rh + h^2}$

and then the line $c_2 = c_1$ at $c_2 = r[1 - r/2(r + h)]$. For $h = r$ these values of c_2 are equal to $r(\sqrt{10} - 3)$ and $3r/4$, respectively. Next, if $c_1 = 0$ one has $c_2 = (r + h)[\sqrt{1 + 2r(r + h)^{-1}} - 1]$ equal to $r(2\sqrt{3} - 3)$ for $h = r$. However, $\Gamma_2^- \cap \{c_2 > c_1\} = \varnothing$ when $a_1 = \infty$.

The assumption $a_1 = \infty$, $a_2 = \infty$ of Theorem 1 is a limiting case of those in Theorems 2 and 3. In contrast to the case without restrictions where all the decisions I, II, and III are used, in the case $a_1 < \infty$ (resp., $a_2 < \infty$) decision I (resp., II) is excluded.

As usual for dynamic programming the optimal control depends on the planning horizon n. Moreover, for n fixed there exist the *stability domains* of cost parameters $(\Gamma_n^-, \Gamma_n^+, \Gamma^- \cap \Delta^k, \Gamma^+ \cap \Delta_l)$ where the optimal behaviour has the same type, that is, determined by the same set of critical levels u_n, v_n, w_n, or $\overline{v}$.

Fortunately, using the asymptotically optimal stationary controls one can reduce the number of stability domains and exclude the dependence on n.

Definition 1. *A control is called* stationary *if it prescribes the same behaviour at each step of the process. It is* asymptotically optimal *if*

$$\lim_{n\to\infty} n^{-1}\widehat{f}_n(x) = \lim_{n\to\infty} n^{-1} f_n(x),$$

where $\widehat{f}_n(x)$ represents the expected n-step costs under this control.

We prove below only the simplest result.

Theorem 4. *If $a_1 = \infty$, $a_2 \le \infty$, and $c_2 > c_1$ it is asymptotically optimal to take $z_1^{(n)} = (\overline{t} - x)^+$, $z_2^{(n)} = 0$ for all n.*

Proof. Denote by $f_n^l(x)$ the expected n-step costs if $\overline{t}$ is applied during the first l steps, whereas the critical levels v_k, $k \le n - l$, optimal under the assumptions made, are used during the other steps.

It is clear that $f_n^n(x) = \widehat{f}_n(x)$ and $f_n^0(x) = f_n(x)$, hence

$$\widehat{f}_n(x) - f_n(x) = \sum_{l=1}^{n} \left(f_n^l(x) - f_n^{l-1}(x)\right). \tag{12.7}$$

Suppose for brevity that $c_1 < r$; that is, v_1 is finite.

Since $v_n \le v_{n+1}$, $n \ge 1$, and $v_n \to \overline{t}$, as $n \to \infty$, one can find, for any $\varepsilon > 0$, such $\widehat{n} = n(\varepsilon)$ that $\overline{t} - \varepsilon < v_n \le \overline{t}$, if $n \ge \widehat{n}$. Furthermore, we have

$$\max_x \left|f_n^l(x) - f_n^{l-1}(x)\right| \le \max_x \left|f_{n-l+1}^1 - f_{n-l+1}^0(x)\right|$$

and

$$f_k^1(x) - f_k^0(x) = \begin{cases} c_1(\overline{t} - v_k) + L(\overline{t}) - L(v_k) + R(v_k), & x < v_k, \\ c_1(\overline{t} - x) + L(\overline{t}) - L(x) + R(x), & x \in [v_k, \overline{t}), \\ 0, & x \ge \overline{t}, \end{cases}$$

where $R(x) = \int_0^\infty (f_{k-1}(\overline{t} - s) - f_{k-1}(x - s))\varphi(s)\,ds$. Obviously, $k - 1 = n - l \ge \widehat{n}$ for $l \le n - \widehat{n}$, therefore

$$\max_x \left|f_k^1(x) - f_k^0(x)\right| \le d\varepsilon \quad \text{with} \quad d = 2(c_1 + \max(r, h))$$

and

$$\sum_{l=1}^{n-\widehat{n}} \left|f_n^l(x) - f_n^{l-1}(x)\right| \leq (n-\widehat{n})d\varepsilon. \tag{12.8}$$

On the other hand,

$$\sum_{l=n-\widehat{n}+1}^{n} \left|f_n^l(x) - f_n^{l-1}(x)\right| \leq \widehat{n}b(x) \tag{12.9}$$

where $b(x) = \max_{k \leq \widehat{n}} \left|f_k^1(x) - f_k^0(x)\right| \leq L(v_1) + d\overline{t} < \infty$, for all x.

It follows immediately from (12.7), (12.8), and (12.9) that

$$n^{-1}(\widehat{f}_n(x) - f_n(x)) \to 0, \qquad \text{as} \quad n \to \infty.$$

To complete the proof we have to verify that there exists, for all x,

$$\lim_{n\to\infty} n^{-1}\widehat{f}_n(x) = c_1\mu + L(\overline{t}), \qquad \mu = \mathsf{E}\xi_k, \qquad k \geq 1. \tag{12.10}$$

This is obvious for $x \leq \overline{t}$, since in this case

$$\widehat{f}_n(x) = c_1(\overline{t} - x) + c_1 \sum_{k=1}^{n-1} \mathsf{E}\xi_k + nL(\overline{t}).$$

Now let $x > \overline{t}$. Then we do not sell (or borrow) during the first ν_x steps; here

$$\nu_x = \inf\left\{k : \sum_{i=1}^{k} \xi_i > x - \overline{t}\right\}.$$

In other words, we wait until the capital falls below the level $\overline{t}$ proceeding after that as in the previous case. Hence,

$$\widehat{f}_n(x) = L(x) + \mathsf{E}\sum_{i=1}^{\nu_x - 1} L\left(x - \sum_{k=1}^{i} \xi_k\right) + c_1\mathsf{E}\left[\zeta_x + \sum_{i=\nu_x+1}^{n-1} \xi_i\right] + \mathsf{E}(n-\nu_x)L(\overline{t});$$

here $\zeta_x = \sum_{i=1}^{\nu_x} \xi_i - (x - \overline{t})$ is the overshot of the level $x - \overline{t}$ by the random walk with jumps ξ_i, $i \geq 1$.

Thus, it is possible to rewrite $\widehat{f}_n(x)$ as follows,

$$\widehat{f}_n(x) = n(c_1\mu + L(\overline{t})) + B(x).$$

Using Wald's identity and renewal processes properties, as well as the fact that $L(\overline{t})$ is the minimum of $L(x)$, it is possible to establish that $|B(x)| < \infty$ for a fixed x. So (12.10) follows immediately.

The same result is valid for $c_1 \geq r$. The calculations being long and tedious are omitted. ∎

Since $\overline{t} = g(r,h)$, with $g(x_1,x_2) = F^{\text{inv}}(x_1/(x_1+x_2))$, it is useful to check its sensitivity with respect to small fluctuations of parameters r and h and perturbations of distribution F. For this purpose we recall some definitions.

Denote by $R = g(x)$ a valuation criterion (objective function, decision made, or optimal control), $x = (x_1, \ldots, x_n)$ being a vector of model parameters. As usual in sensitivity analysis, R is called the system output, $g(\cdot)$ the model, and x_i the ith input parameter (or factor).

At first we apply the local technique, more precisely, differential importance measure (DIM) introduced in Borgonovo and Apostolakis (2001). Let $x^0 = (x_1^0, \dots, x_n^0)$ be the base-case values of parameters, reflecting the decision-maker knowledge of assumptions made. The DIM for parameter x_i, $i = \overline{1,n}$, is defined as follows,

$$D_i(x^0, dx) = g'_{x_i}(x^0)\, dx_i \left(\sum_{j=1}^{n} g'_{x_j}(x^0)\, dx_j \right)^{-1} \quad (= dg_i(x^0)/dg(x^0))$$

if $dg(x^0) \neq 0$. Whence, for uniform parameters changes: $dx_i = u$, $i = \overline{1,n}$, we get

$$D1_i(x^0) = g'_{x_i}(x^0) \Big/ \sum_{j=1}^{n} g'_{x_j}(x^0). \tag{12.11}$$

Theorem 5. *Under assumptions of Theorem 4, DIMs for parameters r and h do not depend on distribution F.*

Proof. Follows immediately from (12.11) and the definition of function g. Since

$$g'_{x_1}(x^0) = \varphi^{-1}(\bar{t}^0)x_2^0/\left(x_1^0 + x_2^0\right)^2, \qquad g'_{x_2}(x^0) = -\varphi^{-1}(\bar{t}^0)x_1^0/\left(x_1^0 + x_2^0\right)^2,$$

it is clear that

$$D1_1(x^0) = \frac{x_2^0}{x_2^0 - x_1^0}, \qquad D1_2(x^0) = -\frac{x_1^0}{x_2^0 - x_1^0} = 1 - D1_1(x^0);$$

that is, they are well defined for $x_1^0 \neq x_2^0$ and do not depend on F. ■

Note that $D1_1(x^0) > 1$, $D1_2(x^0) < 0$ for $x_2^0 > x_1^0$, and $D1_1(x^0) < 0$, $D1_2(x^0) > 1$ for $x_2^0 < x_1^0$; for an illustration see Figure 12.1.

Next use the method proposed in Sobol' (1990) for global sensitivity analysis.

Although the system parameters are some (often unknown) constants it is useful to treat them as r.v.s. Assume that $X = (X_1, \dots, X_n)$ is uniformly distributed in $K^n = [0,1]^n$ and the function $g(x)$, $x \in K^n$, is integrable.

Theorem 6 (Sobol'). *The following decomposition of variance holds for a square integrable random variable $R = g(X)$:*

$$V[R] = \sum_{i=1}^{n} V_i + \sum_{i<j} V_{i,j} + \sum_{i<j<k} V_{i,j,k} + \dots + V_{1,2,\cdots,n}, \tag{12.12}$$

where $V[R] = \int_{K^n} g^2(x)\, dx - g_0^2$ and partial variances are calculated by way of

$$V_{i_1,\dots,i_s} = \int_0^1 \cdots \int_0^1 g^2_{i_1,\dots,i_s}(x_{i_1}, \dots, x_{i_s}) \prod_{k=i_1,\dots,i_s} dx_k. \tag{12.13}$$

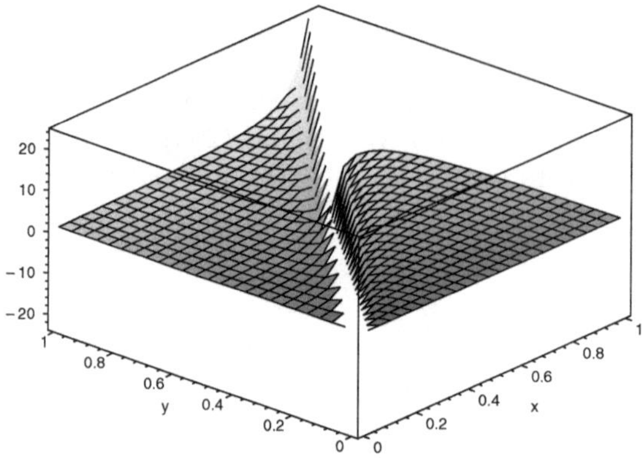

Figure 12.1. Differential importance measure $D1_1(x^0)$ for $\overline{t}$

Here

$$g_0 = \mathsf{E}R = \int_{K^n} g(x)\,dx,$$

$$g_i(x_i) = \int_0^1 \cdots \int_0^1 g(x) \prod_{k \neq i} dx_k - g_0,$$

$$g_{i,j}(x_i, x_j) = \int_0^1 \cdots \int_0^1 g(x) \prod_{k \neq i,j} dx_k - (g_0 + g_i(x_i) + g_j(x_j)),$$

$$\cdots \quad \cdots\cdot$$

Now we can formulate further definitions assuming $V[R] \neq 0$.

Definition 2. *Sensitivity index* $S_{i_1,i_2,\ldots,i_s}$ *for a group of parameters* $(x_{i_1}, x_{i_2}, \cdots, x_{i_s})$, $1 \leq i_1 < i_2 < \cdots < i_s \leq n$, *is given by* $V_{i_1,i_2,\ldots,i_s}/V[R]$, *whereas the sensitivity index of order* s *is* $\sum_{1 \leq i_1 < \cdots < i_s \leq n} S_{i_1,i_2,\ldots,i_s}$.

So, S_i is the first-order contribution of the ith parameter to the output variance, while $S_{i_1,i_2,\ldots,i_s}$ represents the parameter interaction.

Definition 3. *Global sensitivity index* $GI(x_i)$ *of parameter* x_i *is the sum of all indices* $S_{i_1,\ldots,i_s}$, $s \geq 1$, *containing* i,

$$GI(x_i) = \left(V_i + \sum_{j \neq i} V_{i,j} + \cdots + V_{1,2,\ldots,n} \right) / V[R].$$

Thus, $GI(x_i)$ represents the total contribution of parameter x_i to variance of output. That enables us to answer the following questions. Which of the uncertain input factors is so uninfluential that we can safely fix it (them)? If we could eliminatethe uncertainty in one of the input factors, which should we choose to reduce most the variance of output?

Remark 2. Variance decomposition (12.12) is valid (with obvious changes) for any distribution of X.

Turning to our model suppose the ith parameter to be uniformly distributed on $(x_i^0(1-k), x_i^0(1+k))$, $0 < k < 1$, $i = 1, 2$.

Unfortunately, it is impossible to obtain the explicit form of $GI(x_i)$, $i = 1, 2$. However, using Maple 8 it is not difficult to get numerical results. Thus, Figure 12.2 gives the form of global indices as functions of k (for $g(x) = \bar{t}$ and $x_1^0 = x_2^0 = 0.5$). It is easily seen that $IG(x_1)$ (lower curve) decreases, whereas $IG(x_2)$ (upper curve) increases, as k increases, the parameter x_2 being more influential. For $k = 0.25$ one has $IG(x_1) = 0.4949060035$ and $IG(x_2) = 0.5077327062$; that means, even if the relative error in parameters estimation is 25%, the model behaves "almost additively," V_{12} giving only 0.2% of total variance. Moreover, for $k = 0.5$ one has $IG(x_1) = 0.4775592923$ and $IG(x_2) = 0.5334429254$ and V_{12} gives 1.1% of total variance. Hence, parameter interaction is higher for larger errors in their estimation.

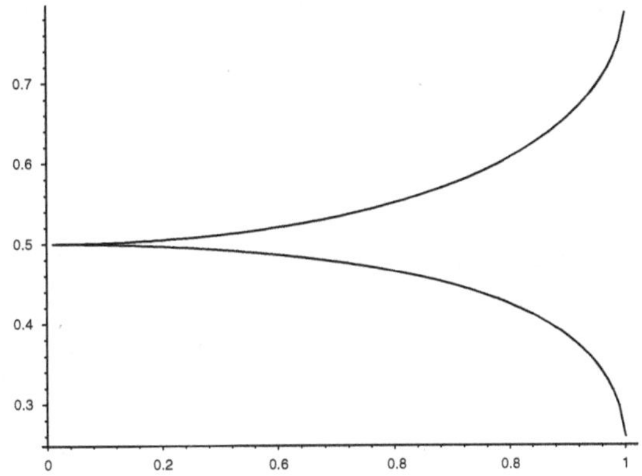

Figure 12.2. Global indices

Finally, we establish that the asymptotically optimal policy based on $\bar{t}$ is stable with respect to small perturbations of distribution F.

Denote by $\bar{t}_k$ the value of $\bar{t}$ corresponding to distribution $F_k(t)$. Moreover, set

$$\gamma(F_k, F) = \sup_t |F_k(t) - F(t)|;$$

that is, γ is the Kolmogorov (or uniform) metric.

Proposition 1. *Let distribution function $F(t)$ be continuous and strictly increasing. Then $\bar{t}_k \to \bar{t}$, provided $\gamma(F_k, F) \to 0$, as $k \to \infty$.*

Proof. According to assumptions $F_k(\bar{t}_k) = F(\bar{t})$ and $|F_k(\bar{t}_k) - F(\bar{t}_k)| \le \gamma(F, F_k)$. Hence $|F(\bar{t}) - F(\bar{t}_k)| \le \gamma(F, F_k)$. That means, $\bar{t}_k \to \bar{t}$, as $k \to \infty$. ∎

Remark 3. This result is also important for construction of asymptotically optimal policies under assumption of no a priori information about distribution F.

12.5 Conclusion

Assuming the underlying distribution laws to be known, we established the optimal controls for a class of discrete-time insurance models and found the stability domains in the parameter space. The notion of asymptotically optimal policy was introduced and such a policy was constructed. Its stability with respect to small fluctuations of parameters and perturbations of distributions was proved.

Thus, we realized the first two steps of the algorithm for dealing with unknown distributions proposed in Bulinskaya (2007a). The third step, that is, construction of the empirical asymptotically optimal policy for our models will be treated elsewhere.

The results pertaining to the case of incomplete information, without restrictions on amounts sold and borrowed, can be found in Bulinskaya (2007b).

It can be shown that in other applications such as inventory theory, queueing, finance, etc., the results obtained for insurance models are of interest and methods employed can be used as well; compare with Bulinskaya (2003).

Acknowledgements. The research was partially supported by RFBR grant 07-01-00362.

References

Bellman, R. (1957). *Dynamic Programming.* Princeton University Press, Princeton, NJ.

Borgonovo, E. and Apostolakis, G.E. (2001). A new importance measure for risk-informed decision-making. *Reliability Engineering and System Safety*, 72:193–212.

Bulinskaya, E. (2003). On a cost approach in insurance. *Review of Applied and Industrial Mathematics*, 10:276–286 (in Russian).

Bulinskaya, E. (2004). Multistep investment policy of insurance company. In *Transactions of XXIV International Seminar on Stability Problems for Stochastic Models*, Jurmala, Latvia, pp. 193–200.

Bulinskaya, E. (2005). Investment policy of insurance company under incomplete information. In *Transactions of XXV International Seminar on Stability Problems for Stochastic Models*, Maiori/Salerno, pp. 71–78.

Bulinskaya, E. (2007a). Sensitivity analysis of some applied probability models. *Pliska Studia Mathematica Bulgarica*, 18:57–90.

Bulinskaya, E. (2007b). Some aspects of decision making under uncertainty. *Journal of Statistical Planning and Inference*, 137:2613–2632.

Dickson, D. and Waters, H. (2004). Some optimal dividends problems. *ASTIN Bulletin*, 34:49–74.

de Finetti, B. (1957). Su un'impostazione alternativa della teoria colletiva del rischio. In *Transactions of the XVth International Congress of Actuaries*, 2:433–443.

Schmidli, H. (1994). Diffusion approximations for a risk process with possibility of borrowing and investment. *Communications in Statistics, Stochastic Models*, 10: 365–388.

Sobol', I.M. (1990). Sensitivity estimates for nonlinear mathematical models. *Matematicheskoe Modelirovanie*, 2:112–118 (in Russian).

13

Central Limit Theorem for Random Fields and Applications

Alexander Bulinski

Department of Mathematics and Mechanics, Moscow State University, Russia

Abstract: A new variant of the CLT is established for random fields defined on $\mathbb{R}^d$ which are strictly stationary, with a finite second moment and weakly dependent (comprising cases of positive or negative association). The summation domains grow in the van Hove sense. At the same time the indices of observations form more and more dense grids in these domains. Thus the effect of combining two scaling procedures is studied. A statistical version of this CLT is also proved. Some stochastic models in radiobiology based on dependent functional subunits are discussed as well.

Keywords and phrases: Random fields, dependence conditions, CLT, stochastic radiobiological models

13.1 Introduction

There is a vast literature devoted to the central limit theorem (CLT) for random fields defined on a lattice $\mathbb{Z}^d$ and to its various applications. The researchers use different dependence and moment conditions and impose some restrictions specifying the summation domains (usually blocks).

We concentrate on strictly stationary real-valued random fields having a finite second moment. Their dependence structure is more general than positive or negative association intensively studied and used in mathematical statistics, reliability theory, percolation theory, and statistical physics (where one applied the classical FKG-inequalities); see, e.g., Bulinski and Shashkin (2007) and references therein.

The random fields under consideration are defined on a space $\mathbb{R}^d$; more exactly, we deal with a sequence of measurable bounded sets $V_n \subset \mathbb{R}^d$ ($n \in \mathbb{N}$) growing to infinity in the van Hove sense as $n \to \infty$. At the same time we consider the so-called infill asymptotics (see Cressie, 1991) when the observations X_t are given for finite sets $U_n = V_n \cap T_n$ where T_n is a grid of points in $\mathbb{R}^d$ with a step $1/\Delta_n$ over each axis and $\Delta_n \to \infty$ as $n \to \infty$. Thus we consider the combination of the effects concerning two scalings.

Theorem 1 below can be viewed as an extension of the beautiful Newman CLT; see Newman (1980). The statistical version of the established CLT develops, in the above

C.H. Skiadas (ed.), *Advances in Data Analysis*, Statistics for Industry and Technology, DOI 10.1007/978-0-8176-4799-5_13,

mentioned setting, the approach proposed for mixing stochastic processes in Peligrad and Shao (1996) and for associated random fields or their modifications in Bulinski and Vronski (1996) and Bulinski (2004); see also Bulinski and Shashkin (2007).

As a domain of applications we mention the interesting stochastic models which arose in radiobiology in the 1990s. We discuss a generalization of these critical volume models using the concept of dependent functional subunits introduced in Bulinski and Khrennikov (2005) to provide the base for further development in this research direction.

13.2 Main results

Let $X = \{X_t, t \in \mathbb{R}^d\}$ be a real-valued wide-sense stationary random field with covariance function $R(u) = \mathrm{cov}(X_t, X_{t+u})$, $u, t \in \mathbb{R}^d$. For positive Δ consider a lattice $T(\Delta) := (\mathbb{Z}/\Delta)^d$, i.e., a collection of points of the form $(j_1/\Delta, \ldots, j_d/\Delta)$ where $j = (j_1, \ldots, j_d) \in \mathbb{Z}^d$. The cardinality of a finite set U is denoted by $|U|$. Let

$$S(U) = \sum_{t \in U} X_t, \quad U \subset \mathbb{R}^d, \quad |U| < \infty.$$

For $\Delta_n > 0$ put $T_n = T(\Delta_n)$, $n \in \mathbb{N}$.

Lemma 1. *Assume that a wide-sense stationary random field X has the covariance function absolutely directly integrable in the Riemann sense and also*

$$\sigma^2 := \int_{\mathbb{R}^d} R(u)du \neq 0. \tag{13.1}$$

Then, for any sets $V_n \to \infty$ in the van Hove sense ($V_n \subset \mathbb{R}^d$) and every sequence $(\Delta_n)_{n \in \mathbb{N}}$ of positive numbers such that $\Delta_n \to \infty$, one has

$$\frac{var\, S(U_n)}{c_n |U_n|} \to 1 \quad as\ n \to \infty \tag{13.2}$$

where $U_n = V_n \cap T_n$ and $c_n = \sigma^2 \Delta_n^d$ $(n \in \mathbb{N})$.

Proof. Set

$$b_n := \sum_{t \in T_n} \mathrm{cov}(X_0, X_t) = \sum_{t \in T_n} R(t).$$

In view of the hypotheses concerning the integrability of the covariance function R we obtain that $b_n \sim c_n$ as $n \to \infty$. For $a > 0$ consider a partition of $\mathbb{R}^d$ by cubes $B_j(a) = B_0(a) + (aj_1, \ldots, aj_d)$, $j = (j_1, \ldots, j_d) \in \mathbb{Z}^d$, where

$$B_0(a) = \{x \in \mathbb{R}^d : 0 < x_k \le a,\ k = 1, \ldots, d\}.$$

Put $J_n^-(a) = \{j \in \mathbb{Z}^d : B_j(a) \subset V_n\}$, $J_n^+(a) = \{j \in \mathbb{Z}^d : B_j(a) \cap V_n \neq \varnothing\}$, and

$$V_n^-(a) = \bigcup_{j \in J_n^-(a)} B_j(a), \quad V_n^+(a) = \bigcup_{j \in J_n^+(a)} B_j(a).$$

Then

$$\operatorname{mes}\left(V_n^-(a)\right) \to \infty \quad \text{and} \quad \operatorname{mes}\left(V_n^+(a)\right)/\operatorname{mes}\left(V_n^-(a)\right) \to 1, \qquad n \to \infty. \tag{13.3}$$

As usual, the Lebesgue measure of a (measurable) set B is denoted by $\operatorname{mes}(B)$. For $U_n^-(a) = V_n^-(a) \cap T_n$ one has

$$A_n := b_n|U_n^-(a)| - \operatorname{var} S\left(U_n^-(a)\right) = \sum_{s \in U_n^-(a),\, t \notin U_n^-(a)} \operatorname{cov}(X_s, X_t).$$

Take $p \in (0, a/2)$ and let $\Lambda_j(a,p) = (u_{j,1}, v_{j,1}] \times \cdots \times (u_{j,d}, v_{j,d}]$ be a cube in $\mathbb{R}^d$ having the same center as $B_j(a)$ and the edge length $a - 2p$, $j \in \mathbb{Z}^d$. Set

$$G_n = \bigcup_{j \in J_n^-(a)} ((B_j(a) \setminus \Lambda_j(a,p)) \cap T_n), \qquad W_n = \bigcup_{j \in J_n^-(a)} (\Lambda_j(a,p) \cap T_n); \tag{13.4}$$

clearly $G_n = G_n(a,p)$ and $W_n = W_n(a,p)$. Then, for all n large enough, $|A_n|\left(b_n|U_n^-(a)|\right)^{-1}$ admits an upper bound

$$\frac{1}{b_n|U_n^-(a)|}\left(\sum_{s \in G_n}\sum_{t \notin U_n^-(a)} |\operatorname{cov}(X_s, X_t)| + \sum_{s \in W_n}\sum_{t \notin U_n^-(a)} |\operatorname{cov}(X_s, X_t)|\right)$$
$$\le \frac{4}{\sigma^2}\left(\frac{pd}{a}\int_{\mathbb{R}^d} |R(u)|du + \int_{\|u\| \ge p} |R(u)|du\right)$$

where $\|\cdot\|$ is a maximal norm in $\mathbb{R}^d$. For any $\varepsilon > 0$ one can choose p large enough and then a large enough to obtain $|A_n|\left(b_n|U_n^-(a)|\right)^{-1} < \varepsilon$ when $n \ge N = N(\varepsilon, p, a)$.

Note that

$$\operatorname{var} S\left(U_n \setminus U_n^-(a)\right) \le |U_n \setminus U_n^-(a)| \sum_{t \in T_n} |R(t)| \le 2|U_n \setminus U_n^-(a)|\Delta_n^d \int_{\mathbb{R}^d} |R(u)|du \tag{13.5}$$

for all n large enough. Let B be a cube (of the form $(u,v]$, $u, v \in \mathbb{R}^d$) with the length of edge $a > 0$. Then it is easily seen that $|B \cap T_n| \sim (a\Delta_n)^d$ as $n \to \infty$. Thus using (13.3), (13.5) we establish the following relation

$$\frac{\operatorname{var} S\left(U_n^-(a)\right)}{b_n|U_n^-(a)|} \to 1, \qquad n \to \infty,$$

and come to (13.2). This completes the proof. □

Remark 1. Relation (13.2) is an extension of the lemma proved by Bolthausen for stationary random field $X = \{X_j, j \in \mathbb{Z}^d\}$ and regularly growing $U_n \subset \mathbb{Z}^d$.

Now we have to generalize the concept of (BL, θ)-dependence introduced for random fields on $\mathbb{Z}^d$ in Bulinski and Suquet (2001) to capture random fields defined on $\mathbb{R}^d$.

Definition 1. *A field $X = \{X_t, t \in \mathbb{R}^d\}$ is called (BL, θ)-dependent if there exists a sequence of positive numbers $\theta_n \searrow 0$ as $n \to \infty$ such that whenever Δ is large enough, then for any finite disjoint sets $I, J \subset T(\Delta)$ and any bounded Lipschitz functions $f : \mathbb{R}^{|I|} \to \mathbb{R}$, $g : \mathbb{R}^{|J|} \to \mathbb{R}$,*

$$|\operatorname{cov}(f(X_s, s \in I), g(X_t, t \in J))| \le \operatorname{Lip}(f)\operatorname{Lip}(g)(|I| \wedge |J|)\Delta^d \theta_r \tag{13.6}$$

where $r = dist(I, J) := \min\{\|s - t\|, s \in I, t \in J\}$ and

$$\mathrm{Lip}(f) := \sup_{x \neq y,\, x,y \in \mathbb{R}^d} \frac{|f(x) - f(y)|}{\sum_{k=1}^{d} |x_k - y_k|}.$$

In Bulinski and Shabanovich (1998) it was shown that, for any positively or negatively associated (PA or NA) random field $X = \{X_t, t \in T\}$ with a finite second moment, for arbitrary finite disjoint sets $I, J \subset T$ and the above-mentioned functions f and g, the left-hand side of (13.6) has the upper bound

$$\mathrm{Lip}(f)\mathrm{Lip}(g) \sum_{s \in I,\, t \in J} |\mathrm{cov}(X_s, X_t)|. \tag{13.7}$$

Thus, in particular, for a wide-sense stationary PA or NA random field X with covariance function absolutely directly integrable in the Riemann sense one can use in (13.6) as θ_r an analogue of the Cox–Grimmett coefficient, namely,

$$\theta_r = 2 \int_{\{u \in \mathbb{R}^d : \|u\| \geq r\}} |R(u)| du, \quad r > 0.$$

There are a number of important stochastic models possessing PA or NA properties or their modifications (see, e.g., Bulinski and Shashkin, 2007 and references therein).

Theorem 1. *Let X be a strictly stationary (BL, θ)-dependent random field with continuous function R satisfying conditions of Lemma* 1. *Then, for any sets $V_n \to \infty$ in the van Hove sense ($V_n \subset \mathbb{R}^d$) and any sequence $(\Delta_n)_{n \in \mathbb{N}}$, $0 < \Delta_n \to \infty$, one has*

$$\frac{S(U_n) - |U_n| \mathsf{E} X_0}{\sqrt{\Delta_n^d |U_n|}} \to N(0, \sigma^2) \quad \textit{in law as } n \to \infty \tag{13.8}$$

where $U_n = V_n \cap T_n$ and σ^2 is defined in (13.1).

Proof. Without loss of generality we can assume that this field X is centered. We have to prove that, for every $\lambda \in \mathbb{R}$,

$$\mathsf{E} \exp\left\{ i\lambda \left(\Delta_n^d |U_n|\right)^{-1/2} S(U_n) \right\} \to \exp\{-\sigma^2 \lambda^2 / 2\} \text{ as } n \to \infty; \tag{13.9}$$

here $i^2 = -1$.

Let $a > 0$ be fixed. Using the same notation as in the proof of Lemma 1, first of all note that

$$\begin{aligned} &|\mathsf{E}\left\{ i\lambda \left(\Delta_n^d |U_n|\right)^{-1/2} S(U_n) \right\} - \mathsf{E} \exp\left\{ i\lambda \left(\Delta_n^d |U_n^-(a)|\right)^{-1/2} S\left(U_n^-(a)\right) \right\}| \\ &\quad \leq |\lambda| \Delta_n^{-d/2} \Big(|U_n|^{-1/2} \mathsf{E} |S(U_n) - S\left(U_n^-(a)\right)| + \delta_n(a) \mathsf{E} |S\left(U_n^-(a)\right)| \Big) \end{aligned}$$

where $\delta_n(a) = ||U_n|^{-1/2} - |U_n^-(a)|^{-1/2}|$. For all n large enough

$$\left(\mathsf{E}|S\left(U_n^-(a)\right)|\right)^2 \leq \mathsf{E}(S(U_n^-(a)))^2 \leq 2\Delta_n^d |U_n^-(a)| \int_{\mathbb{R}^d} |R(u)| du. \tag{13.10}$$

Thus in view of (13.3), (13.5), and (13.10) it suffices to verify that, for every $\lambda \in \mathbb{R}$,

$$\mathsf{E} \exp\left\{ i\lambda \left(\Delta_n^d |U_n^-(a)|\right)^{-1/2} S\left(U_n^-(a)\right) \right\} \to \exp\{-\sigma^2 \lambda^2 / 2\} \quad \text{as } n \to \infty. \tag{13.11}$$

Taking $p \in (0, a/2)$ set $Y_n(a,p) = \left(\Delta_n^d |W_n(a,p)|\right)^{-1/2} S(W_n(a,p))$ where $W_n(a,p)$ appeared in (13.4), $n \in \mathbb{N}$. Thus we are going to deal with normalized sums taken over the sets belonging to separated cubes $\Lambda_j(a,p)$, $j \in J_n^-(a)$.

The same reasoning that was used to prove Lemma 1 leads, for any $\varepsilon > 0$ and arbitrary fixed $\lambda \in \mathbb{R}$, to the estimate

$$\left|\mathsf{E}\exp\left\{i\lambda\left(\Delta_n^d\left|U_n^-(a)\right|\right)^{-1/2} S\left(U_n^-(a)\right)\right\} - \mathsf{E}\exp\{i\lambda Y_n(a,p)\}\right| < \varepsilon$$

provided that p/a is sufficiently small and n is large enough.

For any $L = (v, v+a-2p] \subset \mathbb{R}$ and all n large enough there are $M_n = [(a-2p)\Delta_n]$ points $v_m = z + m\Delta_n^{-1} \in L$, $m = 1, \ldots, M_n$, and $z = q\Delta_n^{-1}$ for some $q \in \mathbb{Z}$ (here $[\cdot]$ stands for the integer part of a number). For each $j \in \mathbb{Z}^d$ let us find a cube $\Gamma_j(a,p) = \prod_{l=1}^d (z_l, z_l + M_n\Delta_n^{-1}\,] \subset \Lambda_j(a,p)$, $(z_1, \ldots, z_d) \in T_n$. Thus $|\Gamma_j(a,p) \cap T_n| = M_n^d$, $j \in \mathbb{Z}^d$. Write $N_n = |J_n^-(a)|$ and enumerate a family of random variables

$$\left\{\left(\Delta_n^d\,|\Gamma_j(a,p) \cap T_n|\right)^{-1/2} S(\Gamma_j(a,p) \cap T_n),\ j \in J_n^-(a)\right\},$$

to obtain a collection $\zeta_{n,1}, \ldots, \zeta_{n,N_n}$ ($N_n = N_n(a, V_n)$, $\zeta_{n,k} = \zeta_{n,k}(a,p,\Delta_n,V_n)$). For any fixed $\lambda \in \mathbb{R}^d$ we have

$$\mathsf{E}\exp\{i\lambda Y_n\} - \mathsf{E}\left\{i\lambda\, N_n^{-1/2} Z_n\right\} \to 0, \qquad n \to \infty,$$

where $Z_n := \sum_{k=1}^{N_n} \zeta_{n,k}$. Due to (13.6) one has, for any $\varepsilon > 0$, $n \in \mathbb{N}$, and $\lambda \in \mathbb{R}$,

$$\left|\,\mathsf{E}\left\{i\lambda\, N_n^{-1/2} Z_n\right\} - \prod_{k=1}^{N_n} \mathsf{E}\exp\left\{i\lambda\, N_n^{-1/2}\zeta_{n,k}\right\}\right|$$
$$\leq \sum_{q=1}^{N_n}\left|\operatorname{cov}\left(\exp\left\{i\lambda N_n^{-1/2}\zeta_{n,q}\right\}, \exp\left\{-i\lambda N_n^{-1/2}\sum_{k=q+1}^{N_n}\zeta_{n,k}\right\}\right)\right| \leq 4\lambda^2\theta_p < \varepsilon$$

if p is large enough. In view of the strict stationarity of X we see that, for every $n \in \mathbb{N}$, the random variables $\zeta_{n,1}, \ldots, \zeta_{n,N_n}$ are identically distributed. Applying the Lindeberg condition to the independent copies of these random variables one finds that $N_n^{-1/2} Z_n$ converges in law as $n \to \infty$ to the Gaussian random variable with mean zero and variance close enough to σ^2 under appropriate choice of a and p. Whence the desired statement follows. □

If we have a sequence $\left(\widehat{\sigma}_n^2(U_n)\right)_{n \in \mathbb{N}}$ of consistent estimates for $\sigma^2 \neq 0$ constructed by means of $X_t, t \in U_n$, then under the conditions of Theorem 1 we can use instead of (13.8) the following *statistical variant* of the CLT:

$$\frac{S(U_n) - n\mathsf{E}X_0}{\widehat{\sigma}(U_n)\Delta_n^{d/2}|U_n|^{1/2}} \to N(0,1) \quad \text{in law as } n \to \infty \tag{13.12}$$

(when $\widehat{\sigma}(U_n) = 0$ we set formally $z/0 := 0$ for $z \in \mathbb{R}$). Thus we obtain a possibility to construct the approximate confidence interval for unknown mean value $\mathsf{E}X_0$.

For a point $t \in T_n$ ($n \in \mathbb{N}$) and any $r > 0$ put

$$K_t(r) = \{s \in T_n : \|s - t\| \le r\}.$$

For a finite set $U_n \subset T_n$ and positive value r_n introduce

$$\widehat{\sigma}(U_n)^2 = \left(\Delta_n^d |U_n|\right)^{-1} \sum_{t \in U_n} |Q_t| \left(\frac{S(Q_t)}{|Q_t|} - \frac{S(U_n)}{|U_n|}\right)^2 \tag{13.13}$$

where $Q_t = K_t(r_n) \cap U_n$ (these Q_t and $\widehat{\sigma}(U_n)$ depend on U_n and r_n), $n \in \mathbb{N}$.

Theorem 2. *Let X be a strictly stationary PA or NA random field with continuous covariance function R satisfying the conditions of Lemma 1. Then, for any sets $V_n \to \infty$ in the van Hove sense ($V_n \subset \mathbb{R}^d$) and any sequence $(\Delta_n)_{n\in\mathbb{N}}$, $0 < \Delta_n \to \infty$, there exists a sequence $(r_n)_{n\in\mathbb{N}}$ of positive numbers such that relation (13.12) holds with $\widehat{\sigma}(U_n)$ defined in (13.13).*

Proof. Due to (13.3) we can find a sequence $(a_n)_{n\in\mathbb{N}}$ such that $0 < a_n \to \infty$ as $n \to \infty$ and

$$\operatorname{mes}\left(V_n^-(a_n)\right) \to \infty, \qquad \operatorname{mes}\left(V_n^+(a_n)\right)/\operatorname{mes}\left(V_n^-(a_n)\right) \to 1, \qquad n \to \infty.$$

At first we consider $U_n^-(a_n)$, $\widehat{\sigma}\left(U_n^-(a_n)\right)$ and to simplify the notation write below U_n instead of $U_n^-(a_n) = V_n^-(a_n) \cap T_n$ and $\widehat{\sigma}_n$ instead of $\widehat{\sigma}\left(U_n^-(a_n)\right)$, $n \in \mathbb{N}$.

Observe that

$$\begin{aligned}
\mathsf{E}|\widehat{\sigma}_n^2 - \sigma^2| &\le \left(\Delta_n^d |U_n|\right)^{-1} \mathsf{E}\left| \sum_{t\in U_n} |Q_t| \left(\left(\frac{S(Q_t)}{|Q_t|} - \frac{S(U_n)}{|U_n|}\right)^2 - \left(\frac{S(Q_t)}{|Q_t|}\right)^2\right)\right| \\
&+ \left(\Delta_n^d |U_n|\right)^{-1} \mathsf{E}\left| \sum_{t\in U_n} |Q_t|^{-1}\left(S(Q_t)^2 - \mathsf{E}(S(Q_t))^2\right)\right| \\
&+ \left| \left(\Delta_n^d |U_n|\right)^{-1} \sum_{t\in U_n} |Q_t|^{-1}\mathsf{E}(S(Q_t))^2 - \sigma^2\right| =: R_{n,1} + R_{n,2} + R_{n,3}.
\end{aligned}$$

For $r > 0$ set $U_n(r) = U_n \setminus (T_n \setminus U_n)^{(r)}$ where $G^{(r)}$ denotes the r-neighbourhood of a finite set $G \subset T_n$; i.e., $G^{(r)} := \{t \in T_n : \min_{s\in G} \|t - s\| < r\}$.

The simplest part is to check that $R_{n,3} \to 0$, $n \to \infty$. Let $r_n < a_n/4$, $n \in \mathbb{N}$. In view of the stationarity of X, one has $\mathsf{E}(S(Q_t(r_n)))^2 = \mathsf{E}(S(K_0(r_n)))^2$ for $t \in U_n(2r_n)$. Consequently,

$$\begin{aligned}
&\left(\Delta_n^d |U_n|\right)^{-1} \sum_{t\in U_n} |Q_t|^{-1}\mathsf{E}(S(Q_t))^2 \\
&\le \frac{\mathsf{E}(S(K_0(r_n)))^2}{\Delta_n^d |K_0(r_n)|} + \left(\Delta_n^d |U_n|\right)^{-1} \sum_{t\in U_n\setminus U_n(2r_n)} |Q_t|^{-1}\mathsf{E}(S(Q_t))^2 =: L_{n,1} + L_{n,2}.
\end{aligned}$$

By Lemma 1 one has $L_{n,1} \to \sigma^2$ as $r_n \to \infty$. For all n large enough

$$L_{n,2} \le 2|U_n|^{-1}|U_n \setminus U_n(2r_n)| \int_{\mathbb{R}^d} |R(u)|du.$$

Applying the arguments used in the proof of Theorem 1 (concerning the concentric cubes) one can infer that $|U_n|^{-1}|U_n \setminus U_n(2r_n)| \le 4dr_n/a_n$ for all n large enough. Now we choose $r_n = o(a_n)$, $n \to \infty$.

For $M > 0$ and $t \in T_n$ ($n \in \mathbb{N}$) write $\xi_t = \left(\Delta_n^d |Q_t|\right)^{-1/2} S(Q_t)$ and introduce the auxiliary random variables $\nu_t = H_M^2(\xi_t)$, $Y_t = \nu_t - \mathsf{E}\nu_t$, $Z_t = \xi_t^2 - \mathsf{E}\xi_t^2 - Y_t$ where

$$H_M(x) = -M\mathbb{I}\{x < -M\} + x\mathbb{I}\{|x| \le M\} + M\mathbb{I}\{x > M\}, \qquad x \in \mathbb{R},$$

and $\mathbb{I}\{A\}$ is the indicator of a set A. Then we may write

$$R_{n,2} \le |U_n|^{-1}\mathsf{E}\left|\sum_{t\in U_n} Y_t\right| + |U_n|^{-1}\mathsf{E}\left|\sum_{t\in U_n} Z_t\right| =: R_{n,2}^{(1)} + R_{n,2}^{(2)}.$$

Furthermore,

$$\left(R_{n,2}^{(1)}\right)^2 \le |U_n|^{-2} \sum_{s,t\in U_n} \operatorname{cov}(\nu_s, \nu_t)$$

$$\le |U_n|^{-2} \sum_{s\in U_n} \sum_{t\in U_n, \|s-t\|\le 3r_n} |\operatorname{cov}(\nu_s, \nu_t)| + |U_n|^{-2} \sum_{s\in U_n} \sum_{t\in U_n, \|s-t\|> 3r_n} |\operatorname{cov}(\nu_s, \nu_t)|.$$

The first summand at the right-hand side of the last inequality admits an upper bound $M^4|U_n|^{-1}(6r_n+1)^d \to 0$ as $r_n = o(a_n)$, $n \to \infty$. To estimate the second summand we use relation (13.7) to obtain the upper bound

$$\begin{aligned} J_n :=& 4M^2\Delta_n^{-d}|U_n|^{-2} \sum_{t\in U_n, \|s-t\|>3r_n} |Q_s|^{-1/2}|Q_t|^{-1/2}|\operatorname{cov}(S(Q_s), S(Q_t))| \\ \le& 4M^2\Delta_n^{-d}|U_n|^{-2} \sum_{t\in U_n, \|s-t\|>3r_n} |Q_s|^{-1/2}|Q_t|^{-1/2}(|Q_s| \wedge |Q_t|) \sum_{r\in T_n, \|r\|\ge r_n} |R(r)|. \end{aligned}$$

Therefore, for any $b > 0$ and all n large enough

$$J_n \le 8M^2 \int_{\|u\|\ge b} |R(u)|du.$$

Consequently, for any fixed $M > 0$, one has $J_n \to 0$ as $n \to \infty$.

For $M > 0$ and all n large enough

$$\begin{aligned} R_{n,2}^{(2)} \le& |U_n|^{-1} \sum_{t\in U_n(2r_n)} \mathsf{E}|Z_t| + |U_n|^{-1} \sum_{t\in U_n\setminus U_n(2r_n)} \mathsf{E}|Z_t| \\ \le& 2\mathsf{E}\left(\frac{S(K_0(r_n))^2}{\Delta_n^d|K_0(r_n)|}\mathbb{I}\left\{\frac{S(K_0(r_n))^2}{\Delta_n^d|K_0(r_n)|} \ge M^2\right\}\right) + 4\frac{|U_n \setminus U_n(2r_n)|}{|U_n|}\int_{\mathbb{R}^d} |R(u)|du. \end{aligned}$$

Theorem 1 implies that a family $\left\{S(K_0(r_n))^2/|K_0(r_n)|, n \in \mathbb{N}\right\}$ is uniformly integrable. Thus $R_{n,2}^{(2)}$ can be made arbitrarily small by an appropriate choice of M.

In view of Lemma 1 for all n large enough one has

$$R_{n,1} \le \Delta_n^{-d}|U_n|^{-3}\mathsf{E}S(U_n)^2 \sum_{t\in U_n} |Q_t| + 2\Delta_n^{-d}|U_n|^{-2}\mathsf{E}\left|S(U_n) \sum_{t\in U_n} S(Q_t)\right|$$

$$\leq 2\sigma^2|U_n|^{-2}\sum_{t\in U_n}|Q_t| + 2\Delta_n^{-d}|U_n|^{-2}\sum_{t\in U_n}\left(\mathsf{E}(S(U_n))^2\mathsf{E}(S(Q_t))^2\right)^{1/2}$$

$$\leq 2\sigma^2|U_n|^{-1}|K_0(r_n)| + 4\left(\sigma^2|U_n|^{-1}|K_0(r_n)|\int_{\mathbb{R}^d}|R(u)|du\right)^{1/2}.$$

Therefore $R_{n,1} + R_{n,2} + R_{n,3} \to 0$ as $n \to \infty$.

To complete the proof for the general case of $V_n \to \infty$ in the van Hove sense one can verify that $\mathsf{E}|\widehat{\sigma}(U_n)^2 - \widehat{\sigma}\left(U_n^-(a_n)\right)^2| \to 0$, as $n \to \infty$, with an appropriate choice of a_n and r_n. □

In fact we have obtained the following more general result.

Theorem 3. *Let the conditions of Theorem 1 be satisfied. Then for any $V_n \to \infty$ in the van Hove sense $(V_n \subset \mathbb{R}^d)$ and any sequence $(\Delta_n)_{n\in\mathbb{N}}$ such that $0 < \Delta_n \to \infty$, there exists a sequence $(r_n)_{n\in\mathbb{N}}$ of positive numbers which satisfies relation* (13.12) *with $\widehat{\sigma}(U_n)$ defined in* (13.13).

Proof. The proof follows the lines used to establish Theorem 2. We have to modify only the estimate $\left(R_{n,1}^{(1)}\right)^2$. Namely,

$$\Delta_n^{-d}|U_n|^{-2}\sum_{s\in U_n}\sum_{t\in U_n:\|s-t\|>3r_n}|\mathrm{cov}(\nu_s,\nu_t)|$$

$$\leq |U_n|^{-2}\sum_{s\in U_n}\sum_{t\in U_n:\|s-t\|>3r_n}4M^2|Q_s|^{-1/2}|Q_t|^{-1/2}(|Q_s|\wedge|Q_t|)\theta_{\|s-t\|-2r_n}$$

$$\leq 4M^2\theta_{r_n} \to 0, \qquad n \to \infty.$$

The proof is complete. ■

Remark 2. Thus if the conditions of Theorem 2 or 3 are met then in the particular important case $V_n = (u_n, v_n]^d$ and $v_n - u_n \to \infty$ one can take $r_n = o(v_n - u_n)$ as $n \to \infty$.

13.3 Applications

There are various stochastic models describing tissue (or organ) response under irradiation. One uses, e.g., the single-hit or multiple-hit or LQ-models for probability that a cell after irradiation of a certain dose will be alive. Due to Withers et al. (1988) the idea of independent functional subunits (FSUs) was introduced for biological modelling. The approach based on this idea and involving the binomial distribution was developed in the papers by Niemierko and Goitein (1993), Jackson et al. (1993), and York et al. (1993); see also Stavrev et al. (2001), Warkentin et al. (2004), and references therein. Namely the tissue (or organ) consists of N functional subunits that behave "statistically independently" under irradiation (one can consider also the irradiation of a part of the tissue or organ). One assumes that there exists "the functional reserve" M, i.e., the number of structural elements that must be damaged to cause a failure in the structure of interest. For tumours it is equal to the total number of clonogens ($M = N$) meaning that all clonogens (tumour cells) should be destroyed to cause the tumour collapse. For "Critical Element" organs $M = 1$ meaning that all FSUs constituting the organ are critical to its normal functioning. It is clear that Tumours and the Critical Element are special (end term) cases of the "Critical Volume" response when $1 \leq M \leq N$.

Let $p_{FSU} = p_{FSU}(\overrightarrow{par}_{FSU}, D)$ be the probability of damaging a FSU when irradiated to dose D and $\overrightarrow{par}_{FSU}$ be the vector of parameters describing the response of the FSU to radiation. For example, in the single-hit model

$$p_{FSU}(\overrightarrow{par}_{FSU}) = (1 - e^{-\alpha D})^{N_0}$$

where N_0 is the number of cells in a subunit (or the number of clonogens) and α describes the radiosensitivity of a cell. Thus here $\overrightarrow{par}_{FSU} = (N_0, \alpha)$ and the model parameters are $N, M, p_{FSU}(\overrightarrow{par}_{FSU}, D)$. Consequently, for the "individual" tissue (or organ) the probability corresponding to the critical volume model is given by the formula

$$P_{ind}(N, M, p_{FSU}(\overrightarrow{par}_{FSU}, D) = \sum_{k=M}^{N} \binom{N}{k} p_{FSU}^k (1 - p_{FSU})^{N-k}.$$

Thus if νN subunits ($1 \leq \nu N \leq N$) are irradiated then (see Stavrev et al., 2001) the following well-known approximation is used for the binomial distribution

$$\sum_{k=M}^{\nu N} \binom{\nu N}{k} p_{FSU}^k (1 - p_{FSU})^{\nu N-k} \approx \Phi\left(\frac{\sqrt{N}(\nu p_{FSU} - \mu_{cr})}{\sqrt{\nu p_{FSU}(1 - p_{FSU})}}\right) \tag{13.14}$$

where Φ is a distribution function of the standard normal random variable and $\mu_{cr} = M/N$.

In Bulinski and Khrennikov (2005) a generalization of the critical volume model was proposed. The idea of the dependent (in particular independent) FSUs was expressed by invoking dependent (mixing) random fields defined on a lattice $\mathbb{Z}^d$. In this chapter we go further. Namely, we consider the dependence structures based on positive or negative association (comprising independent random variables) and dwell on (BL, θ)-dependence. Moreover, here we study the general case of growing in the van Hove sense domains in $\mathbb{R}^d$ with increasing density of grids of observations. The impact of the irradiation on the FSU can be described not only in terms of the indicator functions to take into account the intermediate cases between the killed and alive FSUs. And finally Theorems 2 and 3 show that the approximation of $S(U_n)$ describing the collective effects of the summands' behavior should involve the possible dependence structure of FSUs. It should be also emphasised that the approximation (13.14) will take a different form due to the possible dependence of FSUs. We do not tackle here the problem of nonindividual, say, population response, the problems of nonunifom irradiation. To conclude we indicate a very important problem concerning the accuracy of various approximations. It seems that this problem was not discussed yet in the special biomedical papers. In this regard we refer to Bulinski and Kryzhanovskaya (2006) where the convergence rate of the (vector-valued) statistics with self-normalisation was established for dependent observations defined on subsets of $\mathbb{Z}^d$.

Acknowledgements. The research is partially supported by RFBR grant 07-01-00373a.

References

Bulinski, A. (2004). A statistical version of the central limit theorem for vector-valued random fields. *Math. Notes*, 76:455–464.

Bulinski, A. and Khrennikov, A. (2005). Generalization of the critical volume NTCP model in the radiobiology. Université Pierre et Marie Curie. Paris-6. CNRS U.M.R. 7599. Probabilités et Modèles Aléatoires. Prépublication PMA-977:1–13.

Bulinski, A. and Kryzhanovskaya, A. (2006). Convergence rate in the CLT for vector-valued random fields with self-normalization. *Probab. Math. Statist.*, 26:261–281.

Bulinski, A. and Shabanovich, E. (1998). Asymptotical behaviour for some functionals of positively and negatively dependent random fields. *Fundam. Prikl. Mat.*, 4:479–492 (in Russian).

Bulinski, A. and Shashkin, A. (2007). *Limit Theorems for Associated Random Fields and Related Systems.* World Scientific, Singapore.

Bulinski, A. and Suquet, C. (2001). Normal approxiation for quasi associated random fields. *Stat. Probab. Lett.*, 54:215–226.

Bulinski, A. and Vronski, M. (1996). Statistical variant of the central limit theorem for associated random fields. *Fundam. Prikl. Mat.*, 2:999–1018 (in Russian).

Cressie, N.A.C. (1991). *Statistics for Spatial Data.* Wiley, New York.

Jackson, A., Kutcher, G.J. and York E.D. (1993). Probability of radiation induced complications for normal tissues with parallel architecture subject to non-uniform irradiation. *Med. Phys.*, 20:613–625.

Newman, C.M. (1980). Normal fluctuations and the FKG inequalities. *Commun. Math. Phys.*, 74:119–128.

Niemierko, A. and Goitein, M. (1993). Modeling of normal tissue response to radiation: The critical volume model. *Int. J. Radiat. Oncol., Biol., Phys.*, 25:983–993.

Peligrad, M. and Shao, Q.-M. (1996). Self-normalized central limit theorem for sums of weakly dependent random variables. *Stat. Probab. Lett.*, 26:141–145.

Stavrev, P., Stavreva, N., Niemierko, A. and Goitein M. (2001). Generalization of a model of tissue response to radiation based on the idea of functional subunits and binomial statistics. *Phys. Med. Biol.*, 46:1501–1518.

Warkentin, B., Stavrev, P., Stavreva, N., Field, C. and Gino, B. (2004). A TCP-NTCP estimation module using DVHs and known radiobiological models and parameters sets. *J. Appl. Clin. Med. Phys.*, 5:50–63.

Withers, H.R., Taylor, J.M. and Maciejewski, B. (1988). Treatment volume and tissue tolerance. *Int. J. Radiat. Oncol. Biol. Phys.*, 14:751–759

York, E.D., Kutcher, G.J., Jackson, A. and Ling, C.C. (1993). Probability of radiation-induced complications in normal tissues with parallel architecture under conditions of uniform whole or partial organ irradiation. *Radiother. Oncol.*, 26:226–237.

14

A Berry–Esseen Type Estimate for Dependent Systems on Transitive Graphs

Alexey Shashkin

Department of Mathematics and Mechanics, Moscow State University, Russia

Abstract: Dependent random systems indexed by transitive graphs are studied. The dependence structure generalizes the ideas of positive and negative association. For such random systems the CLT is proved and the rate of convergence is established.

Keywords and phrases: Local dependence, transitive graphs, association, Stein–Tikhomirov techniques.

14.1 Introduction

This chapter is devoted to the proof of the central limit theorem for a dependent random system indexed by points of a graph that is more general than $\mathbb{Z}^d$. The interest in such problems has arisen in connection with interacting particle systems (see Häggström et al., 2000; Doukhan et al., 2008, and references there). The dependence condition we study was considered recently in Doukhan et al. (2008). In contrast with that paper we allow the random variables to be unbounded requiring only that they possess a moment of order higher than two. Also we weaken the condition on stationarity.

Let $G = (V, E)$ be some locally finite graph with countable vertex set; i.e., the degree of any vertex $t \in V$ is finite. Introduce standard metric d on V, agreeing that $d(x, y) = n \in \mathbb{N}$ if and only if there exist pairwise different points $t_0 = x, t_1, \ldots, t_{n-1}, t_n = y \in V$ such that t_{i-1} is connected to t_i $(i = 1, \ldots, n)$, but there is no set of points with the same property having length less than n. We assume that the degree of a vertex is uniformly bounded; i.e., there exists such $\rho > 0$ that $|\{y : d(x, y) = 1\}| \le \rho$. Clearly, in that case the cardinality of a ball of radius r is bounded by ρ^r, $r \in \mathbb{N}$. Denote by $B_r(x)$ the ball of radius r centered at $x \in V$. As usual, $|A|$ is the cardinality of a finite set A.

Recall that a bijection $a : V \to V$ is called an automorphism of the graph G if any two vertices $x, y \in V$ are connected if and only if $a(x)$ is connected to $a(y)$. Suppose that the graph G is transitive; i.e., the group of its automorphisms $Aut(G)$ acts transitively on G. That property means that for any $x, y \in V$, there exists such an automorphism a of G that $a(x) = y$. Note that in this case there exists such $\rho \in \mathbb{N}$ that $|B_1(x)| = \rho$

C.H. Skiadas (ed.), *Advances in Data Analysis*, Statistics for Industry and Technology, DOI 10.1007/978-0-8176-4799-5_14,

for any $x \in V$ (Godsil and Royle, 2001). Clearly, the lattice $\mathbb{Z}^d$ is a transitive graph with automorphisms acting as translations.

As usual, a function $F : \mathbb{R}^n \to \mathbb{R}$ is called Lipschitz if

$$Lip(F) := \sup_{x,y \in \mathbb{R}^n,\, x \neq y} \frac{|F(x) - F(y)|}{|x_1 - y_1| + \cdots + |x_n - y_n|} < \infty.$$

Definition 1. (Bulinski and Suquet, 2001). *Let* $\{X_t, t \in V\}$ *be some system of square-integrable random variables. This system is called* (BL, θ)*-dependent if there exists a nonincreasing sequence* $\theta = (\theta_r)_{r \in \mathbb{N}}$, $\theta_r \to 0$ *as* $r \to \infty$, *such that for any finite disjoint sets* $I, J \subset V$ *with* $\mathrm{dist}(I, J) = r \in \mathbb{N}$, *and any pair of bounded Lipschitz functions* $f : \mathbb{R}^{|I|} \to \mathbb{R}$, $g : \mathbb{R}^{|J|} \to \mathbb{R}$ *one has*

$$|\mathrm{cov}(f(X_i, i \in I), g(X_j, j \in J))| \leq \mathrm{Lip}(f)\mathrm{Lip}(g)(|I| \wedge |J|)\theta_r. \tag{14.1}$$

The assumption that f and g are bounded is inessential and can be avoided via an easy application of the dominated convergence theorem. Recall that the system X is called wide-sense stationary if the group $\mathrm{Aut}(G)$ contains a subgroup H that acts transitively on G and for any different $t, v \in V$ and all $a \in H$ one has $\mathrm{cov}(X_t, X_v) = \mathrm{cov}(X_{a(t)}, X_{a(v)})$, $\mathsf{E}X_t = \mathsf{E}X_{a(t)}$. Stationary associated and negatively associated random systems satisfy the definition (14.1) provided that they satisfy the finite susceptibility condition of Newman; that is, $\sum_{v \in V} \mathrm{cov}(X_t, X_v) \in \mathbb{R}$, $t \in V$. The proof (Bulinski and Shabanovich, 1998) is usually given for $V = \mathbb{Z}^d$, but remains true also on the general graph G. In that case one can take

$$\theta_r = \sum_{v \in V,\, d(v,t) \geq r} |\mathrm{cov}(X_t, X_v)|.$$

Associated random processes and fields and their modifications form an important class which appears with increasing frequency in statistics, statistical physics, random graphs, and random measures theory. An independent random system is always associated. In 1980 Newman (1980) proved the central limit theorem for strictly stationary associated random fields. Since that time a lot of other limit theorems (invariance principles, Berry–Esseen type estimates, laws of the iterated logarithm, etc.) have been established; see Bulinski (1995), Bulinski and Shashkin (2007), and references provided there. There are also examples of random fields that are neither positively nor negatively associated but possess the property (14.1) (Shashkin, 2004; Bulinski and Shashkin, 2007). In particular, they arise in the theory of interacting particle systems indexed by $\mathbb{Z}^d$.

14.2 Main result

Now we formulate the Berry–Esseen type theorem.

Theorem 1. *Suppose that* $X = \{X_t, t \in V\}$ *is a centered* (BL, θ)*-dependent random system such that* $D_s = \sup_{t \in V} \mathsf{E}|X_t|^s < \infty$ *for some* $s > 2$. *Assume also that the* *sequence* (θ_r) *admits the bound* $\theta_r = O(e^{-\lambda r})$ *as* $r \to \infty$, *with some* $\lambda > \log \rho$. *Finally, suppose that there is a class of finite sets* $\mathcal{U}$ *such that for some* $d > 0$ *and all* $U \in \mathcal{U}$ *one has* $\mathsf{D}S(U) \geq d|U|$. *Then there exist such* $C > 0$ *and* $\mu > 0$ *determined by* d, s, ρ, *and* λ *that for any* $U \in \mathcal{U}$ *one has*

$$\sup_{x\in\mathbb{R}} |\mathsf{P}(S(U) \leq x\sqrt{\mathsf{D}S(U)}) - \mathsf{P}(Z \leq x)| \leq C|U|^{-\mu/2},$$

where $Z \sim N(0,1)$. *In particular, if* $D_3 < \infty$ *and* $\lambda > 4\log\rho$, *then one can take*

$$\mu = \varkappa(1+\varkappa)/(9\varkappa^2 + 20\varkappa - 12),$$

where $\varkappa = \lambda/\log\rho$.

Remark 1. Thus, the exponent in the Gaussian approximation estimate tends to $1/9$ when the third moment is finite and λ tends to infinity. If the field is wide-sense stationary and its covariance function is nonnegative (as will be the case if X is associated), then for any finite $W \subset V$ one has $\mathsf{D}S(W) \geq |W|\mathsf{D}X_t$, $t \in V$. So in this case class $\mathcal{U}$ contains all finite sets $U \subset V$.

14.3 Proof

Note that the condition imposed on λ ensures that the series

$$\sigma_0^2(t) := \sum_{t\in V} |\mathrm{cov}(X_t, X_v)|$$

converges uniformly, since this sum does not exceed some positive value multiplied by $\sum_{k=1}^{\infty} \rho^k e^{-\lambda k} < \infty$. In particular, for any finite set $W \subset V$ we have

$$\mathsf{E}S^2(W) \leq \sup_{t\in V} \sigma_0^2(t)|W|. \tag{14.2}$$

The proof adapts the local sectioning (Stein–Tikhomirov) method of the analysis of characteristic functions (Tikhomirov, 1981). For associated random processes and fields it was applied, respectively, in Birkel (1988) and Bulinski (1995, 1996). Let $m = m(U) \in \mathbb{N}$ be some quantity specified later. For $t \in \mathbb{R}$, $l \in \{1, 2\}$, and $j \in U$ denote

$$\begin{aligned}
&\sigma = \sigma(U) = \sqrt{\mathsf{D}S(U)}, && \tau = t/\sigma, && f(t) = \mathsf{E}\exp\{i\tau S(U)\},\\
&U_j^0 = U,\ U_j^l = \{q \in U : d(q,j) > lm\}, && && W_j^l = U_j^{l-1} \setminus U_j^l,\\
&S_j^l = \sigma^{-1} S\left(U_j^l\right), && Z_j^l = S\left(W_j^l\right), && \xi_j^l = \exp\left\{i\tau Z_j^l\right\} - 1.
\end{aligned}$$

By c we denote various positive factors which may depend on the field X, but not on j and m.

We intend to prove that $f(t) \to e^{-t^2/2}$ when $\sigma(U) \to \infty$. To this end we establish a differential equation that $f(t)$ satisfies and make use of the formula of its solution. Write the following differential equation:

$$\begin{aligned}
f'(t) &= i\sigma^{-1}\mathsf{E}X_j \exp\{i\tau S(U)\} \\
&= i\sigma^{-1} f(t) \sum_{j\in U} \mathsf{E}X_j\xi_j^1 + i\sigma^{-1}\sum_{j\in U}\mathsf{E}X_j \exp\left\{itS_j^1\right\} \\
&\quad + i\sigma^{-1}\sum_{j\in U}\mathsf{E}X_j\xi_j^1\left(\exp\left\{itS_j^2\right\} - f(t)\right) + i\sigma^{-1}\sum_{j\in U}\mathsf{E}X_j\xi_j^1\xi_j^2\exp\left\{itS_j^2\right\} \quad (14.3)\\
&=: A(t)f(t) + B_1(t) + B_2(t) + B_3(t).
\end{aligned}$$

The following lemma is a counterpart of Lemmas 1 and 2 in Bulinski (1995). It allows us to estimate $A(t)$, $B_1(t)$, and $B_3(t)$.

Lemma 1. *For any $j \in U$ and all $t \in \mathbb{R}$,*

$$|A(t)+t| \le c|U|\left(|t|\sigma^{-2}\theta_m + (|t|\rho^m)^{s-1}\sigma^{-s}\right), \tag{14.4}$$

$$B_1(t)| \le c|U||t|\sigma^{-2}\theta_m, \tag{14.5}$$

$$\mathsf{E}|X_j\xi_j^1\xi_j^2| \le c|\tau|\rho^m(|\tau|\rho^{m/2} + (|\tau|\theta_m)^{(s-2)/(s-1)}). \tag{14.6}$$

Proof. Let $\alpha_0(x)$ be a function such that $e^{ix} - 1 - x = \alpha_0(x)|x|^{s-1}$ for $x \in \mathbb{R}$. Then one clearly has $|\alpha_0(x)| \le 2$, $x \in \mathbb{R}$. Hence, using that $S(U) = S\left(Z_j^1\right) + S\left(U_j^1\right)$ for all $j \in U$, we have

$$\begin{aligned}
|A(t)+t| &= i\sigma^{-1}\sum_{j\in U}\mathsf{E}X_j i\sigma^{-1}tZ_j^1 + t + i\sigma^{-1}\sum_{j\in U}\mathsf{E}X_j\sigma^{1-s}|t|^{s-1}\left|Z_j^1\right|^{s-1}\alpha\left(\tau Z_j^1\right) \\
&= t\sigma^{-2}\mathrm{cov}(X_j, S\left(U_j^1\right)) + i\sigma^{-s}|t|^{s-1}\sum_{j\in U}\mathsf{E}X_j\left|Z_j^1\right|^{s-1}\alpha\left(\tau Z_j^1\right).
\end{aligned}$$

Now the relation (14.4) follows after applying (14.1) to the first summand and the Hölder inequality to the second one.

The bound (14.5) is the direct consequence of (14.1). Furthermore, take the function $h_M(x) = (x \wedge M) \vee (-M)$, with $M > 0$ selected later, and write

$$\begin{aligned}
\mathsf{E}\left|X_j\xi_j^1\xi_j^2\right| &\le \left(\mathsf{E}\left|\xi_j^1\right|^s\right)^{1/s}\left(\mathsf{E}\left|X_j\xi_j^2\right|^{s/(s-1)}\right)^{(s-1)/s} \\
&\le D_s^{1/s}|\tau|\rho^m\left(\mathsf{E}|X_j|^{s/(s-1)}\mathsf{E}\left|\xi_j^2\right|^{s/(s-1)} + \left|\mathrm{cov}\left(|X_j|^{s/(s-1)}, \left|\xi_j^2\right|^{s/(s-1)}\right)\right|\right)^{(s-1)/s} \\
&\le D_s^{1/s}|\tau|\rho^m\left(\mathsf{E}|X_j|^{s/(s-1)}\mathsf{E}\left|\xi_j^2\right|^{s/(s-1)} + \left|\mathrm{cov}\left(|h_M(X_j)|^{s/(s-1)}, \left|\xi_j^2\right|^{s/(s-1)}\right)\right|\right. \\
&\quad \left. + \left|\mathrm{cov}\left(|X_j|^{s/(s-1)} - |h_M(X_j)|^{s/(s-1)}, \left|\xi_j^2\right|^{s/(s-1)}\right)\right|\right)^{(s-1)/s}.
\end{aligned}$$

Since $\mathrm{Lip}(|h_M(\cdot)|^{s/(s-1)}) = (s/(s-1))M^{1/(s-1)}$, the first covariance can be estimated using (14.1). The second covariance is bounded via the Hölder inequality due to the fact that $|h_M(x) - x| \le |x|\mathbf{I}\{|x| > M\}$. It remains to minimize the obtained bound in M to get (14.6). ■

Lemma 2. *For any $j \in U$ one has*

$$|B_2(t)| \le c|U|\sigma^{-1}\left(|\tau|^2\rho^{3m/2} + |\tau|\rho^m\theta_m + (|\tau|^2\rho^m\theta_m)^{(s-1)/s}\right).$$

Proof. Note that

$$\left|\mathsf{E}X_j\xi_j^1\left(\exp\left\{itS_j^2 - f(t)\right\}\right)\right| \le \left|\operatorname{cov}\left(X_j\xi_j^1, \exp\left\{itS_j^2\right\}\right)\right| + \left|\mathsf{E}X_j\xi_j^1\left(\mathsf{E}\exp\left\{itS_j^2\right\} - \mathsf{E}\exp\{i\sigma^{-1}tS(U)\}\right)\right|. \tag{14.7}$$

To estimate the second summand in the right-hand side of (14.7) use Lemma 1 and note that by relation (14.2) we have

$$\left|\mathsf{E}\exp\left\{itS_j^2\right\} - \mathsf{E}\exp\{i\sigma^{-1}tS(U)\}\right| \le |\tau|\left(\mathsf{E}\left(Z_j^1 + Z_j^2\right)^2\right)^{1/2} \le C|\tau|\rho^m.$$

As for the first summand, again let $M > 0$ be a number picked later and set $h_M(x) = (x \wedge M) \vee (-M)$. Then

$$\left|\operatorname{cov}\left(X_j\xi_j^1, \exp\left\{itS_j^2\right\}\right)\right| \le \left|\operatorname{cov}\left(h_M(X_j)\xi_j^1, \exp\left\{itS_j^2\right\}\right)\right| + \left|\operatorname{cov}\left((X_j - h_M(X_j))\xi_j^1, \exp\left\{itS_j^2\right\}\right)\right| \le (2 + M|\tau|)|\tau|\rho^m\theta_m + 4D_sM^{1-s},$$

where we used (14.1) for the first term and the Markov inequality for the second one. On minimizing the last expression in $M > 0$ we get the lemma.

The final step of the proof is based on Esseen inequality and picking the appropriate values of m (they may depend on t). For simplicity we consider the case $s = 3$, but the proof can be easily transmitted to the case when $s \in (2, 3)$, only x and y below need to be chosen in another way. Denote $g(t) = f(t) - e^{-t^2/2}$. Lemmas 1 and 2 and expansion (14.3) yield a bound for the derivative of g.

Let us take

$$m = \left\lceil \frac{x \log t + y \log \sigma}{\log \rho} \right\rceil$$

where positive numbers x and y are selected later. Recall that by the Esseen inequality for any $T > 0$ one has

$$\sup_{x\in\mathbb{R}} \left|\mathsf{P}(S(U) \le x\sqrt{\mathsf{D}S(U)}) - \mathsf{P}(Z \le x)\right| \le cT^{-1} + \int_{|t|\le T} |t|^{-1}|g(t)|dt.$$

Inserting here the bound for g' and taking the integral over $[-T, T]$ we obtain that

$$\sup_{x\in\mathbb{R}} \left|\mathsf{P}(S(U) \le x\sqrt{\mathsf{D}S(U)}) - \mathsf{P}(Z \le x)\right| \le cT^{-1} + c\Big(T^{2-2\varkappa/(2+\kappa)}\sigma^{-y\kappa} + T^{3+4/(2+\kappa)}\sigma^{2y-1} + T^{4-(\kappa-2)/(\kappa+2)}\sigma^{-1/2-y(\kappa/2-1)} + T^{3+3/(2+\kappa)}\sigma^{3/2y-1} + T^{7/3-4(\kappa-1)/(6+3\kappa)}\sigma^{-1/3-2y(\kappa-1)/3}\Big).$$

Now we take

$$x = \frac{2}{2+\varkappa}, \qquad y = \frac{\varkappa+6}{9\varkappa^2 + 20\varkappa - 12}.$$

Finally, select $T = \sigma^{\mu}$ where μ appears in the statement of theorem. ∎

14.4 Conclusion

Clearly, the argument above allows us to establish the rate of convergence to normal law (in the regular case when $\sigma^2(U) \geq C|U|$). Note that in Doukhan et al. (2008) the central limit theorem is established for the sets $U \subset V$ which tend to infinity in a regular way; that is, $|U| \to \infty$ but $|\partial U|/|U| \to 0$. Here, as usual, $\partial U = \{y \in V : d(x,y) = 1\}$. However, this condition is usually not true in the case of graphs more general than $\mathbb{Z}^d$. For example, in the paper cited above the authors mention the graph $G = (V, E)$ whose vertices are elements of a noncommutative free group with generators $\{g_1, \ldots, g_L\}$ having order 2 ($L > 2$). The vertices $x, y \in V$ are connected by an edge if and only if there exists such $i \in \{1, \ldots, L\}$ that $x = g_i y$. For that graph $|\partial B_n(x)| = L(L-1)^{n-1}$, hence $|B_n(x)| \sim L^n$ and $|\partial B_n(x)|/|B_n(x)|$ does not tend to zero when $n \to \infty$. The theorem given above does not need such a restriction on U.

Acknowledgements. The work was partially supported by the RFBR grant 07-01-00373-a.

References

Birkel, T. (1988). On the convergence rate in central limit theorem for associated processes. *Ann. Probab.*, 16:1685–1698.

Bulinski, A. (1995). Rate of convergence in the central limit theorem for fields of associated random variables. *Theory Probab. Appl.*, 40:136–144.

Bulinski, A. (1996). On the convergence rates in the central limit theorem for positively and negatively dependent random fields. *Probab. Theory Math. Statist.*, 51:3–14.

Bulinski, A.V. and Shabanovich, E. (1998). Asymptotical behaviour for some functionals of positively and negatively dependent random fields. *Fundam. Prikl. Mat.*, 4:479–492.

Bulinski, A. and Shashkin, A. (2007). *Limit Theorems for Associated Random Fields and Related Systems.* World Scienific, Singapore.

Bulinski, A. and Suquet, C. (2001). Normal approximation for quasi associated random fields. *Statist. Probab. Lett.*, 54:215–226.

Doukhan, P., Lang, G., Louhichi, S., and Ycart, B. (2008). A functional central limit theorem for interacting particle systems on transitive graphs. *Markov Proc. Rel. Fields*, 14:79–114.

Godsil, G. and Royle, G.F. (2001). *Algebraic Graph Theory.* Springer, New York.

Häggström, O., Schonmann, R., and Steif, J. (2000). The Ising model on diluted graphs and strong amenability. *Ann. Probab.*, 28:1111–1137.

Newman, C. (1980). Normal fluctuations and the FKG inequalities. *Commun. Math. Phys.*, 74:119–128.

Shashkin, A. (2004). A weak dependence property of a spin system. *Proceedings of the XXIV Int. Symposium on Stability Problems for Stochastic Models*, 30–35.

Tikhomirov, A.N. (1981). On the convergence rate in the central limit theorem for weakly dependent random variables. *Theory Probab. Appl.*, 25:790–809.

15

Critical and Subcritical Branching Symmetric Random Walks on d-Dimensional Lattices

Elena Yarovaya

Department of Mathematics and Mechanics, Moscow State University, Russia

Abstract: We study a symmetric continuous time branching random walk on a d-dimensional lattice with the zero mean and a finite variance of jumps under the assumption that the birth and the death of particles occur at a single lattice point. In the critical and subcritical cases the asymptotic behavior of the survival probability of particles on $\mathbf{Z}^{\mathbf{d}}$ at time t, as $t \to \infty$, is obtained. Conditional limit theorems for the population size are proved. The models of a branching random walk in a spatially inhomogeneous medium could be applied to the study of the long-time behavior of objects in a catalytic environment.

Keywords and phrases: Branching random walks, survival probability, limit theorems, critical case, subcritical case.

15.1 Introduction

Numerous applications of branching processes in various areas of the natural sciences have demonstrated the importance of developing more realistic mathematical models in which the evolutionary processes depend on the structure of a medium. It is well known that the inhomogeneity of a medium plays an essential role in the formation of abnormal properties of particle transport processes. It is worth mentioning that the concept of "strong centers" is used for the interpretation of the intermittency phenomenon in the theory of random media; see Gartner and Molchanov (1990) and Molchanov (1994). Consequently, interest in branching random walks under the assumption that the birth and the death of particles occur at a single lattice point (i.e., the source) has increased. This chapter is devoted to the study of continuous-time branching symmetric random walks on d-dimensional lattices with a reproduction of particles at the origin. The model is of interest mainly in connection with the following two circumstances: the branching medium (i.e., the set of branching characteristics at points of the phase space) is inhomogeneous, and the phase space in which the walk occurs is unbounded (see Bogachev and Yarovaya, 1998a).

C.H. Skiadas (ed.), *Advances in Data Analysis*, Statistics for Industry and Technology, DOI 10.1007/978-0-8176-4799-5_15,

A similar model on a one-dimensional lattice has been studied in the paper by Vatutin et al. (2005). In this model an additional parameter controlling the behavior of a process at a branching source was introduced, but simultaneously the introduction of this parameter destroyed symmetry of an infinitesimal transition matrix of a random walk. As has been shown by Vatutin and Xiong (2007), similar types of branching random walks with a single source are used as approximations of catalytic superprocesses, the theory of which has recently undergone active development in papers by Fleischmann and Le Gall (1995) and Greven et al. (1999).

One of the main problems in such models is the study of the evolution of populations of particles. The offspring reproduction intensity at the origin exerts essential influence on the asymptotic behavior of a branching random walk. In connection with this, the definition of criticality for a branching random walk is introduced. Particular attention in this chapter is paid to critical and subcritical processes. Then conditions of reaching a critical regime and their dependence on the lattice dimension are formulated and analyzed.

15.2 Description of a branching random walk

We consider a branching random walk on $\mathbf{Z^d}$ with a single source. The population of individuals is initiated at time $t = 0$ by a single particle. Being outside the origin the particle performs a continuous time random walk with infinitesimal transition matrix $A = \| a(x, y) \|_{x,y \in \mathbf{Z^d}}$. The random walk is assumed to be symmetric, homogeneous, irreducible, and having the zero mean and a finite variance of jumps: $a(x, y) = a(y, x)$, $a(x, y) = a(0, y - x) = a(y - x)$ with $a(x) \geq 0$, $x \neq 0$, $a(0) < 0$, $\sum_x a(x) = 0$, and $\sum_{x \in \mathbf{Z^d}} x^2 a(x) < \infty$. In particular, this class includes the simple symmetric random walk defined by $a(x, y) = a(0)/2d$ for $|y - x| = 1, a(x, x) = -a(0)$, and $a(x, y) = 0$ otherwise. The branching mechanism at the source is governed by the infinitesimal generating function $f(u) := \sum_{n=0}^{\infty} b_n u^n$ $(0 \leq u \leq 1)$, where $b_n \geq 0$ for $n \neq 1$, $b_1 < 0$, and $\sum_n b_n = 0$. It is supposed that the particle spends at the origin an exponentially distributed time with parameter $-(a(0) + b_1)$ and then either jumps to a point $y \in \mathbf{Z^d}$ (distinct from the origin) or dies producing just before the death a random number of offspring. The newborn particles behave independently and stochastically in the same way as the parent individual. This model was first introduced (for the case of a simple symmetric random walk without the death of particles: $b_0 = 0$) by Yarovaya (1991).

Under these conditions, the random walk transition probabilities $p(t, x, y)$ satisfy the system of differential-difference equations (Kolmogorov's backward equations)

$$\frac{\partial p}{\partial t} = Ap, \qquad p(0, x, y) = \delta_y(x), \tag{15.1}$$

where the (linear) operator A acts with respect to the variable x in accordance with the following rule,

$$Ap(t, x, y) := \sum_{x'} a(x, x') p(t, x', y),$$

while y is treated as a parameter. By Schur's test (see, e.g., Halmos, 1982), A is a bounded operator on $l^q(\mathbf{Z^d})$ for any $1 \le q \le \infty$.

Set

$$\phi(\theta) = \sum_x a(x,0)e^{i(x,\theta)}, \qquad \theta \in [-\pi, \pi]^d,$$

then the long-time asymptotics of the transition probability is given by

$$p(t,x,y) \sim \gamma_d \cdot t^{-d/2}, \qquad t \to \infty, \tag{15.2}$$

where $\gamma_d = \sqrt{(2\pi)^d \left|\det \phi''_{\theta\theta}(0)\right|}$ is a constant depending on the space dimension. The main tool in proving (15.2) is the Fourier transform with respect to the space variable x. Namely, for the function $\tilde{p}(t,\theta,y) := \sum_{x \in \mathbf{Z^d}} p(t,x,y)e^{i(x,\theta)}$ defined for $\theta \in [-\pi,\pi]^d$ where $[-\pi,\pi]^d$ is the d-dimensional cube, the Cauchy problem (15.1) can be rewritten in the form

$$\frac{\partial \tilde{p}}{\partial t} = \phi(\theta)\tilde{p}(t,\theta,y), \qquad \tilde{p}(0,\theta,y) = e^{i(\theta,y)}.$$

Hence, $\tilde{p}(t,\theta,y) = e^{\phi(\theta)t}e^{i(\theta,y)}$, and by applying the inverse Fourier transform we obtain the representation

$$p(t,x,y) = \frac{1}{(2\pi)^d} \int\limits_{[-\pi,\pi]^d} e^{\phi(\theta)t + i(\theta, y-x)}\, d\theta, \qquad t \ge 0, \quad x,y \in \mathbf{Z^d}. \tag{15.3}$$

Note that by the symmetry of the matrix A the function (15.2) is real-valued and symmetric:

$$\phi(\theta) = \sum_x a(x,0)\cos(x,\theta), \qquad \theta \in [-\pi,\pi]^d.$$

Furthermore, the function ϕ is twice continuously differentiable and has a unique non-degenerate maximum $\phi_{\max} = \phi(0) = 0$; see Yarovaya (2007). Thus, (15.3) is the Laplace integral and its asymptotics has the form (15.2), which is obtained as in the book by Fedoruk (1987).

By setting $x = y = 0$ in (15.3) we can write

$$p(t,0,0) = \frac{1}{(2\pi)^d} \int\limits_{[-\pi,\pi]^d} e^{\phi(\theta)t}\, d\theta, \qquad t \ge 0.$$

This implies the monotonicity of the transition probability $p(t,0,0)$.

Denote the Green function of the random walk by

$$G^d_\lambda(x,y) := \int_0^\infty e^{-\lambda t} p(t,x,y)\, dt,$$

where the right-hand side is the Laplace transform of $p(t,x,y)$ with respect to t. Also put $\beta_c := 1/G^d_0(0,0)$; then $\beta_c = 0$ for $d = 1,2$ and $\beta_c > 0$ for $d \ge 3$.

15.3 Definition of criticality for branching random walks

Suppose that $f^{(r)}(u)|_{u=1} < \infty$ for all $r \in \mathbf{N}$, where $\beta := f^{(1)}(u)|_{u=1} = f'(1)$ and $f^{(2)}(u)|_{u=1} := f''(1)$. The point β_c is critical since the asymptotic behavior of the branching random walk is essentially different for $\beta > \beta_c$, $\beta = \beta_c$, and $\beta < \beta_c$; see Bogachev and Yarovaya (1998a) and Albeverio et al. (1998). Let us introduce the definition of criticality for a continuous time branching random walk.

Definition 1. *If $\beta < \beta_c$ then the infinitesimal offspring generating function $f(u)$ of a continuous time branching random walk on $\mathbf{Z^d}$ is subcritical, if $\beta = \beta_c$ then the function $f(u)$ is critical, and if $\beta > \beta_c$ then the function $f(u)$ is supercritical.*

Hence, in dimensions $d = 1, 2$ the average number of offspring of a particle in a source should be equal to 1 ($\beta = \beta_c = 0$) to reach a critical regime. Thus, if $d = 1, 2$ then the definition of a critical branching random walk is equivalent to the definition of a critical branching process at the origin (i.e., $f'(1) = 0$).

If $d \geq 3$ then the average number of offspring should exceed 1 to reach a critical regime ($\beta = \beta_c > 0$), and simultaneously with increase of the dimension of a lattice an average number of offspring should also increase to reach a critical regime. Therefore, the subcritical case in dimensions ($d \geq 3$) is possible even without the death of particles at the source ($b_0 = 0$).

Let $\mu_t(y)$ be the number of particles at a point $y \in \mathbf{Z^d}$; then $\mu_t := \sum_y \mu_t(y)$ is the total number of particles at time $t > 0$.

Definition 2. *If μ_t vanishes at some finite time t then a branching random walk is called extinct.*

We denote by $P_x(\mu_t = 0)$ the probability of extinction at time t. The survival probability of the particles' population on $\mathbf{Z^d}$ at time t is equal to

$$Q(t, x) := 1 - P_x(\mu_t = 0) = P_x(\mu_t > 0).$$

Set $m_n(t, x) := E_x \mu_t^n$, $n \in \mathbf{N}$, where E_x denotes the mathematical expectation under the condition $\mu_0(\cdot) = \delta_x(\cdot)$. The long-time asymptotics of all the moments for the μ_t has been studied in Bogachev and Yarovaya (1998a). If $\beta > \beta_c$ then in the sense of convergence of the moments the random variable μ_t has a limit distribution, as $t \to \infty$, (see Bogachev and Yarovaya, 1998b) under the normalization $e^{-\lambda_0 t}$ where exponent λ_0 is determined by the equation

$$\beta G^d_{\lambda_0}(0, 0) = 1.$$

In the case $\beta \leq \beta_c$, the growth of the moments for μ_t appears to be irregular with respect to n; see Yarovaya (2007). This means that the behavior of the random variable μ_t, as $t \to \infty$, substantially differs from the behavior of the moments. For that reason the asymptotic behavior of the survival probability $Q(t, x)$ of the particles on $\mathbf{Z^d}$ at time t is of great importance. The aim of the present chapter is to study the asymptotic behavior of the survival probabilities $Q(t, x)$ and to establish conditional limit theorems for μ_t in critical and subcritical cases.

15.4 Main equations

Denote the generating function of the total number of particles μ_t at time $t > 0$ by

$$F(z;t,x) := E_x e^{-z\mu_t}, \qquad z \geq 0.$$

From this definition it follows that

$$F(z;t,x) = P_x\{\mu_t = 0\} + \sum_{n=1}^{\infty} P_x\{\mu_t = n\} e^{-zn}.$$

If $z = \infty$, we set $F(\infty;t,x) := P_x\{\mu_t = 0\} = 1 - Q(t,x)$.

Lemma 1. *The generating function $F(z;t,x)$ for every $0 \leq z \leq \infty$ is continuously differentiable with respect to t uniformly with respect to $x, y \in \mathbf{Z^d}$. It satisfies the inequalities $0 \leq F(z;t,x) \leq 1$ and the following evolution equation*

$$\frac{\partial F(z;t,x)}{\partial t} = (AF(z;t,\cdot))(x) + \delta_0(x) f(F(z;t,x)),$$

with the initial condition $F(z;0,x) = e^{-z}$.

The inequalities $0 \leq F(z;t,x) \leq 1$ follow from the definition of the generating function. The continuous differentiability of the function $F(z;t,x)$ with respect to t can be proved by standard methods of the analysis of evolution of the system on an interval $(t, t+h)$ by using the Markov property of the process; see Yarovaya (2007).

Corollary 1. *The generating function $F(z;t,x)$ satisfies the following integral equation,*

$$F(z;t,x) = e^{-z} + \int_0^t p(t-s,x,0) f(F(z,s,0))\, ds. \tag{15.4}$$

Since $Q(t,x) = 1 - F(\infty,t,x)$, then (15.4) implies the following proposition.

Corollary 2. *The survival probability of the process $Q(t,x)$ at time t for $x \in \mathbf{Z^d}$ satisfies the following integral equation*

$$Q(t,x) = 1 - \int_0^t p(t-s,x,0) f(1 - Q(s,0)) ds. \tag{15.5}$$

For an arbitrary dimension d and an arbitrary regime the function $F(z,t,x)$ is nondecreasing with respect to t for every z and x, and simultaneously the survival probability of the process $Q(t,x)$ is nonincreasing with respect to t for every x. This statement is a corollary of known theorems on positive solutions of differential equations with off-diagonal positive right-hand part in Banach spaces.

Let us reduce auxiliary results about the infinitesimal generating function $f(u)$ in the critical and subcritical cases $\beta \leq \beta_c$ to the barest essentials:

Lemma 2. *If $\beta \leq \beta_c$ then $f^{(r)}(u) = \sum_{n=r}^{\infty} (n!/(n-r)!) b_n u^{n-r}$ $(r \geq 0)$, where the series converges for all $u \in [0,1]$. Furthermore, for $d = 1, 2$ and $u \to 1$,*

$$f(u) = \begin{cases} \frac{f''(1)}{2}(1-u)^2 + o((1-u)^2) & \text{for } \beta = \beta_c, \\ -\beta(1-u) + o(1-u) & \text{for } \beta < \beta_c. \end{cases} \tag{15.6}$$

Proof. From the condition $\sum_n nb_n \leq \beta_c$ it follows that $\sum_{n=2}^{\infty} nb_n \leq \beta_c - b_1$. Hence, the series with positive elements in the left-hand side converges. Therefore, the series $\sum_{n=0}^{\infty} b_n$ absolutely converges. Then the function $f(u) = \sum_{n=0}^{\infty} b_n u^n$ has the radius of convergence not smaller then 1, and so it converges as $u \in [0,1]$. Then, as it is known from the theory of analytical functions, for every $r \geq 1$ the series $\sum_{n=r}^{\infty} (n!/(n-r)!) b_n u^{n-r}$ also has the radius of convergence not smaller then 1. Convergence of this series follows from the condition $f^{(r)}(u)|_{u=1} < \infty$.

If $\sum_n b_n = 0$ and $\beta = \sum_n nb_n$ then $b_0 = -b_1 - \sum_{n=2}^{\infty} b_n = -\beta + \sum_{n=2}^{\infty} nb_n - \sum_{n=2}^{\infty} b_n = -\beta + \sum_{n=2}^{\infty} (n-1) b_n$, and so

$$f(u) = -\beta(1-u) + \sum_{n=2}^{\infty} b_n \left(u^n - nu + n - 1\right).$$

If $d = 1, 2$ then $\beta = \beta_c = 0$ and the function $f(u)$ can be presented in the form $f(u) = \sum_{n=2}^{\infty} b_n g_n(u)$, where $g_n(u) = u^n - nu + n - 1$. According to Taylor's formula

$$g_n(u) = g_n(1) + g_n'(1)(u-1) + \frac{1}{2!} g_n''(1)(u-1)^2 + \frac{1}{3!} g_n^{(3)}(\theta_n(u))(u-1)^3, \quad 0 \leq \theta_n(u) \leq 1.$$

Here

$$g_n(1) = g_n'(1) = 0, \qquad g_n''(1) = n(n-1),$$

and

$$g_n^{(3)}(\theta_n(u)) = n(n-1)(n-2)\theta_n^3(u) \leq n(n-1)(n-2).$$

Thus,

$$f(u) = \sum_{n=2}^{\infty} b_n g_n(u) = \left(\sum_{n=2}^{\infty} \frac{n(n-1)}{2} b_n \right) (1-u)^2 + f_*(u) = \frac{f''(1)}{2} (1-u)^2 + f_*(u), \tag{15.7}$$

where

$$|f_*(u)| \leq \sum_{n=2}^{\infty} \frac{n(n-1)(n-2)}{3!} b_n (1-u)^3 = \frac{f^{(3)}(1)}{3!} (1-u)^3. \tag{15.8}$$

From the relations (15.7) and (15.8) the proof of Lemma 2 follows for $\beta = \beta_c$ in dimensions $d = 1, 2$. The statement of Lemma 2 for $\beta < \beta_c$ in dimensions $d = 1, 2$ is proved similarly.

15.5 Asymptotic behavior of survival probabilities

Some results for the critical case were obtained for the branching symmetric random walk in Yarovaya (2005).

Lemma 3. *If $\beta \leq \beta_c$ then for every $z > 0$ and $x \in \mathbf{Z^d}$*

$\lim_{t\to\infty} F(z,t,x) = 1$,	$\lim_{t\to\infty} Q(t,x) = 0$	*for* $d = 1, 2$,
$\lim_{t\to\infty} F(z,t,x) = 1 - c_d(z,x)$,	$\lim_{t\to\infty} Q(t,x) = c_d(x)$	*for* $d \geq 3$,

where $\lim_{z\to\infty} c_d(z,x) = c_d(x) > 0$ *and* $c_d(z,x)$ *is the least nonnegative root of the equation*

$$\frac{1 - c_d(z,x) - e^{-z}}{G_0^d(x,0)} = f(1 - c_d(z,0)). \tag{15.9}$$

Lemma 3 can be derived directly from equations (15.4) and (15.5) by using the facts that the function $F(z,t,x)$ is nondecreasing with respect to t for every $z > 0$ and $x \in \mathbf{Z^d}$ and hence the survival probability of the process $Q(t,x)$ is nonincreasing with respect to t for every $x \in \mathbf{Z^d}$.

Theorem 1. *If* $\beta \le \beta_c$ *then the survival probabilities* $Q(t,x)$ *have the following asymptotics, as* $t \to \infty$,

$$Q(t,x) \sim \begin{cases} C_d(x)v(t), & \text{if } \beta = \beta_c, \\ K_d(x)u(t), & \text{if } \beta < \beta_c, \end{cases}$$

where the functions v *and* u *are of the form:*

$$\begin{array}{lll} v(t) = t^{-1/4}, & u(t) = t^{-1/2} & \text{for } \ d = 1, \\ v(t) = (\ln t)^{-1/2}, & u(t) = (\ln t)^{-1} & \text{for } \ d = 2, \\ v(t) \equiv 1, & u(t) \equiv 1 & \text{for } \ d \ge 3, \end{array}$$

and

$$\begin{array}{lll} C_1(x) = \sqrt{2}\left(f''(1)\gamma_1\pi\right)^{-1/2}, & K_1(x) = \left(-\beta\gamma_1\pi\right)^{-1} & \text{for } \ d = 1, \\ C_2(x) = \sqrt{2}\left(f''(1)\gamma_2\right)^{-1/2}, & K_2(x) = \left(-\beta\gamma_2\right)^{-1} & \text{for } \ d = 2. \end{array}$$

Both the functions $C_d(x)$ *and* $K_d(x)$ *for every* $x \in \mathbf{Z^d}$ $(d \ge 3)$ *are determined by*

$$1 - \beta_c G_0^d(x,0)\left(1 - c_d(0)\right),$$

where $c_d(0)$ *is the least nonnegative root of the equation*

$$\beta_c(1 - c_d(0)) = f(1 - c_d(0)). \tag{15.10}$$

15.6 Limit theorems

Here we establish conditional limit theorems for the total number of particles μ_t existing on $\mathbf{Z^d}$ at time t using the asymptotics of the survival probabilities of the process $Q(t,x)$ at time t.

Theorem 2. *Let* $\beta = \beta_c$ *and* $d = 1, 2$; *then for any* $z > 0$ *and* $x \in \mathbf{Z^d}$,

$$\lim_{t\to\infty} E_x\left[e^{-z\mu_t} | \mu_t > 0\right] = 1 - \sqrt{1 - e^{-z}}.$$

Theorem 3. *Let* $\beta < \beta_c$ *and* $d = 1, 2$; *then for any* $z > 0$ *and* $x \in \mathbf{Z^d}$,

$$\lim_{t\to\infty} E_x\left[e^{-z\mu_t} | \mu_t > 0\right] = e^{-z}.$$

Theorem 4. *Let $\beta \le \beta_c$ and $d \ge 3$; then for any $z > 0$ and $x \in \mathbf{Z^d}$,*

$$\lim_{t\to\infty} E_x\left[e^{-z\mu_t}|\mu_t > 0\right] = \frac{(1-\beta_c G_0(x,0))\,e^{-z} + \beta_c G_0(x,0)\,(c_d(0) - c_d(z,0))}{1-\beta_c G_0(x,0)\,(1-c_d(0))}.$$

Furthermore,

$$\lim_{t\to\infty} E_x\left[e^{-z\mu_t}\right] = \left(1 - \frac{G_0(x,0)}{G_0(0,0)}\right) e^{-z} + \frac{G_0(x,0)}{G_0(0,0)}\,(1-c_d(z,0)),$$

where $c_d(z,0)$ and $c_d(0)$ are the least nonnegative roots of equations (15.9) and (15.10), respectively.

Theorems 1–4 can be proved by applying the Laplace transform with respect to λ to equations (15.4) and (15.5). Then we study the asymptotics of the Laplace transforms of the functions $F(z,t,x)$ and $Q(t,x)$, as $\lambda \to 0$, using the asymptotic representation for the Laplace transforms of the Green functions $G_\lambda^d(x,0)$. The asymptotic behavior of $F(z,t,x)$ and $Q(t,x)$, as $t \to \infty$, can then be derived from the Tauberian theorems; see, e.g., Feller (1971). It should be noted that these theorems require the monotonicity of the originals $F(z,t,x)$ and $Q(t,x)$ in t. Since $P_x\{\mu_t = 0|\,\mu_t > 0\} = 0$ then

$$\begin{aligned} E_x[e^{-z\mu_t}|\,\mu_t > 0] &= \sum_{n=1}^{\infty} e^{-zn} P_x\{\mu_t = n|\,\mu_t > 0\} \\ &= P_x^{-1}\{\mu_t > 0\} \sum_{n=1}^{\infty} e^{-zn} P_x\{\mu_t = n\} \\ &= P_x^{-1}\{\mu_t > 0\}(F(z,t,x) - P_x\{\mu_t = 0\}). \end{aligned}$$

Whence

$$E_x[e^{-z\mu_t}|\,\mu_t > 0] = \frac{F(z,t,x) - F(\infty,t,x)}{1-F(\infty,t,x)} = 1 - \frac{1-F(z,t,x)}{Q(t,x)}. \tag{15.11}$$

Using the asymptotic behavior of the functions $F(z,t,x)$ and $Q(t,x)$, as $t \to \infty$, the statements of the theorems are proved. If $d \ge 3$ then the proof of Theorem 4 follows from the statement of Lemma 3.

Below we present the detailed proof of Theorems 1, 2, and 3 for dimensions $d = 1, 2$ in critical and subcritical cases.

15.7 Proof of theorems for dimensions $d = 1, 2$ in critical and subcritical cases

Equation (15.5) for the survival probability $Q(t,x)$ at the point $x = 0$ has the following form

$$1 - Q(t,0) = \int_0^t p(t-s,0,0) f(1-Q(s,0))\,ds. \tag{15.12}$$

Hence, by applying the Laplace transform to (15.12) we get the representation

$$\widehat{1-Q} = G_{\lambda}^{d}(0,0) f\widehat{(1-Q)}, \tag{15.13}$$

where the left part of (15.13) is the Laplace transform of the extinction probability $P_0\{\mu_t = 0\} = 1 - Q(t,0)$. By Lemma 3, if $\beta \le \beta_c$ then the extinction probability $P_0\{\mu_t = 0\}$ tends to 1, as $t \to \infty$. Therefore, the asymptotics of the Laplace transform of the extinction probability, as $\lambda \to 0$, has the following form, $\widehat{1-Q} \sim (1/\lambda)$. With the help of the representation (15.2) of the monotone decreasing transition probability $p(t,0,0)$ and the Tauberian theorem (see Feller, 1971), it can be shown that the asymptotics of the Green function $G_{\lambda}^{d}(0,0)$, as $\lambda \to 0$, has the following form

$$G_{\lambda}^{d}(0,0) \sim \begin{cases} \gamma_1 \sqrt{\pi} \lambda^{-(1/2)} & \text{for } d = 1, \\ \gamma_2 \ln\left(\frac{1}{\lambda}\right) & \text{for } d = 2. \end{cases} \tag{15.14}$$

From (15.13) and (15.14) we deduce that

$$f\widehat{(1-Q)} \sim \begin{cases} \left(\gamma_1 \sqrt{\pi}\right)^{-1} \lambda^{-(1/2)} & \text{for } d = 1, \\ (\gamma_2 \lambda)^{-1} \ln^{-1}\left(\frac{1}{\lambda}\right) & \text{for } d = 2. \end{cases}$$

Hence, due to the monotonicity of the function $f(1-Q)$ and the Tauberian theorem "for densities" (see Feller, 1971), we obtain the following asymptotic equality as $t \to \infty$,

$$f(1-Q) \sim \begin{cases} \left(\gamma_1 \pi\right)^{-1} t^{-(1/2)} & \text{for } d = 1, \\ (\gamma_2)^{-1} \ln^{-1} t & \text{for } d = 2. \end{cases} \tag{15.15}$$

If $d = 1, 2$ and $\beta \le \beta_c = 0$ then by Lemma 3 the survival probability $Q(t,0) \to 0$, as $t \to \infty$, and $1 - Q(t,0) \to 1$, as $t \to \infty$, respectively. Thus, by Lemma 2 the function

$$f(1-Q) = \sum_{n=0}^{\infty} b_n (1-Q)^n$$

has the representation (15.6). If $\beta = \beta_c$ then having used (15.15) we obtain, as $t \to \infty$,

$$Q^2(t,0) \sim \begin{cases} 2(f''(1)\gamma_1 \pi)^{-1} t^{-(1/2)} & \text{for } d = 1, \\ 2(f''(1)\gamma_2 \ln t)^{-1} & \text{for } d = 2. \end{cases}$$

Hence, when $x = 0$, we get the statement of Theorem 1 for the critical case in low dimensions:

$$Q(t,0) \sim \begin{cases} \sqrt{2}(f''(1)\gamma_1 \pi)^{-(1/2)} t^{-(1/4)} & \text{for } d = 1, \\ \sqrt{2}(f''(1)\gamma_2 \ln t)^{-(1/2)} & \text{for } d = 2. \end{cases} \tag{15.16}$$

In the same manner, using asymptotic equalities (15.15) and the representation (15.6) for $\beta < \beta_c$ we get in the subcritical case that

$$Q(t,0) \sim \begin{cases} \left(-\beta \gamma_1 \pi \sqrt{t}\right)^{-1} & \text{for } d = 1, \\ \left(-\beta \gamma_2 \ln t\right)^{-1} & \text{for } d = 2, \end{cases} \tag{15.17}$$

as $t \to \infty$.

Since the first term of the asymptotic expansion of transition probabilities (15.2) does not depend on x as $t \to \infty$ then the first term of the asymptotic expansion of its Laplace transform $G^d_\lambda(x,0)$ for $\lambda \to 0$ also does not depend on x. Therefore, the first term of the asymptotic expansion of survival probabilities $Q(t,x)$ does not depend on x as $t \to \infty$ and has the representation (15.16) for the critical case and (15.17) for the subcritical case. This immediately implies the assertion of Theorem 1 in the critical and subcritical cases for dimensions $d = 1,2$ and every $x \in \mathbf{Z^d}$.

The Laplace transform of integral equation (15.4) at the point $x = 0$ has the following form

$$\widehat{F} = \frac{e^{-z}}{\lambda} + G^d_\lambda(0,0)\widehat{f(F)}.$$

By Lemma 3 if $\beta \le \beta_c$ and $d = 1,2$ then the function $F(z,t,0)$ tends to 1 as $t \to \infty$. So the asymptotics of the Laplace transform of the function $F(z,t,0)$ when $\lambda \to 0$ has the form $\widehat{F} \sim 1/\lambda$. Hence we get for $d = 1,2$:

$$\widehat{f(F)} \sim \left(1 - e^{-z}\right)\left(\lambda G^d_\lambda(0,0)\right)^{-1}, \quad \text{as } \lambda \to 0.$$

On the other hand, from (15.13) we have the following representation for $\widehat{f(1-Q)}$:

$$\widehat{f(1-Q)} \sim \left(\lambda G^d_\lambda(0,0)\right)^{-1}, \quad \text{as } \lambda \to 0.$$

Thus,

$$\widehat{f(F)} \sim \left(1 - e^{-z}\right)\widehat{f(1-Q)}, \quad \text{as } \lambda \to 0. \tag{15.18}$$

Applying the Tauberian theorem (see Feller, 1971) the monotonicity of the function $F(t,z,0)$ in t required by this theorem and the representations (15.15), as $t \to \infty$, we obtain for $\beta \le \beta_c$ and $d = 1,2$

$$f(F(z,t,0)) \sim (1 - e^{-z}) f(1 - Q(t,0)).$$

By Lemma 3 if $d = 1,2$ then $F(t,z,0) \to 1$, as $t \to \infty$, for every $z > 0$. Thus, the representation (15.6) for F takes the form

$$f(F) = \begin{cases} \frac{f''(1)}{2}(1-F)^2 + o((1-F)^2) & \text{for } \beta = \beta_c, \\ -\beta(1-F) + o(1-F) & \text{for } \beta < \beta_c. \end{cases}$$

From (15.18) we get the following representation, as $t \to \infty$,

$$(1 - e^z) f(1 - Q(t,0)) \sim \begin{cases} \frac{f''(1)}{2}(1 - F(z,t,0))^2 & \text{for } \beta = \beta_c, \\ -\beta(1 - F(z,t,0)) & \text{for } \beta < \beta_c, \end{cases}$$

from which one can show that

$$1 - F(z,t,0) \sim \begin{cases} \sqrt{(1-e^z)Q(t,0)} & \text{for } \beta = \beta_c, \\ (1-e^z)Q(t,0) & \text{for } \beta < \beta_c, \end{cases} \tag{15.19}$$

as $t \to \infty$.

Now, using (15.19) we have for $d = 1, 2$ and $x = 0$

$$\lim_{t\to\infty} \frac{1 - F(z,t,0)}{Q(t,0)} = \begin{cases} \sqrt{1-e^z} & \text{for } \beta = \beta_c, \\ 1 - e^z & \text{for } \beta < \beta_c. \end{cases}$$

Since the first term of the asymptotic expansion of the Laplace transform for $G_\lambda^d(x,0)$, as $\lambda \to 0$, does not depend on x so the first term of the asymptotic expansion of the function $F(z,t,x)$ as $t \to \infty$ does not depend on x. Let us note that this fact is valid only in the case when $d = 1, 2$. Hence, if $d = 1, 2$ then for every $x \in \mathbf{Z^d}$

$$\lim_{t\to\infty} \frac{1 - F(z,t,x)}{Q(t,x)} = \begin{cases} \sqrt{1-e^z} & \text{for } \beta = \beta_c, \\ 1 - e^z & \text{for } \beta < \beta_c. \end{cases}$$

By (15.11) we immediately obtain the statements of Theorems 2 and 3.

15.8 Conclusions

In conclusion we formulate some results for the simple model where we can find the explicit solutions of equations (15.9) and (15.10).

Theorem 5. *Let $\beta = \beta_c$, $d \geq 3$, and $f(u) = b_0 + b_1 u + b_2 u^2$; then for any $z > 0$ and $x \in \mathbf{Z^d}$,*

$$\lim_{t\to\infty} E_x\left[e^{-z\mu_t} | \mu_t > 0\right] = \frac{\left(1 - \beta_c G_0^d(x,0)\right) e^{-z} + \beta_c G_0^d(x,0)\sqrt{\frac{\beta_c}{b_2}}\left(1 - \sqrt{1-e^{-z}}\right)}{1 - \beta_c G_0^d(x,0)\left(1 - \sqrt{\frac{\beta_c}{b_2}}\right)}.$$

Furthermore,

$$\lim_{t\to\infty} E_x\left[e^{-z\mu_t}\right] = \left(1 - \frac{G_0(x,0)}{G_0(0,0)}\right) e^{-z} + \frac{G_0(x,0)}{G_0(0,0)}\left(1 - \sqrt{(1-e^z)\frac{\beta_c}{b_2}}\right).$$

Proof. For every finite n by Lemma 2 we have the representation

$$f(u) = -\beta(1-u) + \frac{f^{(2)}(1)}{2}(1-u)^2 + \cdots + (-1)^n b_n (1-u)^n.$$

Therefore, by applying this representation for $n = 2$ to the right-hand side of equation (15.10) we get

$$\beta_c(1 - c_d(0)) = -\beta c_d(0) + \frac{f''(1)}{2} c_d^2(0).$$

Putting $\beta = \beta_c$ in this equation we have

$$\beta_c = \frac{f''(1)}{2} c_d^2(0). \tag{15.20}$$

The positive solution of (15.20) has the form $c_d(0) = \sqrt{\beta_c/b_2}$. In the same manner, the solution $c_d(z,0)$ of equation (15.9) is obtained by $c_d(z,0) = \sqrt{1-e^{-z}}c_d(0)$ for every $z > 0$. Therefore, the statements of Theorem 5 follow immediately from Theorem 4.

Note that for the particle starting from the origin at time $t = 0$ we get the following corollary from Theorem 5 valid for all dimensions d.

Corollary 3. *Let $\beta = \beta_c$; then for any $z > 0$ on $\mathbf{Z}^d$,*

$$\lim_{t\to\infty} E_0\left[e^{-z\mu_t}|\mu_t > 0\right] = 1 - \sqrt{1-e^{-z}}.$$

Acknowledgements. This work is supported by the RFBR grant 07-01-00362.

References

Albeverio, S., Bogachev, L., Yarovaya, E. (1998). Asymptotics of branching symmetric random walk on the lattice with a single source. *Comptes Rendus de l'Academie des Sciences, Paris, Series I*, 326:975–980.

Bogachev, L., Yarovaya, E. (1998a). Moment analysis of a branching random walk on a lattice with a single source. *Doklady Akademii Nauk*, 363:439–442.

Bogachev, L., Yarovaya, E. (1998b). A limit theorem for supercritical branching random walk on $\mathbf{Z}^{\mathbf{d}}$ with a single source. *Russian Mathematical Surveys*, 53:1086–1088.

Fedoruk, M. (1987). *Asymptotics: Integrals and Series.* Nauka, Moscow (in Russian).

Feller, W. (1971). *An Introduction to Probability Theory and Its Applications.* Wiley, New York, 2nd ed., volume 2.

Fleischmann, K., Le Gall, J. (1995). A new approach to the single point catalytic super-Brownian motion. *Probability Theory and Related Fields*, 102:63–82.

Gartner, J., Molchanov, S., (1990). Parabolic problems for the Anderson model.I: Intermittency and related topics. *Communications in Mathematical Physics*, 132:613–655.

Greven, A., Klenke, A., Wakolbinger A. (1999). The long-time behavior of branching random walk in a catalytic medium. *Electronic Journal of Probability*, 4:1–80.

Halmos, P. (1982). *Hilbert Space Problem Book.* Springer-Verlag, New York, 3rd ed.

Molchanov, S. (1994). Lectures on random media. *Lecture Notes in Mathematics*, 1581:242–411.

Vatutin V. A., Topchiĭ, V. A., Yarovaya, E. B. (2005). Catalytic branching random walks and queueing systems with a random number of independently operating servers. *Theory Probability and Mathematical Statistics*, 69:1–15.

Vatutin, V., Xiong, J. (2007). Limit theorems for a particle system of single point catalytic branching random walks. *Acta Mathematica Sinica*, 23:997–1012.

Yarovaya, E. (1991). Use of the spectral methods to study branching processes with diffusion in a noncompact phase space. *Theoretical and Mathematical Physics*, 88:25–30.

Yarovaya, E. (2005). A limit theorem for critical branching random walk on $\mathbf{Z}^{\mathbf{d}}$ with a single source. *Russian Mathematical Surveys*, 60:175–176.

Yarovaya, E. (2007). *Branching Random Walks in Inhomogeneous Medium.* MSU, Moscow (in Russian).

Part IV

Bioinformatics and Markov Chains

16

Finite Markov Chain Embedding for the Exact Distribution of Patterns in a Set of Random Sequences

Juliette Martin,[1] Leslie Regad,[2] Anne-Claude Camproux,[2] and Grégory Nuel[3]

[1] Unité Mathématique Informatique et Génome UR1077, INRA, Jouy-en-Josas F-78350, France
Equipe de Bioinformatique Génomique et Moléculaire, INSERM UMR-S726/Université Denis Diderot Paris 7, Paris F-75005, France
Université de Lyon, Lyon, France; Université Lyon 1; IFR 128; CNRS, UMR 5086; IBCP, Institut de Biologie et Chime des Protéines, 7 passage du Vercors, Lyon F-69367, France (e-mail: `juliette.martin@ibcp.fr`)

[2] Equipe de Bioinformatique Génomique et Moléculaire, INSERM UMR-S726/Université Denis Diderot Paris 7, Paris F-75005, France
MTI, Inserm UMR-S 973; Université Denis Diderot Paris 7, Paris F-75205, France (e-mail: `leslie.regad@univ-paris-diderot.fr`, `anne-claude.camproux@univ-paris-diderot.fr`)

[3] CNRS, Paris, France; MAP5 UMR CNRS 8145, Laboratory of Applied Mathematics, Department of Mathematics and Computer Science, Université Paris Descartes, Paris F-75006, France (e-mail: `gregory.nuel@parisdescartes.fr`)

Abstract: Patterns with "unusual" frequencies are new functional candidate patterns. Their identification is usually achieved by considering an homogeneous m-order Markov model ($m \geqslant 1$) of the sequence, allowing the computation of p-values. For practical reasons, stationarity of the model is often assumed. This approximation can result in some artifacts especially when a large set of small sequences is considered. In this work, an exact method, able to take into account both nonstationarity and fragmentary structure of sequences, is applied on a simulated and a real set of sequences. This illustrates that pattern statistics can be very sensitive to the stationary assumption.

Keywords and phrases: stationary distribution, pattern Markov chain, biological patterns, finite Markov chain embedding.

16.1 Introduction

It is well known that selection can affect the frequencies of functional patterns in biological sequences (DNA, proteins, etc.). It is hence natural to search for new functional

Juliette Martin and Leslie Regad equally contributed Anne-Claude Camproux is corresponding author

C.H. Skiadas (ed.), *Advances in Data Analysis*, Statistics for Industry and Technology, DOI 10.1007/978-0-8176-4799-5_16,

candidate patterns among patterns with "unusual" frequencies. This is usually achieved by considering an homogeneous m-order Markov model ($m \geqslant 1$) of sequence allowing pattern p-value computation. For practical reasons, stationarity of the model is often assumed.[1] As the marginal distribution of the usual encountered Markov model quickly converges toward its stationary distribution, this assumption is false only for a small portion of the sequence. The error done is usually harmless when considering long biological sequences such as DNA. But things could be quite different when large sets of short sequences are considered. For instance, considering a large set of protein sequences (of some hundred residues) can result in an accumulation of side effects and lead to erroneous conclusions. In this work, an exact method, able to take into account both nonstationarity and fragmentary structures of sequences, is proposed. First, mathematical notions as patterns of Markov chain and finite Markov chain embedding needed to obtain exact p-value computations, are introduced. Then, the exact approach is used and compared with two classical approximations (considering one single sequence and stationarity hypothesis) on a simulated and a real set of sequences. This illustrates that pattern statistics can be very sensitive to these approximations and that our exact method allows us to improve pattern extraction by avoiding artifacts on particular sets of biological sequences.

16.2 Methods

16.2.1 Notations

We consider a set of r sequences over the finite alphabet $\mathcal{A}$. For $1 \leqslant j \leqslant r$ we denote by $x^j = x_1^j \cdots x_i^j \cdots x_{\ell_j}^j$ the jth sequence. Let us now consider a pattern on $\mathcal{A}$ and denote by n^j its number of occurrences in the sequence x^j. The question then is: how to associate an overrepresentation (or underrepresentation) p-value to these observations.

The classical framework to answer that question consists in studying the distribution of N^j, the random number of pattern occurrences in the random sequence X^j assuming that all sequences are independently drawn according to an homogeneous order m Markov model. For each sequence, it is hence possible to compute $p_j = P(N^j \geqslant n^j)$[2] and one can combine all these probabilities to get a global overrepresentation p-value: $\prod_{j=1}^r p_j$.

If this approach is quite natural, it is seldom used in practice for two reasons: first, most available methods to compute the p_j are asymptotic approximations which are hence not suitable on short sequences; second, one may not be interested in taking into account local fluctuations in each sequence and prefer rather to consider some more global statistic. This is why we usually compute

$$p = \mathbb{P}(\underbrace{N^1 + \cdots + N^j + \cdots + N^r}_{N} \geqslant \underbrace{n^1 + \cdots n^j + \cdots n^j}_{n})$$

instead of the p_j product.

As the computation of p remains a difficult task, two approximations are commonly done in order to ease it:

---

[1] The starting distribution is the stationary distribution

[2] This concerns the overrepresented case. Replace all $\geqslant$ by $\leqslant$ in the underrepresented case.

Type I: If N' is defined as the number of pattern occurrences on a single sequence of length $\ell = \ell_1 + \cdots + \ell_r$, then $p' = \mathbb{P}(N' \geqslant n) \simeq p$.

Type II: It is also classical to assume both ergodicity[3] and stationarity of the considered Markov chains. If ergodicity is in general a harmless assumption, stationarity is rarely achieved in practice.

As we show later with simulations and applications, the more numerous and short are the considered sequences the more erroneous are the results obtained through both these approximations. Let us now see how to overcome this problem.

16.2.2 Pattern Markov chains

Theorem 1. *Let $X = X_1 \cdots X_\ell$ be an order m Markov chain[4] over the finite alphabet $\mathcal{A}$ and let $\mathcal{W}$ be a pattern on this alphabet. It is then possible to build an order 1 Markov chain $Y = Y_m \cdots Y_\ell$ over the finite state space $\mathcal{Q}$ such as*

$$\forall m \leqslant i \leqslant \ell \quad \mathcal{W} \text{ ends in position } i \text{ in } X \iff Y_i \in \mathcal{F}$$

where $\mathcal{F} \subset \mathcal{Q}$. A Markov chain Y having these properties is called a Pattern Markov Chain (PMC).

Proof. The proof is constructive and uses results from pattern matching theory. The technique consists in building a deterministic finite state automaton that recognizes the language, Nuel (2008) $\mathcal{L} = \mathcal{A}^*\mathcal{W}$ from which the PMC is derived. Check Nuel (2006), Nuel (2008) and Nuel and Prum (2007) for more details.

Thanks to Theorem 1 and without loss of generality we consider now the following framework: for all $1 \leqslant j \leqslant r$, N^j is the random number of occurrences of $\mathcal{F}$ in the PMC $Y^j = Y^j_m \cdots Y^j_{\ell_j}$; we denote by ν^j_m the distribution of Y^j_m[5] and by Π its transition matrix.

16.2.3 Exact computations

Moments

Proposition 1. *For all $1 \leqslant j \leqslant r$ we have:*

$$\mathbb{E}[N^j] = \sum_{f \in \mathcal{F}} E^j(f) \quad \text{with} \quad E^j(f) = \sum_{i=m}^{\ell_j} \nu^j_m \Pi^{i-m} e_f^T \forall f \in \mathcal{F}$$

and

$$\mathbb{V}[N^j] = \sum_{f,f' \in \mathcal{F}} \mathbb{I}_{a=b} \times E^j(f) + C^j(f,f') + C^j(f',f)$$

with

$$C^j(f,f') = \sum_{i=m}^{\ell_j - 1} \left(\nu^j_m \Pi^{i-m} e_f^T\right) \sum_{j=i+1}^{\ell_j} \left(e_f \Pi^{j-i} e_{f'}^T\right)$$

where e_q is the indicatrix row-vector of q for all $q \in \mathcal{Q}$.

---

[3] The marginal distribution converges towards the stationary distribution.

[4] Homogeneous or heterogeneous Markov chain.

[5] For computations using the stationary assumption, simply build ν^j_m from the stationary distribution of the Markov chain X^j. Please note that, even in this case, ν^j_m is in general *not* the stationary distribution of Y^j.

Proof. We simply use the decomposition $N^j = \sum_{i=m}^{\ell} \mathbb{I}_{Y_i^j \in \mathcal{F}}$ and the fact that $\nu_m^j \Pi^{i-1}$ is the marginal distribution of Y_i^j.

We can use Proposition 1 to compute the first two moments of N over a sequence of length ℓ with complexities $O(\ell)$ in space and time. Both complexities can be reduced to $O(\log \ell)$ using the convergence of Y towards its stationary distribution. See Nuel (2006), Nuel (2008) and Nuel and Prum (2007) for more details.

As we assume that our r sequences are independent, it is hence possible to compute the exact expectation and variance of $N = N^1 + \cdots + N^r$ by computing and summing each individual one.

Finite Markov Chain embedding

The PMC introduced above allows us to monitor step by step the formation of the next pattern occurrence but does not have memory of the previous ones. In order to study the distribution of N we hence need to keep track of the accumulated number of pattern occurrences. This is exactly the purpose of the new Markov chain we introduce here.

For all $c \in \mathbb{N}$ and for all $1 \leqslant j \leqslant r$ we define the Finite Markov Chain Embedding (FMCE, first proposed by Fu and Koutras (1994) in the field of pattern statistics) $Z^j = Z_m^j \cdots Z_{\ell_j}^j$ by:

$$Z_i^j = \begin{cases} \left(Y_i^j, N_i^j\right) & \text{if } N_i^j < c \\ c_+ & \text{if } N_i^j \geqslant c \end{cases}$$

where N_i^j is the number of occurrences of $\mathcal{F}$ in $Y_m^j \cdots Y_i^j$. We denote by T its transition matrix which is naturally defined by blocks, using the transition matrix $\Pi = P + Q$, where Q contains all transitions ending in $\mathcal{F}$ and P all other ones:

$$T\big((q,n),(q',n')\big) = \begin{cases} P & \text{if } n' = n \\ Q & \text{if } n' = n+1 \text{ and } n' < c \\ 0 & \text{else} \end{cases}$$

Proposition 2. *For all $c \geqslant 1$ and $1 \leqslant j \leqslant r$ we have*[6]

$$\mathbb{P}(N^1 + \cdots + N^j \geqslant c) = M_m^j T^{\ell_j - m} E_{c+}^T$$

where M_m^j is defined by recurrence for all $(q,n) \in \mathcal{Q} \times \{0,\ldots,c-1\}$:

$$\begin{cases} M_m^1(q,n) = \mathbb{I}_{n=0} \times \nu_m^1(q) & \text{and} \quad M_m^1(c+) = 0 \\ M_m^j(q,n) = M_m^{j-1} T^{l_j - m} E_n^T \times \nu_m^j(q) & \text{and} \quad M_m^j(c+) = M_m^{j-1} T^{l_j - m} E_{c+}^T \end{cases}$$

where E_n (resp., E_{c+}) is the indicatrix row-vector of $(\mathcal{Q}, n)$ (resp., of $c+$).

Proof. As we start with 0 occurrence of the pattern, the starting distribution of Z_m^1 is necessarily concentrated on the $n = 0$ block, which explains the expression of M_m^1. At the end of the first sequence, the distribution of $Z_{\ell_1}^1$ is given by $M_m^1 T^{\ell_1 - m}$ and can

[6] This result holds with the event $N^1 + \cdots + N^j < c$ when replacing E_{c+} by $\sum_{n<c} E_n$.

hence be used to get the distribution of N. At the beginning of the second sequence, Y_m^2 is distributed according to ν_m^2. We hence first compute the weight of each block using the distribution of $Z_{\ell_1}^1$ and then distribute this weight within the block according to ν_m^2. We simply repeat the process until we reach the end of the last sequence.

Using Proposition 2 and the computational algorithms and techniques proposed in Nuel (2008) it is hence possible to get the exact value of $\mathbb{P}(N \geqslant n)$ in $O(n)$ in space and $O(n \times \ell)$ in time.

All these methods have been implemented in the Statistics for Patterns (SPatt) software which is freely available.[7]

16.3 Data

16.3.1 Simulated data

To illustrate the effect of type I and II approximations, sequences are simulated over the alphabet $\mathcal{A} = \{\mathtt{A}, \mathtt{B}\}$ with initial distribution $\mu_0 = \begin{pmatrix} 0.0 \; 1.0 \end{pmatrix}$ and transition matrix

$$\pi = \begin{pmatrix} 0.7 & 0.3 \\ 0.4 & 0.6 \end{pmatrix}$$

Numerical convergence towards the stationary distribution is achieved after six steps. Different datasets are characterized by different sequence lengths uniformly distributed in the range $[L - 50\%, L + 50\%]$. The number of sequences, N, is chosen so as to obtain equivalent dataset sizes: N varies from 1 to 1000 and L from $10,000$ to 10.

16.3.2 Real data

Exon databank: 139,416 exons of *Caenorhabditis elegans* genome are obtained from the query-oriented data management system Biomart (Durinck et al., 2005). The exon average size is 221.61, 5th percentile = 57 and 95th percentile = 598.

Protein databank: A protein databank of 1,100 protein sequences presenting less than 50% of sequence identity is used. The protein sequences are translated into a simplified alphabet $\mathcal{A} = \{\mathtt{M}, \mathtt{O}\}$: M is methionine and O regroups other amino acids, as it is well known that the majority of protein sequences begin with a methionine. The sequence average size is 237 residues and 5th percentile = 66 and 95th percentile = 516.

Protein loop databank: A set of 3,152 tridimensional protein structures presenting less than 50% of sequence identity is considered. Each structure is simplified into a string sequence using the structural alphabet HMM-27 (Camproux et al., 2004;

[7] http://stat.genopole.cnrs.fr/spatt.

Regad et al., 2008). HMM-27 is a library of 27 structural letters which are 4 residue prototype fragments. From the simplified protein structures, 34,267 simplified loops[8] are extracted (Regad et al., 2006). The average size of the simplified loops is 8.72 structural letters, 5th percentile = 4, and 95th percentile = 20.

16.4 Results and discussion

In this section, we compare the exact approach with type I (considering a single sequence) and type II approximations (assuming stationarity) under an homogeneous order 1 Markov model. Over- and underrepresentations are defined using a 5% threshold including the Bonferonni correction. Global measures of the approximation effects are: False Positive Rate (FPR), False Negative Rate (FNR), and Kendall's tau (Myles and Douglas, 1973) using exact computations as reference.

16.4.1 Simulation study

We study the statistics of three-letter patterns on trials of 5,000 datasets. A p-value under H_0 being, by definition, uniformly distributed between 0.0 and 1.0, the resulting Z-score[9] should be distributed according to an $\mathcal{N}(0,1)$ distribution in our simulations. Using type I approximation, we observe that Z-scores of all patterns are strongly biased towards negative values when the dataset contains several sequences (data not shown). On a global point of view, it results in the identification of many false positive underrepresented patterns. Table 16.1 reports the effect of the sequence number on this FPR. On a dataset with only 10 sequences, 22.9% of underrepresented patterns are false positives. This proportion rapidly increases with the sequence number: with 100 sequences, nearly all underrepresented patterns are artifacts. Type I approximation is thus clearly inappropriate when dealing with several sequences.

Table 16.1. FPR in underrepresented patterns using type I approximation. N is the number of sequences

N	1	10	20	40	50	100
FPR	0	22.9	40.2	69.6	80.7	95.9

Type II approximation is studied in more detail. Results are shown in Figure 16.1. Using the exact approach, all Z-scores are $\mathcal{N}(0,1)$ distributed whatever the sequence length. Type II approximation results in non normal distributions on short sequences. As shown in Figure 16.1a, the effect depends on the considered pattern: Z-scores of `BAA` are biased towards positive values and those of `ABA` are biased towards negative values. This is related to the difference between the actual initial distribution $\begin{pmatrix}0 & 1\end{pmatrix}$

---

8 Unlike helices and strands, loops are nonperiodic regions in protein structures and they are less conserved during evolution.

9 The Z-score is given by $(\mathbb{E}(N)-N_{\text{obs}})/\sqrt{\mathbb{V}(N)}$ where $\mathbb{E}(N)$ and $\sqrt{\mathbb{V}(N)}$ denote, respectively, the expectation and standard deviation of the pattern occurrence, and N_{obs} denotes its observed number of occurrences.

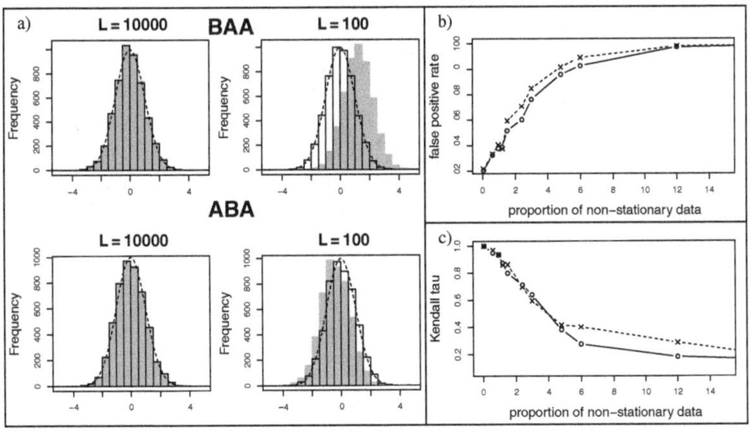

Figure 16.1. Effect of type II approximation on pattern statistics. (**a**) Z-score distributions of two patterns, BAA and ABA. Dashed curved: normal distribution, black histograms: exact Z-scores, gray histograms: type II Z-scores. (**b**) FPR as a function of the proportion of the dataset that is not stationary. Dashed line with crosses: FPR for overrepresentation, plain line with circles: FPR for underrepresentation. (**c**) Kendall tau correlation of the 200 most extreme Z-scores as a function of the proportion of the dataset that is not stationary. Dashed line with crosses: tau obtained on the 200 higher Z-scores, plain line with circles: tau obtained on the 200 lower Z-scores

and the stationary distribution $\begin{pmatrix}0.57 & 0.43\end{pmatrix}$. Patterns starting with `B` are falsely seen as overrepresented because `B` stationary frequency is 0.43 but all sequences start with a `B`. Then, on short sequences, there is a risk of falsely concluding that pattern `BAA` is overrepresented.

The evolution of FPR over all patterns is shown in Figure 16.1b. It can be seen that over- and underrepresented FPR have similar evolution when the sequence length decreases. When only 1.5% of the dataset is not stationary ($L = 400$), about 40% (resp., 30%) of overrepresented (resp., underrepresented) scores are false positives. The last analysis, shown in Figure 16.1c, concerns the pattern ranking. The Kendall's tau correlation coefficient rapidly decreases when sequences are short. This illustrates that not only the scores become falsely significant, but the pattern ranks also become rapidly meaningless.

16.4.2 Illustrations on biological sequences

The simulations are made using a two-letter alphabet. Such an alphabet can be observed in biology. For example, protein sequences can be translated into an alphabet $\mathcal{A} = \{\mathtt{M}, \mathtt{O}\}$. In this case, initial distribution (0.44 0.55) clearly differs from the stationary distribution (0.02 0.98). On this simple real case, errors occur when considering six-letter patterns. Indeed, among the four overrepresented patterns obtained with type I approximation, two are false positives (FPR = 0.5) (cf. Table 16.2). Two false positives are counted among six overrepresented patterns (FPR = 0.33) extracted by type II approximation (cf. Table 16.2). Moreover, the type I and type II Kendall's tau are 0.67, so even ranks are badly conserved (cf. Table 16.2).

Another example, maybe more realistic, is exon sequences which are portions of DNA ($\mathcal{A} = \{\mathtt{A}, \mathtt{C}, \mathtt{G}, \mathtt{T}\}$). Statistics of three-letter patterns are computed in the exon

Table 16.2. Overrepresentation results of statistic computation in the biological data

Databank	Approximation	FPR	FNR	Kendall's tau
Protein sequence	type I	0.5	0.033	0.667
	type II	0.333	0	0.667
Exon sequence	type I	0	0.064	0.939
	type II	0.028	0	0.946
Loop structure	type II	0.126	0.004	0.73

databank. Even if the difference between the initial distribution (0.49 0.13 0.12 0.25) and stationary distribution (0.30 0.21 0.28 0.21) seems not very high, type I and type II approximations lead to some errors. When type II approximation is used, 2.8% of overrepresented patterns are false positives and no false negatives are observed. Using type I approximation, 6.4% of overrepresented patterns are false negatives and no false positives are observed. The Kendall's tau for the two approximations is close to 1, so the pattern ranks are mostly the same.

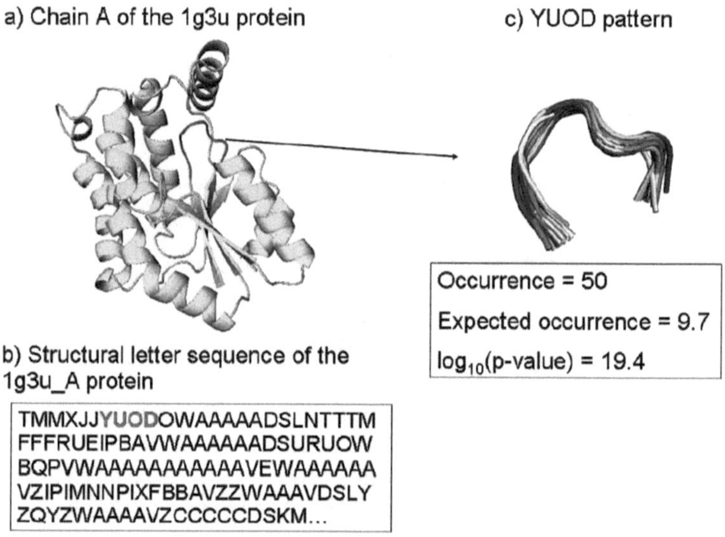

Figure 16.2. Illustration of an overrepresented pattern YUOD extracted from simplified loops. (**a**) The tridimensional structure of the protein 1g3uA (PDB code). (**b**) The series of structural letters obtained after translation of the protein 1g3uA into the structural alphabet space. (**c**) The statistic of YUOD pattern, and the superposition of fragments corresponding to this pattern

Our last example deals with structural motifs in protein loop structures. We suppose that a structural pattern is overrepresented because its structure was conserved during evolution, suggesting a biological functional implication. In order to test that hypothesis, we used a simplification of loop structures as strings. In that case, the alphabet cardinality is 27. We compute statistics of 16,977 four-letter patterns. When type II approximation is used, 12.6% of the four-letter overrepresented patterns are false positives and 4.0% of nonsignificant patterns are false negatives. The Kendall's tau is 0.73 indicating that ranks are also affected by this approximation (cf. Table 16.2).

Moreover, among the 50 most overrepresented patterns, 12% (= 6 patterns) are ranked below 50 using type II approximation. On these 50 patterns, the Kendall's tau (0.64) is lower than the global one.

Figure 16.2 presents an illustration of the overrepresented pattern YUOD. This pattern is extracted from the series of letters corresponding to the tridimensional structure of the protein 1g3uA (Figure 16.2a). The structure superposition of the YUOD different occurrence shows that these fragments correspond to the same shape.

16.5 Conclusion

In this chapter we present a new method allowing to fully take into account the specificity of a large set of sequences when computing pattern statistics. For exact p-value computations, the method adapts classical results on finite Markov chain embedding to our particular framework. One should note that while the method is proposed for an homogeneous Markov model, it also can be easily extended to heterogeneous ones (with no further computational cost).

On simulated data, we show that the more numerous and short are the considered sequences the more erroneous are the results obtained through both these approximations, even with a small alphabet and short patterns. On real data, also, classical approximations lead to many errors. In particular, the identification of exceptional structural patterns within loops is strongly affected by the used method. Indeed, both statistical values and pattern ranking are modified by usual approximations. It is of prime importance since it has been shown that extraction of overrepresented structural patterns in protein loops is a new promising direction for loop analysis and mining (Regad et al., 2006). Current development deals with the particular characteristics of these patterns.

References

Camproux, A. C., Gautier, R. and Tufféry, T. (2004). A hidden Markov model derivated structural alphabet for proteins. *J. Mol. Biol.*, 339: 561–605.

Durinck, S., Moreau, Y., Kasprzyk, A., Davis, S., De Moor, B., Brazma, A. and Huber, W. (2005). BioMart and Bioconductor: a powerful link between biological databases and microarray data analysis. *Bioinformatics*, 21, 16: 3439–3440.

Fu, J. C. and Koutras, M. V. (1994). Distribution theory of runs: A Markov chain approach. *J. Am. Statist. Assoc.* 89: 1050–1058.

Myles, H. and Douglas, A. W. (1973). *Nonparametric Statistical Inference.* John Wiley & Sons: 185–194.

Nuel, G. (2006). Effective p-value computations using Finite Markov Chain Imbedding (FMCI): Application to local score and to pattern statistics. *Algo. Mol. Biol.*, 1(5).

Nuel, G. (2008). Pattern Markov chains: optimal Markov chain embedding through deterministic finite automata. *J. Appl. Prob.*, 45: 226–243.

Nuel, G. and Prum, B. (2007). *Analyse statistique des séquences biologiques: modélisation markovienne, alignements et motifs*. Hermes, Paris, in Press.

Regad, L., Guyon, F., Maupetit, J., Tufféry, P. and Camproux, A. C. (2008). A hidden Markov model applied to the protein 3D structure analysis. *Comput. Statist. Data Anal.*, 52: 3198–3207.

Regad, L., Martin, J. and Camproux, A. C. (2006). Identification of non random motifs in loops using a structural alphabet. *Proceedings of IEEE Symposium on Computational Intelligence in Bioinformatics and Computational*: 92–100.

17

On the Convergence of the Discrete-Time Homogeneous Markov Chain

I. Kipouridis[1] and G. Tsaklidis[2]

[1] Technological Institution of West Macedonia, Department of General Sciences, Koila Kozanis, Greece
[2] Department of Mathematics, Aristotle University of Thessaloniki, Thessaloniki, Greece

Abstract: The evolution of a discrete-time Markov Chain (MC) is determined by the evolution equation $\mathbf{p}^T(t) = \mathbf{p}^T(t-1) \cdot \mathbf{P}$, where $\mathbf{p}(t)$ stands for the stochastic state vector at time t, $t \in \mathbb{N}$, $\mathbf{P}$ interprets the stochastic transition matrix of the MC, and the superscript ^{T} denotes transposition of the respective column vector (or matrix). The present chapter examines under which conditions concerning the stochastic matrix $\mathbf{P}$, a set of stochastic vectors, $\{\mathbf{p}(t-1)\}$, representing a hypersphere on the set of the attainable structures of the MC, is transformed into a stochastic set $\{\mathbf{p}(t)\}$ also representing a hypersphere of the MC. The results concerning the form of the transition matrix $\mathbf{P}$ are derived by means of the product $\mathbf{P}\mathbf{P}^T$. The set of the matrices $\mathbf{P}$ turns out to be a subset of the set of the doubly stochastic matrices.

Keywords and phrases: Discrete-time homogeneous Markov chains, discrete-time homogeneous Markov systems

17.1 Introduction

Basic results associated with Markov chains (MCs) in discrete or continuous time, concern among other topics the variation of the probability state vectors $\mathbf{p}(t) = (p_i(t))$, where $p_i(t)$ represents the probability of the chain to possess state i at time t. The evolution of a MC is usually examined by studying the evolution and the asymptotic behaviour of the probability state vectors $\mathbf{p}(t)$.

In Section 17.2 we present the evolution equation of a MC. In Section 17.3 we evaluate the equation of the image of a hypersphere of the MC under the one-step transformation expressed by $\mathbf{p}^T(t) = \mathbf{p}^T(t-1) \cdot \mathbf{P}$, where $\mathbf{P}$ stands for the transition matrix of the system, and we examine the kind of the respective hypersurface. The motivation for the study of this problem is to investigate by means of the MC's transition probabilities, the variability of the distance of the stochastic structures $\mathbf{p}(t)$, $t = 0, 1, \ldots$ (considered also as points of $\mathbb{R}^n$) from any structure point. Especially, if the points $\mathbf{p}(t-1)$ belong to a hypersphere whose center is the stability point of the

C.H. Skiadas (ed.), *Advances in Data Analysis*, Statistics for Industry and Technology, DOI 10.1007/978-0-8176-4799-5_17,

(convergent) MC, then the shape of the deformed hypersphere $\{\mathbf{p}(t)\}$, at time t, obviously characterizes the evolution of the MC and the way it approaches the stability point. Thus, if the image $\{\mathbf{p}(t)\}$ is also a hypersphere, then all the initial points $\mathbf{p}(t-1)$ tend to the limit as $t \to \infty$, in a somewhat uniform way, in the sense that their new distances from the center-stability point remain equal; in other words, every sequence $\{\mathbf{p}(t) : \mathbf{p}^T(t) = \mathbf{p}^T(t-1) \cdot \mathbf{P}, \quad t = 1, 2, ...\}$ converges to the chain's stability point by the same rate of convergence, i.e., independently of the direction from which the associated trajectory $\{\mathbf{p}(t),\, t = 0, 1, ...\}$ approaches the stability point.

In Section 17.4, the equation which represents the image of the hypersphere derived in Section 17.3, is given in matrix notation.

In Section 17.5, we pay attention to the problem of Section 17.3 by deriving conditions for a hypersphere of $\mathbb{R}^{n-1}$ to be the image of a hypersphere under the stochastic transformation $\mathbf{p}^T(t) = \mathbf{p}^T(t-1) \cdot \mathbf{P}$. These conditions concern the transition matrix $\mathbf{P}$ of the MC by means of the product $\mathbf{PP}^T$. As a byproduct of our analysis, results are derived concerning the matrix equation $\mathbf{PP}^T = a\mathbf{I} + b\mathbf{J}$, with $a > b \geq 0$, where $\mathbf{J}$ is a matrix with its main diagonal entries equal to 0, and all the other entries equal to 1.

17.2 The homogeneous Markov chain in discrete time

For a discrete-time MC let t, $t = 0, 1, ...$, be the time variable and $\mathrm{S} = \{1, 2, ..., n\}$ the state space of the chain. Denote by p_{ij} the one-step (conditional) transition probability of moving from state i to state j, and by $\mathbf{P} = (p_{ij})$ the respective transition matrix. Also, denote by $p_i(t)$ the probability that the MC is in state i at time t, and by $\mathbf{p}(t) = (p_i(t))$, $i = 1, 2, ..., n$, the (column) probability state vector. It is known that (Iosifescu, 1980)

$$\mathbf{p}^T(t) = \mathbf{p}^T(t-1) \cdot \mathbf{P}. \tag{2.1}$$

The results which we present in Sections 17.3–17.5 can be applied (among other topics) directly to the homogeneous Markov systems, whose evolution equation is exactly (2.1) (Bartholomew, 1982). These systems are met in manpower planning (Gani, 1963), Bartholomew (1982), Tsaklidis (1994), and Vassiliou (1982, 1997)), in demography (Bartholomew, 1982), biology (Patoucheas and Stamou, 1993), or in order to describe the patients' flows and costs in a hospital (McClean et al., 1998), Taylor et al. (2000)) etc.

17.3 The equation of the image of a hypersphere under the transformation (2.1)

We consider the general case, where the matrix $\mathbf{P}$ is a nonsingular $n \times n$ stochastic matrix. In order to simplify the notation of the previously mentioned equation $\mathbf{p}^T(t) = \mathbf{p}^T(t-1)\,\mathbf{P}$, we adopt the notation $\mathbf{y} = \mathbf{p}(t)$ and $\mathbf{x} = \mathbf{p}(t-1)$, and thus we have

$$\mathbf{y}^T = \mathbf{x}^T \cdot \mathbf{P}. \tag{3.1}$$

Under the assumption that $\mathbf{P}$ is a nonsingular matrix, equation (3.1) leads to

$$\mathbf{x}^T = \mathbf{y}^T \cdot \mathbf{Q}, \tag{3.2}$$

where $\mathbf{Q} = (q_{ij}) = \mathbf{P}^{-1}$, or, written in analytical form,

$$x_i = \sum_{j=1}^{n} y_j q_{ji}, \qquad i = 1, 2, ..., n. \tag{3.3}$$

Let $\mathbf{x} = (x_1, x_2, ..., x_n)^T$ represent the radius vector of an arbitrary point X of the stochastic simplex $\mathrm{S}^{n-1}_{\text{simplex}}$; that is, $X \in \{(x_1, x_2, ..., x_n) \in \mathbb{R}^n \backslash x_1 + x_2 + \cdots + x_n = 1, x_1, ..., x_n \geqslant 0\}$. As the coordinate system of reference we consider the Cartesian coordinate system $\{\mathrm{O}, \mathbf{e}_1, \mathbf{e}_2, ..., \mathbf{e}_n\}$. Now, assume that X also belongs to a hypersphere with radius R and center at some point $\mathrm{K} \in \mathrm{S}^{n-1}_{\text{simplex}}$ with radius vector $\mathbf{k} = (k_1, k_2, ..., k_n)^T$. Then the coordinates of X satisfy the equation

$$(x_1 - k_1)^2 + (x_2 - k_2)^2 + \cdots + (x_n - k_n)^2 = R^2. \tag{3.4}$$

We evaluate the equation of the hypersurface to which the image $\mathbf{y} = (y_1, y_2, ..., y_n)^T$ of $\mathbf{x}$ belongs, under the transformation (3.1). From (3.4) we get that

$$\sum_{i=1}^{n} x_i^2 - 2\sum_{i=1}^{n} k_i x_i + \sum_{i=1}^{n} k_i^2 = R^2.$$

Then, due to (3.3), we have

$$\sum_{i=1}^{n} \left(\sum_{j=1}^{n} y_j q_{ji} \right)^2 - 2\sum_{i=1}^{n} k_i \sum_{j=1}^{n} y_j q_{ji} + \sum_{i=1}^{n} k_i^2 = R^2,$$

and

$$\sum_{i=1}^{n}\sum_{j=1}^{n}\sum_{s=1}^{n} y_j q_{ji} y_s q_{si} - 2\sum_{j=1}^{n} y_j \sum_{i=1}^{n} k_i q_{ji} + \|\mathbf{k}\|^2 = R^2,$$

or

$$\sum_{j=1}^{n}\sum_{s=1}^{n} y_j y_s \sum_{i=1}^{n} q_{ji} q_{si} - 2\sum_{j=1}^{n} y_j \sum_{i=1}^{n} k_i q_{ji} + \|\mathbf{k}\|^2 = R^2,$$

where $\|\mathbf{k}\| = \sqrt{(k_1)^2 + (k_2)^2 + \cdots + (k_n)^2}$ represents the Euclidean norm of $\mathbf{k}$.

Now, denote by $\mathbf{q}_i^T$ the ith row of the matrix $\mathbf{Q}$, to get

$$\sum_{j=1}^{n}\sum_{s=1}^{n} y_j y_s \langle \mathbf{q}_j, \mathbf{q}_s \rangle + 2\sum_{j=1}^{n} y_j \langle \mathbf{q}_j, -\mathbf{k} \rangle + \|\mathbf{k}\|^2 = R^2,$$

and by isolating in every sum the term y_n there follows

$$\sum_{j=1}^{n-1}\sum_{s=1}^{n-1} y_j y_s \langle \mathbf{q}_j, \mathbf{q}_s \rangle + 2y_n \sum_{s=1}^{n-1} y_s \langle \mathbf{q}_n, \mathbf{q}_s \rangle + y_n^2 \langle \mathbf{q}_n, \mathbf{q}_n \rangle$$
$$+ 2\sum_{j=1}^{n-1} y_j \langle \mathbf{q}_j, -\mathbf{k} \rangle + 2y_n \langle \mathbf{q}_n, -\mathbf{k} \rangle + \|\mathbf{k}\|^2 = R^2. \tag{3.5}$$

Denote by $\mathbf{1}$ the $n \times 1$ vector of 1s, and note that $\mathbf{y}^T\mathbf{1} = \mathbf{x}^T\mathbf{P}\mathbf{1} = \mathbf{x}^T\mathbf{1} = 1$; that is, $y_1 + y_2 + \cdots + y_n = 1$. Then $y_n = 1 - y_1 - y_2 \cdots - y_{n-1}$, and by substituting y_n into (3.5) we have,

$$\sum_{j=1}^{n-1}\sum_{s=1}^{n-1} y_j y_s \langle \mathbf{q}_j, \mathbf{q}_s \rangle + 2\left(1 - \sum_{j=1}^{n-1} y_j\right) \sum_{s=1}^{n-1} y_s \langle \mathbf{q}_n, \mathbf{q}_s \rangle + \left(1 - \sum_{j=1}^{n-1} y_j\right)^2 \langle \mathbf{q}_n, \mathbf{q}_n \rangle$$
$$+ 2\sum_{j=1}^{n-1} y_j \langle \mathbf{q}_j, -\mathbf{k} \rangle + 2\left(1 - \sum_{j=1}^{n-1} y_j\right) \langle \mathbf{q}_n, -\mathbf{k} \rangle + \|\mathbf{k}\|^2 = R^2,$$

or

$$\sum_{j=1}^{n-1}\sum_{s=1}^{n-1} y_j y_s \langle \mathbf{q}_j, \mathbf{q}_s \rangle + 2\sum_{s=1}^{n-1} y_s \langle \mathbf{q}_n, \mathbf{q}_s \rangle - 2\sum_{j=1}^{n-1} y_j \sum_{s=1}^{n-1} y_s \langle \mathbf{q}_n, \mathbf{q}_s \rangle + \left(1 - 2\sum_{j=1}^{n-1} y_j\right.$$
$$\left. + \left(\sum_{j=1}^{n-1} y_j\right)^2\right) \langle \mathbf{q}_n, \mathbf{q}_n \rangle + 2\sum_{j=1}^{n-1} y_j \langle \mathbf{q}_j, -\mathbf{k} \rangle + 2\langle \mathbf{q}_n, -\mathbf{k} \rangle$$
$$- 2\sum_{j=1}^{n-1} y_j \langle \mathbf{q}_n, -\mathbf{k} \rangle + \|\mathbf{k}\|^2 = R^2,$$

or

$$\sum_{j=1}^{n-1}\sum_{s=1}^{n-1} y_j y_s \langle \mathbf{q}_j, \mathbf{q}_s \rangle + 2\sum_{j=1}^{n-1}\sum_{s=1}^{n-1} y_s y_j \langle -\mathbf{q}_n, \mathbf{q}_s \rangle + 2\sum_{s=1}^{n-1} y_s \langle \mathbf{q}_n, \mathbf{q}_s \rangle$$
$$+ \langle \mathbf{q}_n, \mathbf{q}_n \rangle - 2\sum_{j=1}^{n-1} y_j \langle \mathbf{q}_n, \mathbf{q}_n \rangle + \sum_{j=1}^{n-1}\sum_{s=1}^{n-1} y_j y_s \langle \mathbf{q}_n, \mathbf{q}_n \rangle$$
$$+ 2\sum_{j=1}^{n-1} y_j \langle \mathbf{q}_j, -\mathbf{k} \rangle + 2\langle \mathbf{q}_n, -\mathbf{k} \rangle - 2\sum_{j=1}^{n-1} y_j \langle \mathbf{q}_n, -\mathbf{k} \rangle + \|\mathbf{k}\|^2 = R^2.$$

By renaming some indices, last equation becomes

$$\underbrace{\sum_{j=1}^{n-1}\sum_{s=1}^{n-1} y_j y_s \langle \mathbf{q}_j, \mathbf{q}_s \rangle + 2\sum_{j=1}^{n-1}\sum_{s=1}^{n-1} y_s y_j \langle -\mathbf{q}_n, \mathbf{q}_s \rangle + \sum_{j=1}^{n-1}\sum_{s=1}^{n-1} y_j y_s \langle \mathbf{q}_n, \mathbf{q}_n \rangle}$$

$$+\underbrace{2\sum_{s=1}^{n-1} y_s \langle \mathbf{q}_n, \mathbf{q}_s \rangle + 2\sum_{s=1}^{n-1} y_s \langle \mathbf{q}_s, -\mathbf{k} \rangle - 2\sum_{s=1}^{n-1} y_s \langle \mathbf{q}_n, \mathbf{q}_n \rangle - 2\sum_{s=1}^{n-1} y_s \langle \mathbf{q}_n, -\mathbf{k} \rangle}$$

$$+\underbrace{\langle \mathbf{q}_n, \mathbf{q}_n \rangle + 2\langle \mathbf{q}_n, -\mathbf{k} \rangle + \|\mathbf{k}\|^2} = R^2,$$

and finally

$$\sum_{j=1}^{n-1}\sum_{s=1}^{n-1} y_j y_s \langle \mathbf{q}_j - \mathbf{q}_n, \mathbf{q}_s - \mathbf{q}_n \rangle + 2\sum_{s=1}^{n-1} y_s \langle \mathbf{q}_s - \mathbf{q}_n, \mathbf{q}_n - \mathbf{k} \rangle + \|\mathbf{q}_n - \mathbf{k}\|^2 = R^2, \quad (3.6)$$

where $0 \leq y_i \leq 1$, $i = 1, 2, ..., n-1$, and $y_1 + y_2 + \cdots + y_{n-1} \leq 1$.

17.4 Representation of equation (3.6) in matrix form

In order to identify the kind of hypersurface expressed by equation (3.6), we observe that the coefficient of $y_i y_j$ is $\alpha_{ij} = \langle \mathbf{q}_i - \mathbf{q}_n, \mathbf{q}_j - \mathbf{q}_n \rangle$, $i, j = 1, 2, ..., n-1$. Moreover, it is obvious that $\alpha_{ij} = \alpha_{ji}$, so the matrix $\mathbf{A} = [\alpha_{ij}]$ of the second-order hypersurface (3.6) is symmetric. Then,

$$\mathbf{A} = [\alpha_{ij}] = \overset{(n-1)\times n}{\begin{bmatrix} \mathbf{q}_1^T - \mathbf{q}_n^T \\ \mathbf{q}_2^T - \mathbf{q}_n^T \\ \cdots\cdots\cdots \\ \mathbf{q}_{n-1}^T - \mathbf{q}_n^T \end{bmatrix}} \cdot \overset{n\times(n-1)}{\begin{bmatrix} \mathbf{q}_1 - \mathbf{q}_n & \mathbf{q}_2 - \mathbf{q}_n & \cdots & \mathbf{q}_{n-1} - \mathbf{q}_n \end{bmatrix}}.$$

Thus $\mathbf{A}$ is representable in the form $\mathbf{A} = \mathbf{C}\mathbf{C}^T$, with

$$\mathbf{C} = [c_{ij}] = \begin{bmatrix} \mathbf{q}_1^T - \mathbf{q}_n^T \\ \mathbf{q}_2^T - \mathbf{q}_n^T \\ \cdots\cdots\cdots \\ \mathbf{q}_{n-1}^T - \mathbf{q}_n^T \end{bmatrix} \in M_{n-1,n},$$

where $M_{s,r}$ denotes the set of $s \times r$ matrices ($s, r \in \mathbb{N}^+$). Especially for $s = r$, we use the notation M_s.

Now, let $\mathbf{y}_s = (y_1, y_2, ..., y_{n-1})^T$, $\mathbf{r} = \mathbf{q}_n - \mathbf{k}$, and $\mathbf{B} = \mathbf{C} \cdot \mathbf{r}$. Then, (3.6) can be written as

$$\mathbf{y}_s^T \mathbf{A} \mathbf{y}_s + 2\mathbf{B}^T \mathbf{y}_s + \|\mathbf{r}\|^2 = R^2,$$

or

$$\mathbf{y}_s^T \mathbf{C}\mathbf{C}^T \mathbf{y}_s + 2\mathbf{r}^T \mathbf{C}^T \mathbf{y}_s + \mathbf{r}^T \mathbf{r} = R^2, \quad (4.1)$$

and, since $\mathbf{r}^T\mathbf{C}^T\mathbf{y}_s = \mathbf{y}_s^T\mathbf{C}\mathbf{r}$, we get

$$\left(\mathbf{C}^T\mathbf{y}_s + \mathbf{r}\right)^T \cdot \left(\mathbf{C}^T\mathbf{y}_s + \mathbf{r}\right) = R^2,$$

where the vectors $\mathbf{y}_s$ are substochastic.

The augmented matrix $\mathbf{A}_\varepsilon = \mathbf{A}_\varepsilon(R)$ of $\mathbf{A}$ is

$$\mathbf{A}_\varepsilon = [(\alpha_\varepsilon)_{ij}] = \overset{n\times(n+1)}{\begin{bmatrix} \mathbf{q}_n^T - \mathbf{k}^T & R \\ \mathbf{q}_1^T - \mathbf{q}_n^T & 0 \\ \mathbf{q}_2^T - \mathbf{q}_n^T & 0 \\ \cdots\cdots\cdots & \cdots \\ \mathbf{q}_{n-1}^T - \mathbf{q}_n^T & 0 \end{bmatrix}} \cdot \overset{(n+1)\times n}{\begin{bmatrix} \mathbf{q}_n - \mathbf{k} & \mathbf{q}_1 - \mathbf{q}_n & \mathbf{q}_2 - \mathbf{q}_n & \cdots & \mathbf{q}_{n-1} - \mathbf{q}_n \\ -R & 0 & 0 & \cdots & 0 \end{bmatrix}},$$

or

$$\mathbf{A}_\varepsilon = \begin{bmatrix} \langle \mathbf{q}_n - \mathbf{k}, \mathbf{q}_n - \mathbf{k}\rangle - R^2 & \langle \mathbf{q}_n - \mathbf{k}, \mathbf{q}_1 - \mathbf{q}_n\rangle & \cdots & \langle \mathbf{q}_n - \mathbf{k}, \mathbf{q}_{n-1} - \mathbf{q}_n\rangle \\ \langle \mathbf{q}_n - \mathbf{k}, \mathbf{q}_1 - \mathbf{q}_n\rangle & \langle \mathbf{q}_1 - \mathbf{q}_n, \mathbf{q}_1 - \mathbf{q}_n\rangle & \cdots & \langle \mathbf{q}_1 - \mathbf{q}_n, \mathbf{q}_{n-1} - \mathbf{q}_n\rangle \\ \cdots\cdots\cdots & \cdots\cdots\cdots & \cdots & \cdots\cdots\cdots \\ \langle \mathbf{q}_n - \mathbf{k}, \mathbf{q}_{n-1} - \mathbf{q}_n\rangle & \langle \mathbf{q}_1 - \mathbf{q}_n, \mathbf{q}_{n-1} - \mathbf{q}_n\rangle & \cdots & \langle \mathbf{q}_{n-1} - \mathbf{q}_n, \mathbf{q}_{n-1} - \mathbf{q}_n\rangle \end{bmatrix}.$$

Thus,

$$\mathbf{A}_\varepsilon = \mathbf{A}_\varepsilon(R) = \begin{bmatrix} (\alpha_\varepsilon)_{11} & \mathbf{b}^T \\ \mathbf{b} & \mathbf{A} \end{bmatrix}, \tag{4.2}$$

where $(\alpha_\varepsilon)_{11} = \langle \mathbf{q}_n - \mathbf{k}, \mathbf{q}_n - \mathbf{k}\rangle - R^2$, and

$$\mathbf{b}^T = \left(\langle \mathbf{q}_n - \mathbf{k}, \mathbf{q}_1 - \mathbf{q}_n\rangle, \ldots, \langle \mathbf{q}_n - \mathbf{k}, \mathbf{q}_{n-1} - \mathbf{q}_n\rangle\right).$$

Proposition 4.1. *For the matrix* $\mathbf{A}$ *of the hypersurface* (3.6) *and for the corresponding augmented matrix* $\mathbf{A}_\varepsilon$ *it holds that*

$$\operatorname{rank} \mathbf{A} = n - 1$$

and

$$\operatorname{rank} \mathbf{A}_\varepsilon(R) = \begin{cases} n-1 & \text{when } R = 0 \\ n & \text{when } R \neq 0 \end{cases},$$

and

$$\det \mathbf{A}_\varepsilon(R) = -R^2 \det \mathbf{A}.$$

Proof. In order to evaluate the rank of the matrix $\mathbf{A}$, we consider $\mathbf{A}$ in the above-mentioned form $\mathbf{A} = \mathbf{C}\mathbf{C}^T$, where

$$\mathbf{C} = [c_{ij}] = \begin{bmatrix} \mathbf{q}_1^T - \mathbf{q}_n^T \\ \mathbf{q}_2^T - \mathbf{q}_n^T \\ \cdots\cdots\cdots \\ \mathbf{q}_{n-1}^T - \mathbf{q}_n^T \end{bmatrix} \in M_{n-1,n}.$$

Under the assumption that $\det \mathbf{P} \neq 0$, we get $\det \mathbf{Q} \neq 0$, and so the rows $\mathbf{q}_i^T$ of the matrix $\mathbf{Q}$ are linearly independent. Thus the rows of the matrix $\mathbf{C}$ are linearly independent and consequently rank$\mathbf{C} = n-1$. Since rank$(\mathbf{C}\mathbf{C}^T) =$ rank $\mathbf{C}$, we conclude that

$$\text{rank}\mathbf{A} = \text{rank}(\mathbf{C}\mathbf{C}^T) = \text{rank}\mathbf{C} = n-1.$$

Hence the $(n-1) \times (n-1)$ matrix $\mathbf{A}$ is nonsingular.

Taking into consideration the form of the augmented matrix $\mathbf{A}_\varepsilon$ given in (4.2) we get

$$\det \mathbf{A}_\varepsilon(R) = (\alpha_\varepsilon)_{11} \det \mathbf{A} - \mathbf{b}^T(\text{Adj}(\mathbf{A}))\mathbf{b},$$

or

$$\det \mathbf{A}_\varepsilon(R) = (\langle \mathbf{q}_n - \mathbf{k}, \mathbf{q}_n - \mathbf{k}\rangle) \cdot \det \mathbf{A} - R^2 \det \mathbf{A} - \mathbf{b}^T(\text{Adj}(\mathbf{A}))\mathbf{b}. \tag{4.3}$$

Let

$$\mathbf{C}_\varepsilon(R) = \begin{bmatrix} \mathbf{q}_n^T - \mathbf{k}^T & R \\ \mathbf{q}_1^T - \mathbf{q}_n^T & 0 \\ \dots\dots\dots & \dots \\ \mathbf{q}_{n-1}^T - \mathbf{q}_n^T & 0 \end{bmatrix} \in M_{n,n+1}.$$

Then $\mathbf{A}_\varepsilon(R)$ can be written in the form

$$\mathbf{A}_\varepsilon(R) = \mathbf{C}_\varepsilon(R) \cdot \mathbf{C}_\varepsilon^T(-R),$$

and especially for $R = 0$ it becomes

$$\mathbf{A}_\varepsilon(0) = \mathbf{C}_\varepsilon(0) \cdot \mathbf{C}_\varepsilon^T(0).$$

Let $\mathbf{k}_0 = (k_{01}, k_{02}, ..., k_{0n})^T$ be the radius vector of the image K_0 of the point K under the transformation (3.1). Then $\mathbf{k}_0$ satisfies the equation $\mathbf{k}_0^T = \mathbf{k}^T\mathbf{P}$, and thus $\mathbf{k}^T = \mathbf{k}_0^T\mathbf{Q}$. Since the point K_0 belongs to the stochastic $(n-1)$-simplex, we get

$$\begin{aligned}
&- k_{01}\left(\mathbf{q}_1^T - \mathbf{q}_n^T\right) - k_{02}\left(\mathbf{q}_2^T - \mathbf{q}_n^T\right) - \cdots - k_{0n-1}\left(\mathbf{q}_{n-1}^T - \mathbf{q}_n^T\right) \\
&= -k_{01}\mathbf{q}_1^T - k_{02}\mathbf{q}_2^T - \cdots - k_{0n-1}\mathbf{q}_{n-1}^T + (k_{01} + k_{02} + \cdots + k_{0n-1})\mathbf{q}_n^T \\
&= -k_{01}\mathbf{q}_1^T - k_{02}\mathbf{q}_2^T - \cdots - k_{0n-1}\mathbf{q}_{n-1}^T + (1 - k_{0n})\mathbf{q}_n^T \\
&= -k_{01}\mathbf{q}_1^T - k_{02}\mathbf{q}_2^T - \cdots - k_{0n-1}\mathbf{q}_{n-1}^T - k_{0n}\mathbf{q}_n^T + \mathbf{q}_n^T \\
&= \mathbf{q}_n^T - \mathbf{k}^T.
\end{aligned}$$

Thus $\mathbf{q}_n^T - \mathbf{k}^T$ is a linear combination of the $n-1$ linearly independent vectors $\mathbf{q}_1^T - \mathbf{q}_n^T$, $\mathbf{q}_2^T - \mathbf{q}_n^T$,...,$\mathbf{q}_{n-1}^T - \mathbf{q}_n^T$. So:

(1) If $R = 0$, then rank$(\mathbf{C}_\varepsilon(0)) = n-1$, and

$$\text{rank}\mathbf{A}_\varepsilon(0) = \text{rank}(\mathbf{C}_\varepsilon(0) \cdot \mathbf{C}_\varepsilon^T(0)) = n-1,$$

hence

$$\det \mathbf{A}_\varepsilon(0) = 0.$$

Then,

$$\det \mathbf{A}_\varepsilon(0) = (\langle \mathbf{q}_n - \mathbf{k}, \mathbf{q}_n - \mathbf{k}\rangle) \det \mathbf{A} - \mathbf{b}^T(\text{Adj}(\mathbf{A}))\mathbf{b} = 0,$$

and consequently

$$(\langle \mathbf{q}_n - \mathbf{k}, \mathbf{q}_n - \mathbf{k} \rangle) \det \mathbf{A} = \mathbf{b}^T (\mathrm{Adj}(\mathbf{A}))\mathbf{b}.$$

Now, due to (4.3), we arrive at

$$\det \mathbf{A}_\varepsilon(R) = -R^2 \det \mathbf{A}.$$

(2) If $R \neq 0$, and since $\mathbf{A}$ is nonsingular, $\det \mathbf{A}_\varepsilon(R) = -R^2 \det \mathbf{A} \neq 0$. Hence

$$\mathrm{rank}\mathbf{A}_\varepsilon(R) = n. \qquad \square$$

Remark 4.1. (a) Taking into account Proposition 4.1, we conclude that equation (3.6) defines a second-order hypersurface with one center.
(b) In the special case $R = 0$, the hypersphere given by equation (3.4) becomes the single point K, and consequently its image via equation (3.1) is the single point K_0.

By using the singular value decomposition of the matrix $\mathbf{C}$, we have $\mathbf{C} = \mathbf{V\Sigma W}^T$, where $\mathbf{V} \in M_{n-1}$, $\mathbf{W} \in M_n$ are unitary matrices and $\mathbf{\Sigma} = (\sigma_{ij})$ is an $(n-1) \times n$ matrix with $\sigma_{ij} = 0,\ i \neq j,\ \ i = 1, 2, ..., n-1,\ \ j = 1, 2, ..., n$, and there are exactly $n - 1$ nonzero (positive) elements at the positions (i, i), $i = 1, 2, ..., n - 1$. The main diagonal entries of $\mathbf{\Sigma}$ are $\sigma_{ii} = \sqrt{\lambda_i}$, where λ_i, $i = 1, 2, ..., n - 1$, are the eigenvalues of the (positive definite) matrix $\mathbf{CC}^T$. Now, via (4.1), we have

$$\mathbf{y}_s^T \mathbf{V\Sigma W}^T (\mathbf{V\Sigma W}^T)^T \mathbf{y}_s + 2\mathbf{r}^T (\mathbf{V\Sigma W}^T)^T \mathbf{y}_s + \|\mathbf{r}\|^2 = R^2,$$

or

$$\mathbf{y}_s^T \mathbf{V\Sigma W}^T \mathbf{W\Sigma}^T \mathbf{V}^T \mathbf{y}_s + 2\mathbf{r}^T \mathbf{W\Sigma}^T \mathbf{V}^T \mathbf{y}_s + \mathbf{r}^T\mathbf{r} = R^2,$$

or

$$\mathbf{y}_s^T \mathbf{V\Sigma\Sigma}^T \mathbf{V}^T \mathbf{y}_s + 2\mathbf{r}^T \mathbf{W\Sigma}^T \mathbf{V}^T \mathbf{y}_s + \mathbf{r}^T\mathbf{r} = R^2,$$

or

$$\mathbf{y}_s^T \mathbf{V\Sigma\Sigma}^T \mathbf{V}^T \mathbf{y}_s + 2\mathbf{r}^T \mathbf{W\Sigma}^T \mathbf{V}^T \mathbf{y}_s + \mathbf{r}^T\mathbf{WW}^T\mathbf{r} = R^2.$$

Then

$$\left(\mathbf{\Sigma}^T \mathbf{V}^T \mathbf{y}_s + \mathbf{W}^T\mathbf{r}\right)^T \left(\mathbf{\Sigma}^T \mathbf{V}^T \mathbf{y}_s + \mathbf{W}^T\mathbf{r}\right) = R^2,$$

and by setting $\mathbf{z} = \mathbf{V}^T \mathbf{y}_s$ we get

$$\mathbf{z}^T \mathbf{\Sigma\Sigma}^T \mathbf{z} + 2\mathbf{r}^T \mathbf{W\Sigma}^T \mathbf{z} + \mathbf{r}^T \mathbf{WW}^T \mathbf{r} = R^2,$$

or, in analytical form,

$$\sum_{i=1}^{n-1} \lambda_i z_i^2 + 2\sum_{i=1}^{n} \left(\sum_{k=1}^{n} r_k w_{ki}\right)\left(\sum_{l=1}^{n-1} \sigma_{il}^T z_l\right) + \sum_{i=1}^{n}\left(\sum_{k=1}^{n} r_k w_{ki}\right)^2, \tag{4.4}$$

where $\mathbf{z} = (z_i)$, $i = 1, 2, ..., n - 1$, and $\mathbf{W} = (w_{ij})$, $i, j = 1, 2, ..., n$.

In order to achieve a more useful form of equation (4.4), the following lemma is proved.

Lemma 4.1. *The nth column,* $\mathbf{w}_n$*, of the unitary matrix* $\mathbf{W} = (\mathbf{w}_1, \mathbf{w}_2, ..., \mathbf{w}_n)$ *is the vector*

$$\left(\frac{1}{\sqrt{n}}, \frac{1}{\sqrt{n}}, ..., \frac{1}{\sqrt{n}}\right)^T.$$

Proof. Since

$$\mathbf{C} = \mathbf{V\Sigma W}^T,$$

then

$$\mathbf{C1} = \mathbf{V\Sigma W}^T\mathbf{1},$$

or

$$\mathbf{0} = \mathbf{V\Sigma W}^T\mathbf{1}.$$

Furthermore, taking into account that $\mathbf{V} \in M_{n-1}$ is a unitary matrix, and thus invertible, we get

$$\mathbf{\Sigma W}^T\mathbf{1} = \mathbf{0}.$$

Hence,

$$\sum_{j=1}^{n} \sqrt{\lambda_i} w_{ij}^T = 0, \qquad i = 1, 2, ..., n-1,$$

or

$$\sqrt{\lambda_i} \cdot \sum_{j=1}^{n} w_{ij}^T = 0, \qquad i = 1, 2, ..., n-1,$$

or

$$\sum_{j=1}^{n} w_{ji} = 0, \qquad i = 1, 2, ..., n-1.$$

Thus, the sum of the elements of each one of the $n-1$ columns $\mathbf{w}_1, \mathbf{w}_2, \ldots, \mathbf{w}_{n-1}$ of the matrix $\mathbf{W}$ is equal to zero. By combining this result with the fact that $\mathbf{W}$ is a unitary matrix, it results that

$$\mathbf{w}_n = \left(\frac{1}{\sqrt{n}}, \frac{1}{\sqrt{n}}, ..., \frac{1}{\sqrt{n}}\right)^T. \qquad \square$$

According to the previous lemma, and taking into account that $\mathbf{r}^T\mathbf{1} = (\mathbf{q}_n - \mathbf{k})^T\mathbf{1} = 0,$ we have

$$\sum_{k=1}^{n} r_k w_{kn} = 0.$$

Moreover, since $\sigma_{ij} = 0$ for $i \neq j$, while $\sigma_{ii} = \sqrt{\lambda_i}$ for $i = 1, 2, ..., n-1$, the equation of the hypersurface (4.4) becomes

$$\sum_{i=1}^{n-1} \lambda_i z_i^2 + 2\sum_{i=1}^{n-1} \left(\sum_{k=1}^{n} r_k w_{ki}\right) \left(\sqrt{\lambda_i} z_i\right) + \sum_{i=1}^{n-1} \left(\sum_{k=1}^{n} r_k w_{ki}\right)^2 = R^2,$$

or

$$\sum_{i=1}^{n-1}\left(\sqrt{\lambda_i}z_i+\sum_{k=1}^{n}r_k w_{ki}\right)^2=R^2. \tag{4.5}$$

Relation (4.5) is the equation of the image of the intersection of hypersphere (3.4) with the hyperplane $x_1+x_2+\cdots+x_n=1$, by means of the transformation (3.1), and it represents a hyperellipsoid of $\mathbb{R}^{n-1}$.

17.5 Conditions for a hypersphere of $\mathbb{R}^{n-1}$ to be the image of a hypersphere under the stochastic transformation $\mathbf{p}^T(t)=\mathbf{p}^T(t-1)\cdot\mathbf{P}$

Let us express the state vector $\mathbf{p}(t)$ of a HMS in the form

$$\mathbf{p}^T(t)=\mathbf{k}^T(t)+\varepsilon^T(t), \tag{5.1}$$

where $\mathbf{k}(t)$ is the radius vector of some point K(t) of the stochastic $(n-1)$-simplex $\mathrm{S}^{n-1}_{\text{simplex}}$ moving according to (2.1); i.e., $\mathbf{k}^T(t)=\mathbf{k}^T(t-1)\cdot\mathbf{P}$. Then, we have

$$\mathbf{p}^T(t)\cdot\mathbf{1}=\mathbf{k}^T(t)\cdot\mathbf{1}+\varepsilon^T(t)\cdot\mathbf{1},$$

thus

$$\sum_{i=1}^{n}\varepsilon_i(t)=0, \qquad t\in\mathbb{N}. \tag{5.2}$$

By combining the relation expressing the evolution of the HMS

$$\mathbf{p}^T(t)=\mathbf{p}^T(t-1)\cdot\mathbf{P} \tag{5.3}$$

with (5.1), we get

$$\mathbf{k}^T(t)+\varepsilon^T(t)=(\mathbf{k}^T(t-1)+\varepsilon^T(t-1))\cdot\mathbf{P}$$

and since $\mathbf{k}^T(t)=\mathbf{k}^T(t-1)\cdot\mathbf{P}$, we are led to

$$\varepsilon^T(t)=\varepsilon^T(t-1)\cdot\mathbf{P}. \tag{5.4}$$

Now, assume that the (stochastic) state vectors $\mathbf{p}^T(t)$ of the transformation (5.3) satisfy the equation

$$(\mathbf{p}^T(t)-\mathbf{k}^T(t))\cdot(\mathbf{p}^T(t)-\mathbf{k}^T(t))^T=R^2; \tag{5.5}$$

i.e., the corresponding points lay on the surface of a hypersphere of $\mathrm{S}^{n-1}_{\text{simplex}}$ with radius R, $R\in\mathbb{R}^+$, and center at some point K of radius vector $\mathbf{k}(t)$.

In order to find the equation of the hypersurface to which the points with radius vector $\mathbf{p}(t-1)$ belonged before being transformed (according to (5.3)) to the points $\mathbf{p}(t)$ of the hypersphere (5.5), we combine equations (5.1) and (5.5) to get

$$\varepsilon^T(t) \cdot \varepsilon(t) = R^2, \tag{5.6}$$

where $\sum_{i=1}^{n} \varepsilon_i(t) = 0$ (by (5.2)).

In what follows, while referring to equation (5.6) we still assume the validity of (5.2). By combining equations (5.4) and (5.6), we have

$$\varepsilon^T(t-1) \cdot \mathbf{PP}^T \cdot \varepsilon(t-1) = R^2. \tag{5.7}$$

The symmetric matrix $\mathbf{PP}^T$ is positive definite, and consequently it has positive eigenvalues. So it obeys the Jordan canonical form

$$\mathbf{PP}^T = \mathbf{V\Lambda V}^T, \tag{5.8}$$

where $\mathbf{V}$ is a unitary matrix and $\mathbf{\Lambda} = \text{diag}\{\lambda_1, \lambda_2, ..., \lambda_n\}$, with $\lambda_i \in \mathbb{R}^+$, $i = 1, 2, \ldots, n$. Consequently (5.7) can be written as

$$\varepsilon^T(t-1) \cdot \mathbf{V\Lambda V}^T \cdot \varepsilon(t-1) = R^2.$$

Now, let $\mathbf{V}^T = \left(v_{ij}^T\right)$ and $\mathbf{z} = \mathbf{V}^T \cdot \varepsilon(t-1)$. Then, the latter equation becomes

$$\mathbf{z}^T \mathbf{\Lambda z} = R^2,$$

or

$$\sum_{i=1}^{n} \lambda_i z_i^2 = R^2, \tag{5.9}$$

where $z_i = \sum_{j=1}^{n} v_{ij}^T \varepsilon_j(t-1)$, $i = 1, 2, ..., n$. Written in analytical form, (5.9) becomes

$$\sum_{i=1}^{n} \lambda_i \left(\sum_{j=1}^{n} v_{ij}^T \varepsilon_j(t-1) \right)^2 = R^2,$$

or

$$\sum_{i=1}^{n} \lambda_i \left(\sum_{j=1}^{n} v_{ij}^T \varepsilon_j(t-1) \right) \cdot \left(\sum_{j=1}^{n} v_{ij}^T \varepsilon_j(t-1) \right) = R^2,$$

or

$$\sum_{i=1}^{n} \sum_{j=1}^{n} \sum_{s=1}^{n} \lambda_i v_{is}^T v_{ij}^T \varepsilon_s(t-1) \varepsilon_j(t-1) = R^2,$$

or

$$\sum_{j=1}^{n} \sum_{s=1}^{n} \varepsilon_s(t-1) \varepsilon_j(t-1) \sum_{i=1}^{n} \lambda_i v_{is}^T v_{ij}^T = R^2. \tag{5.10}$$

Equation (5.9) and its equivalent form (5.10) represent a hyperellipsoid of $\mathbb{R}^n$ and they express the analytical equation of a hypersurface the image of which, via (5.4), is a hypersphere.

Since (5.10) was derived from (5.7) by decomposing the matrix product $\mathbf{PP}^T$, we focus attention on the matrix $\mathbf{PP}^T$, aware that $\mathbf{P}$ is a stochastic matrix. The following lemma and proposition lead us to a necessary condition for (5.10) to represent a hypersphere.

Lemma 5.1. *The eigenvalues and the singular values of* $\mathbf{PP}^T$ *are identical, and if* $\mathbf{P}$ *is a doubly stochastic matrix, its eigenvalue* 1 *is also an eigenvalue (singular value) of* $\mathbf{PP}^T$.

Proof. Since the symmetric matrix $\mathbf{PP}^T$ is the product of a matrix with its transposed matrix, then it is positive semidefinite. In the case that $\mathbf{P}$ is nonsingular, then $\mathbf{PP}^T$ is positive definite. By using Takagi's factorization theorem, it appears that the eigenvalues and the singular values of $\mathbf{PP}^T$ are identical. If $\mathbf{P}$ is doubly stochastic, then

$$\mathbf{P}^T\mathbf{1} = \mathbf{1},$$

hence

$$\mathbf{PP}^T\mathbf{1} = \mathbf{P1} = 1 \cdot \mathbf{1}.$$

Based on Lemma 5.1 we can prove the following proposition.

Proposition 5.1. *Let* $\mathbf{P}$ *be an* $n \times n$ *nonsingular doubly stochastic matrix. If the eigenvalues* $\lambda_2, \lambda_3, \ldots, \lambda_n$ *of the matrix* $\mathbf{PP}^T$ *satisfy the condition*

$$\lambda_2 = \lambda_3 = \cdots = \lambda_n = \lambda, \qquad \lambda \in (0, 1],$$

while $\lambda_1 = 1$, *then* (5.10) *represents the equation of a hypersphere with radius* $R/\sqrt{\lambda}$.

Proof. Taking into account Lemma 5.1 and assuming – without loss of generality – that $\lambda_1 = 1$, we derive from (5.10)

$$\sum_{j=1}^{n}\sum_{s=1}^{n} \varepsilon_s(t-1)\varepsilon_j(t-1)v_{1s}^T v_{1j}^T + \sum_{j=1}^{n}\sum_{s=1}^{n} \varepsilon_s(t-1)\varepsilon_j(t-1)\sum_{i=2}^{n}\lambda_i v_{is}^T v_{ij}^T = R^2,$$

or

$$\left(\sum_{s=1}^{n} \varepsilon_s(t-1)v_{1s}^T\right)^2 + \sum_{j=1}^{n}\sum_{s=1}^{n} \varepsilon_s(t-1)\varepsilon_j(t-1)\sum_{i=2}^{n}\lambda_i v_{is}^T v_{ij}^T = R^2. \tag{5.11}$$

We have by (5.8) (which we used to prove (5.10) and (5.11)), that $\mathbf{PP}^T\mathbf{V} = \mathbf{V\Lambda}$; that is, the columns of the matrix $\mathbf{V}$ are right eigenvectors of $\mathbf{PP}^T$. Let us denote by $\mathbf{v}_1$ the first column of $\mathbf{V}$, and note that $\mathbf{PP}^T\mathbf{v}_1 = \lambda_1\mathbf{v}_1 = \mathbf{v}_1$. Then taking into account the form of the right eigenvector of $\mathbf{PP}^T$ (according to the proof of Lemma 5.1), in combination with the fact that $\mathbf{V}$ is a unitary matrix, we conclude that

$$\mathbf{v}_1 = \left(\frac{1}{\sqrt{n}}, \frac{1}{\sqrt{n}}, \ldots, \frac{1}{\sqrt{n}}\right)^T.$$

Then,

$$\sum_{s=1}^{n} \varepsilon_s(t-1)v_{1s}^T = 0,$$

and (5.11) becomes

$$\sum_{j=1}^{n}\sum_{s=1}^{n}\varepsilon_s(t-1)\varepsilon_j(t-1)\sum_{i=2}^{n}\lambda_i v_{ij}^T v_{is}^T = R^2. \tag{5.12}$$

Now, since $\lambda_i = \lambda$ for $i = 2, 3, ..., n$, (5.12) yields

$$\lambda\sum_{j=1}^{n}\sum_{s=1}^{n}\varepsilon_s(t-1)\varepsilon_j(t-1)\sum_{i=2}^{n} v_{ij}^T v_{is}^T = R^2,$$

or

$$\lambda\sum_{i=2}^{n}\sum_{j=1}^{n}\sum_{s=1}^{n}\varepsilon_s(t-1)\varepsilon_j(t-1) v_{ij}^T v_{is}^T = R^2,$$

or

$$\lambda\sum_{i=2}^{n}\left(\sum_{j=1}^{n}\varepsilon_j(t-1)v_{ij}^T\right)^2 = R^2. \tag{5.13}$$

By combining the equalities $z_i = \sum_{j=1}^{n} v_{ij}^T\varepsilon_j(t-1)$, $i = 1, 2, ..., n$, with (5.13), we arrive at

$$\lambda\sum_{i=2}^{n} z_i^2 = R^2. \tag{5.14}$$

Equation (5.14) represents an orthogonal transformation of (5.6) and (since the coordinate system of reference is considered to be Cartesian, then) it represents a hypersphere of $\mathbb{R}^{n-1}$. Now, by means of (5.2), we have from (5.13) that

$$\lambda\sum_{i=2}^{n}\left(\sum_{j=1}^{n-1}\varepsilon_j(t-1)v_{ij}^T - \sum_{j=1}^{n-1}\varepsilon_j(t-1)v_{in}^T\right)^2 = R^2,$$

or

$$\sum_{j=1}^{n-1}\sum_{s=1}^{n-1}\varepsilon_s(t-1)\varepsilon_j(t-1)\sum_{i=2}^{n}\left(v_{ij}^T - v_{in}^T\right)\left(v_{is}^T - v_{in}^T\right) = \frac{R^2}{\lambda}.$$

Thus, when the image of a hypersurface via the transformation $\mathbf{p}^T(t) = \mathbf{p}^T(\text{t-1})\cdot \mathbf{P}$ is a hypersphere of the stochastic $(n-1)$-simplex $\mathrm{S}_{\text{simplex}}^{n-1}$ with radius R, and the transition matrix $\mathbf{P}$ is a nonsingular doubly stochastic matrix having its $n-1$ singular values $\lambda_2,...,\lambda_n$ equal to some λ (where $\lambda \in (0,1]$ and $\lambda_1 = 1$), then the initial hypersurface is a hypersphere of $\mathrm{S}_{\text{simplex}}^{n-1}$ with radius $R = R/\sqrt{\lambda}$. □

Remark 5.1. In the special case that $\lambda = 1$ (in Proposition 5.1), the symmetric matrix $\mathbf{PP}^T$ has all its eigenvalues equal to 1, thus $\mathbf{PP}^T = \mathbf{I}$. Hence the transition matrix $\mathbf{P}$ is a permutation matrix. In other words, $\mathbf{P}$ is a periodic stochastic matrix of period n.

In order to derive more inferences from the equation (5.10) we focus more attentively on the matrix product $\mathbf{PP}^T$. So we observe that $\mathbf{PP}^T = (\langle \mathbf{p}_i, \mathbf{p}_j\rangle)$, i, j $= 1, 2, ..., n$, where $\mathbf{p}_i$ are the row-vectors of the transition matrix $\mathbf{P}$. Since the elements of $\mathbf{PP}^T$ represent inner products of the vectors $\mathbf{p}_i$, $i = 1, 2, ..., n$, we formulate the following lemma.

Lemma 5.2. *Assume that the nonnegative vectors* $\mathbf{p}_i \in \mathbb{R}^n$, $i = 1, 2, ..., n$, *satisfy the relation*

$$\langle \mathbf{p}_i, \mathbf{p}_i \rangle = a > b = \langle \mathbf{p}_i, \mathbf{p}_j \rangle, \qquad i, j = 1, 2, ..., n, \qquad i \neq j.$$

Then the vectors $\mathbf{p}_i$, $i = 1, 2, ..., n$, *are linearly independent.*

Proof. Let

$$k_1\mathbf{p}_1 + k_2\mathbf{p}_2 + \cdots + k_n\mathbf{p}_n = \mathbf{0}, \qquad k_1, k_2, ..., k_n \in \mathbb{R}.$$

By multiplying this equation by the nonnegative vectors $\mathbf{p}_i$, i $= 1, 2, ..., n$, we get the linear system

$$k_1 \langle \mathbf{p}_1, \mathbf{p}_i \rangle + k_2 \langle \mathbf{p}_2, \mathbf{p}_i \rangle + \cdots + k_n \langle \mathbf{p}_n, \mathbf{p}_i \rangle = 0, \qquad i = 1, 2, ..., n, \tag{5.15}$$

with coefficient matrix

$$\mathbf{K} = (\langle \mathbf{p}_i, \mathbf{p}_j \rangle) = \begin{bmatrix} a & b & .. & b \\ b & a & .. & b \\ ... & .. & .. & .. \\ b & b & .. & a \end{bmatrix} \in M_n(\mathbb{R}).$$

Then, by Graybill (1983, p.204),

$$\det \mathbf{K} = (a + (n-1)b)(a-b)^{n-1}.$$

Since $a > b \geq 0$, $\det \mathbf{K}$ cannot be equal to zero, thus the system (5.15) has the unique solution $k_1 = k_2 = \cdots = k_n = 0$, and as a result the vectors $\mathbf{p}_i$, $i = 1, 2, ..., n$, are linearly independent. □

Moreover, referring to the matrix $\mathbf{PP}^T$ and equation (5.10) the following lemma is established.

Lemma 5.3. *A necessary condition for the surface of a hypersphere of* $S^{n-1}_{simplex}$ *to be via* (5.4) *the image of a hypersphere of* $S^{n-1}_{simplex}$, *is that the transition matrix* $\mathbf{P}$ *satisfies a relation of the form*

$$\mathbf{PP}^T = (\langle \mathbf{p}_i, \mathbf{p}_j \rangle) = a\mathbf{I} + b\mathbf{J}, \quad \text{with } a > b \geq 0, \tag{5.16}$$

where $\mathbf{J}$ *is an* $n \times n$ *matrix with its main diagonal entries equal to* 0, *and all the other entries equal to* 1. *Then, if we denote by* c_t *the radius of the hypersphere at time* t, *it results that*

$$c_{t-1} = \frac{c_t}{\alpha - b}.$$

Proof. By hypothesis

$$\varepsilon^T(t)\varepsilon(t) = c_t, \tag{5.17}$$

which, due to (5.4), yields

$$\varepsilon^T(t-1)\mathbf{PP}^T\varepsilon(t-1) = c_t. \tag{5.18}$$

If the transition matrix $\mathbf{P}$ satisfies (5.16) (i.e., the rows $\mathbf{p}_i$, $i = 1, 2, ..., n$, of $\mathbf{P}$ satisfy $\langle \mathbf{p}_i, \mathbf{p}_i \rangle = a$, $\langle \mathbf{p}_i, \mathbf{p}_j \rangle = b$ for $i \neq j$), then from (5.18) we get

$$\varepsilon^T(t-1)(\alpha \mathbf{I} + b\mathbf{J})\varepsilon(t-1) = c_t,$$

or

$$\varepsilon^T(t-1)(\alpha \mathbf{I}\varepsilon(t-1) + b\mathbf{J}\varepsilon(t-1)) = c_t,$$

or

$$\varepsilon^T(t-1)(\alpha\varepsilon(t-1) - b\varepsilon(t-1)) = c_t,$$

or

$$(\alpha - b)\varepsilon^T(t-1)\varepsilon(t-1) = c_t,$$

or

$$\varepsilon^T(t-1)\varepsilon(t-1) = \frac{c_t}{\alpha - b}. \tag{5.19}$$

By setting $c_{t-1} = c_t/(\alpha - b)$, we arrive at the formula $\varepsilon^T(t-1)\varepsilon(t-1) = c_{t-1}$, thus the hypersphere expressed by (5.17) is the image of the hypersphere (5.19) via the transformation (5.4). □

Remark 5.2. The condition $\alpha > b$ of Lemmas 5.2 and 5.3 is not particularly restrictive. Since $\langle \mathbf{p}_i, \mathbf{p}_i \rangle = a$, for every $i \in \{1, 2, ..., n\}$, then

$$\alpha = \langle \mathbf{p}_i, \mathbf{p}_i \rangle = |\mathbf{p}_i|^2 \geq |\mathbf{p}_i|\,|\mathbf{p}_j| \cdot \cos(\phi_{ij}) = b, \qquad i, j = 1, 2, ..., n,$$

where ϕ_{ij} stands for the angle of the vectors $\mathbf{p}_i$ and $\mathbf{p}_j$. Thus, the conditions $\alpha > b$ and $\alpha \neq b$ are equivalent. Moreover, by Lemma 5.2, the rows of the matrix $\mathbf{P}$ are linearly independent and consequently $\mathbf{P}$ is nonsingular.

Based upon the Remark 5.2, we derive the following proposition.

Proposition 5.2. *Let* $\mathbf{P}$ *be an* $n \times n$ *stochastic matrix for which equation* (5.16) *holds.*

(i) If $a \neq b$, *then* $\mathbf{P}$ *is a nonsingular, doubly stochastic matrix and* $a + (n-1)b = 1$.
(ii) If $a = b$, *then* $\mathbf{P}$ *is a stable matrix and*

$$a = \frac{\sum_{j=1}^{n} \left(\sum_{i=1}^{n} p_{ij} \right)^2}{n^2}.$$

Proof. (i) From (5.16) we infer $\langle \mathbf{p}_i, \mathbf{p}_i \rangle = a$ and $\langle \mathbf{p}_i, \mathbf{p}_j \rangle = b$, $i, j = 1, 2, ..., n$, with $i \neq j$. Moreover, by Remark 5.2, the condition $a \neq b$ is equivalent to $a > b$, and consequently the vectors $\mathbf{p}_i$, $i = 1, 2, ..., n$, are linearly independent (by Lemma 5.2). Thus, $\det \mathbf{P} \neq 0$ and consequently $\mathbf{Q} = \mathbf{P}^{-1}$ exists.
By multiplying (5.16) from the left side with the matrix $\mathbf{Q} = \mathbf{P}^{-1}$, we get

$$\mathbf{Q} \cdot \mathbf{P}\mathbf{P}^T = \mathbf{Q} \cdot (a\mathbf{I} + b\mathbf{J})$$

or

$$\mathbf{P}^T = a\mathbf{Q} + b\mathbf{Q}\mathbf{J}.$$

Then

$$\mathbf{P}^T\mathbf{1} = a\mathbf{Q1} + b\mathbf{QJ1}.$$

Since $\mathbf{P1} = \mathbf{1}$ then $\mathbf{QP1} = \mathbf{Q1}$, and $\mathbf{1} = \mathbf{Q1}$. Hence,

$$\mathbf{P}^T\mathbf{1} = (a + (n-1)b)\mathbf{1}; \tag{5.20}$$

that is, the sum of the elements of each column of $\mathbf{P}$ is constant, equal to $a+(n-1)b$. From the relation $\mathbf{y}^T = \mathbf{x}^T\mathbf{P}$ we have for $\mathbf{x} = (1/n)\mathbf{1}$ that

$$\mathbf{y}^T = \left(\frac{1}{n}\right)\mathbf{1}^T\mathbf{P}$$

or

$$\mathbf{y}^T = \left(\frac{1}{n}\right)(\mathbf{P}^T\mathbf{1})^T.$$

Then, based on (5.20) we get

$$\mathbf{y}^T = \left(\frac{1}{n}\right)(a + (n-1)b)\mathbf{1}^T,$$

and consequently

$$\mathbf{y}^T\mathbf{1} = 1 = a + (n-1)b.$$

Thus, $\mathbf{P}$ is doubly stochastic.

(ii) By assuming that $\langle \mathbf{p}_i, \mathbf{p}_i \rangle = \langle \mathbf{p}_i, \mathbf{p}_j \rangle = a$, $i, j = 1, 2, ..., n$, we have that $|\mathbf{p}_i|^2 = |\mathbf{p}_i| \cdot |\mathbf{p}_j| \cdot \cos(\phi_{ij}) = a$ for $i, j = 1, 2, ..., n$, where ϕ_{ij} stands for the angle formed by the vectors $\mathbf{p}_i$ and $\mathbf{p}_j$. Since $|\mathbf{p}_i|^2 = a$, for $i = 1, 2, \ldots, n$, we conclude that $\cos(\phi_{ij}) = 1$, so the rows of $\mathbf{P}$ are parallel vectors. Since two stochastic parallel vectors are identical, we conclude that the rows of $\mathbf{P}$ are equal. Moreover, from (5.16) we have

$$\mathbf{PP}^T = a(\mathbf{I} + \mathbf{J}).$$

By multiplying the latter equation from the left and the right side with the vectors $\mathbf{1}^T$ and $\mathbf{1}$, respectively, we get

$$\mathbf{1}^T\mathbf{PP}^T\mathbf{1} = \mathbf{1}^T a(\mathbf{I} + \mathbf{J})\mathbf{1}$$

or

$$\sum_{j=1}^{n}\left(\sum_{i=1}^{n} p_{ij}\right)^2 = an^2,$$

or

$$a = \frac{\sum_{j=1}^{n}\left(\sum_{i=1}^{n} p_{ij}\right)^2}{n^2}.$$

□

Remark 5.3. Proposition 5.2 is not valid in the inverse direction; i.e., the assumption that the stochastic matrix $\mathbf{P}$ is doubly stochastic is necessary for (5.16) to be valid, but not sufficient. For example, the stochastic matrix

$$\mathbf{P} = \begin{bmatrix} 0.4 & 0.2 & 0.4 \\ 0.4 & 0.5 & 0.1 \\ 0.2 & 0.3 & 0.5 \end{bmatrix},$$

is doubly stochastic, but

$$\mathbf{P}\mathbf{P}^T = \begin{bmatrix} 0.36 & 0.3 & 0.34 \\ 0.3 & 0.42 & 0.28 \\ 0.34 & 0.28 & 0.38 \end{bmatrix},$$

is not of the form (5.16). So the doubly stochastic matrices $\mathbf{P}$ which satisfy (5.16) constitute a subclass of the class of the doubly stochastic matrices. Moreover, note that the class of the matrices $\mathbf{P}\mathbf{P}^T$ of the form (5.16) is single-parametric because of the condition $a + (n-1)b = 1$ proved in Proposition 5.2 (i).

Lemma 5.4. *If an $n \times n$ stochastic matrix $\mathbf{P}$ satisfies formula (5.16), with $a > b \geq 0$, $n > 1$, then we have*

$$a = \frac{1 + (n-1)(\det \mathbf{P})^{2/(n-1)}}{n}, \tag{5.21}$$

and

$$b = \frac{1 - (\det \mathbf{P})^{2/(n-1)}}{n}. \tag{5.22}$$

Proof. It is known (Graybill, 1983) that

$$\det(a\mathbf{I} + b\mathbf{J}) = (a + (n-1)b)(a-b)^{n-1}.$$

Since $a + (n-1)b = 1$ (by Proposition 5.2(i)), we get by (5.16),

$$\det(\mathbf{P}\mathbf{P}^T) = (\det \mathbf{P})^2 = \det(a\mathbf{I} + b\mathbf{J}) = (a-b)^{n-1},$$

and using again the result $a + (n-1)b = 1$, (5.21) and (5.22) follow readily. □

We can perceive that the last two relations are valid also if $a = b$, since thereupon it results by Proposition 5.2(ii) that $\mathbf{P}$ is a stable matrix and consequently $\det \mathbf{P} = 0$. Moreover they are valid in the case $a - b = 1$, since then, in combination with the relation $a + (n-1)b = 1$, we get $\mathbf{P}\mathbf{P}^T = \mathbf{I}$ and therefore $(\det \mathbf{P})^2 = 1$.

Lemma 5.5. *Sufficient and necessary conditions for an $n \times n$ stochastic matrix $\mathbf{P}$ to satisfy* 5.16, *that is,*

$$\mathbf{P}\mathbf{P}^T = (\langle \mathbf{p}_i, \mathbf{p}_j \rangle) = a\mathbf{I} + b\mathbf{J}, \quad \textit{with } a > b \geq 0,$$

is for $\mathbf{P}$ to be nonsingular, doubly stochastic, and the product $\mathbf{P}\mathbf{P}^T$ to have the eigenvalues $\lambda_2, \lambda_3, \ldots, \lambda_n$ equal to some $\lambda \in (0,1]$, while $\lambda_1 = 1$.

Proof. (i) Necessary conditions:

By Proposition 5.2(i), $\mathbf{P}$ is nonsingular and doubly stochastic. It is also known (Graybill, 1983) that

$$\det(\mathbf{P}\mathbf{P}^T - \lambda\mathbf{I}) = (a + (n-1)b - \lambda)(a - b - \lambda)^{n-1},$$

and since $a + (n-1)b = 1$, the eigenvalues of $\mathbf{P}\mathbf{P}^T$ are $\lambda_1 = 1$, and $\lambda_2 = \lambda_3 = \cdots = \lambda_n = 1 - nb = \lambda$. Now, using (5.22), we get $\lambda = (\det \mathbf{P})^{2/(n-1)}$ and consequently $\lambda \in (0,1]$.

(ii) Sufficient conditions:
Let $\mathbf{P}$ be an $n \times n$ nonsingular, doubly stochastic matrix and $\mathbf{PP}^T$ have the eigenvalues $\lambda_1 = 1$ and $\lambda_2 = \lambda_3 = \cdots = \lambda_n = \lambda \in (0,1])$. It is known by Lemma 5.1 that $\mathbf{PP}^T$ has 1 as an eigenvalue and that its eigenvalues are identical with its singular values. Now, consider the Jordan canonical form (5.8) of $\mathbf{PP}^T$; i.e.,

$$\mathbf{PP}^T = \mathbf{V\Lambda V}^T,$$

where

$$\mathbf{\Lambda} = \begin{bmatrix} 1 & 0 & .. & 0 \\ 0 & \lambda & .. & 0 \\ ... & .. & .. & .. \\ 0 & 0 & .. & \lambda \end{bmatrix}.$$

Since $\mathbf{PP}^T$ is symmetric, the matrix $\mathbf{V} = (v_{ij})$ can be chosen to be unitary; that is, $\mathbf{V}^{-1} = \mathbf{V}^T$. Then

$$\mathbf{PP}^T = \mathbf{V\Lambda V}^T = \begin{bmatrix} v_{11} & v_{12} & .. & v_{1n} \\ v_{21} & v_{22} & .. & v_{2n} \\ ... & .. & .. & .. \\ v_{n1} & v_{n2} & .. & v_{nn} \end{bmatrix} \cdot \begin{bmatrix} 1 & 0 & .. & 0 \\ 0 & \lambda & .. & 0 \\ ... & .. & .. & .. \\ 0 & 0 & .. & \lambda \end{bmatrix} \cdot \begin{bmatrix} v_{11} & v_{21} & .. & v_{n1} \\ v_{12} & v_{22} & .. & v_{n2} \\ ... & .. & .. & .. \\ v_{1n} & v_{2n} & .. & v_{nn} \end{bmatrix},$$

or

$$\mathbf{PP}^T = \mathbf{V\Lambda V}^T = \begin{bmatrix} v_{11} & \lambda v_{12} & .. & \lambda v_{1n} \\ v_{21} & \lambda v_{22} & .. & \lambda v_{2n} \\ ... & .. & .. & .. \\ v_{n1} & \lambda v_{n2} & .. & \lambda v_{nn} \end{bmatrix} \cdot \begin{bmatrix} v_{11} & v_{21} & .. & v_{n1} \\ v_{12} & v_{22} & .. & v_{n2} \\ ... & .. & .. & .. \\ v_{1n} & v_{2n} & .. & v_{nn} \end{bmatrix},$$

or

$$\mathbf{PP}^T = \begin{bmatrix} v_{11}^2 + \lambda\sum_{i=2}^n v_{1i}^2 & v_{11}v_{21} + \lambda\sum_{i=2}^n v_{1i}v_{2i} & .. & v_{11}v_{n1} + \lambda\sum_{i=2}^n v_{1i}v_{ni} \\ v_{11}v_{21} + \lambda\sum_{i=2}^n v_{1i}v_{2i} & v_{21}^2 + \lambda\sum_{i=2}^n v_{2i}^2 & .. & v_{21}v_{n1} + \lambda\sum_{i=2}^n v_{2i}v_{ni} \\ ... & .. & .. & .. \\ v_{11}v_{n1} + \lambda\sum_{i=2}^n v_{1i}v_{ni} & v_{21}v_{n1} + \lambda\sum_{i=2}^n v_{2i}v_{ni} & .. & v_{n1}^2 + \lambda\sum_{i=2}^n v_{ni}^2 \end{bmatrix},$$

or

$$\mathbf{PP}^T$$
$$= \begin{bmatrix} (1-\lambda)v_{11}^2 + \lambda\sum_{i=1}^n v_{1i}^2 & (1-\lambda)v_{11}v_{21} + \lambda\sum_{i=1}^n v_{1i}v_{2i} & .. & (1-\lambda)v_{11}v_{n1} + \lambda\sum_{i=1}^n v_{1i}v_{ni} \\ (1-\lambda)v_{11}v_{21} + \lambda\sum_{i=1}^n v_{1i}v_{2i} & (1-\lambda)v_{21}^2 + \lambda\sum_{i=1}^n v_{2i}^2 & .. & (1-\lambda)v_{21}v_{n1} + \lambda\sum_{i=1}^n v_{2i}v_{ni} \\ ... & .. & .. & .. \\ (1-\lambda)v_{11}v_{n1} + \lambda\sum_{i=1}^n v_{1i}v_{ni} & (1-\lambda)v_{21}v_{n1} + \lambda\sum_{i=1}^n v_{2i}v_{ni} & .. & (1-\lambda)v_{n1}^2 + \lambda\sum_{i=1}^n v_{ni}^2 \end{bmatrix}.$$

Since $\mathbf{V}^{-1} = \mathbf{V}^T = \left(v_{ij}^T\right)$ or equivalently $\mathbf{VV}^T = \mathbf{I}$, we have

$$\sum_{i=1}^n v_{ki}v_{is}^T = \sum_{i=1}^n v_{ki}v_{si} = \delta_{ks},$$

thus

$$\mathbf{PP}^T = \begin{bmatrix} (1-\lambda)v_{11}^2+\lambda & (1-\lambda)v_{11}v_{21} & .. & (1-\lambda)v_{11}v_{n1} \\ (1-\lambda)v_{11}v_{21} & (1-\lambda)v_{21}^2+\lambda & .. & (1-\lambda)v_{21}v_{n1} \\ ... & .. & .. & .. \\ (1-\lambda)v_{11}v_{n1} & (1-\lambda)v_{21}v_{n1} & .. & (1-\lambda)v_{n1}^2+\lambda \end{bmatrix}.$$

Given that the vector $\mathbf{v}_1 = (v_{11}, v_{21}, ..., v_{n1})^T$ is a left and right eigenvector of $\mathbf{PP}^T$ for the eigenvalue 1, and $\mathbf{V}$ is a unitary matrix, we conclude that

$$\mathbf{v}_1 = \left(\frac{1}{\sqrt{n}}, \frac{1}{\sqrt{n}}, ..., \frac{1}{\sqrt{n}}\right)^T$$

(as stated in the proof of Proposition 5.1), therefore

$$\mathbf{PP}^T = \begin{bmatrix} (1-\lambda)\frac{1}{n}+\lambda & (1-\lambda)\frac{1}{n} & .. & (1-\lambda)\frac{1}{n} \\ (1-\lambda)\frac{1}{n} & (1-\lambda)\frac{1}{n}+\lambda & .. & (1-\lambda)\frac{1}{n} \\ ... & .. & .. & .. \\ (1-\lambda)\frac{1}{n} & (1-\lambda)\frac{1}{n} & .. & (1-\lambda)\frac{1}{n}+\lambda \end{bmatrix},$$

or

$$\mathbf{PP}^T = \left((1-\lambda)\frac{1}{n}+\lambda\right)\mathbf{I} + \left((1-\lambda)\frac{1}{n}\right)\mathbf{J}.$$

Thus, for $a = (1-\lambda)(1/n)+\lambda$ and $b = (1-\lambda)(1/n)$ we have $\mathbf{PP}^T = (\langle \mathbf{p}_i, \mathbf{p}_j \rangle) = a\mathbf{I} + b\mathbf{J}$, with $a > b \geq 0$. □

A numerical example. Let a homogeneous discrete-time MC with state space $S = \{1,\ 2,\ 3,\ 4\}$, transition matrix

$$\mathbf{P} = \begin{bmatrix} 0.161452 & 0.113666 & 0.584123 & 0.140759 \\ 0.541301 & 0.299853 & 0.126677 & 0.032169 \\ 0.013071 & 0.543429 & 0.166215 & 0.277285 \\ 0.284176 & 0.043052 & 0.122985 & 0.549787 \end{bmatrix},$$

and consider the equation of the hypersphere

$$(x_1 - 0.2)^2 + (x_2 - 0.3)^2 + (x_3 - 0.4)^2 + (x_4 - 0.1)^2 = 0.0144. \tag{5.23}$$

For $x_i \geq 0$, $i = 1, ..., 4$, the radius vectors of the points (x_1, x_2, x_3, x_4) in (5.23), are assumed to represent probability state vectors of the MC at some time point $t-1$, $t \in \mathbb{N}^+$. We are interested in the form of the image, via $\mathbf{P}$, of the set of the points (x_1, x_2, x_3, x_4) lying on the hypersphere (5.23). We note that $\mathbf{P}$ is doubly stochastic and that

$$\mathbf{PP}^T = \begin{bmatrix} 0.4 & 0.2 & 0.2 & 0.2 \\ 0.2 & 0.4 & 0.2 & 0.2 \\ 0.2 & 0.2 & 0.4 & 0.2 \\ 0.2 & 0.2 & 0.2 & 0.4 \end{bmatrix} = 0.4\mathbf{I} + 0.2\mathbf{J};$$

i.e., $\mathbf{PP}^T$ obeys the form $a\mathbf{I} + b\mathbf{J}$, $a > b \geq 0$.

The image via $\mathbf{P}$ of the hypersphere (5.23) is

$$4x_1^2 + 4x_2^2 + 4x_3^2 + 4x_4^2 - 2x_1x_2 - 2x_1x_3 - 2x_1x_4 - 2x_2x_3 - 2x_2x_4 - 2x_3x_4 - 0.283267x_1 - 1.34366x_2 - 0.336121x_3 - 0.0369534x_4 + 0.2856 = 0. \quad (5.24)$$

and the points (x_1, x_2, x_3, x_4) in (5.24), now express for $x_i \geq 0$, $i = 1, ..., 4$, points whose radius vectors are the state vectors of the MC at time t. Equation (5.24) represents, by Proposition 5.1 and Lemma 5.5, a hypersphere. This fact can be easily verified, since (5.24) can be written as

$$(x_1 - 0.228327)^2 + (x_2 - 0.334366)^2 + (x_3 - 0.233612)^2 + (x_4 - 0.203695)^2 = 0.00288.$$

Acknowledgements. This research was partly supported by the Archimedes program of the Technological Institution of West Macedonia, Department of General Sciences, Koila Kozanis, Greece

References

Bartholomew, D.J. (1982). *Stochastic Models for Social Processes*, 3rd ed., Wiley, New York.

Gani, J. (1963). Formulae for projecting enrolments and degrees awarded in universities, *J. R. Stat. Soc., Ser. A.* 126, 400–409.

Graybill, F.A. (1983). *Matrices with Applications in Statistics*, 2nd ed., Wadsworth, Belmont, CA.

Iosifescu, M. (1980). *Finite Markov Processes and Their Applications,* Wiley, New York.

McClean, S.I., B. McAlea, and Millard, P. (1998). Using a Markov reward model to estimate spend-down costs for a geriatric department, *J. Oper. Res. Soc.* 10, 1021–1025.

Patoucheas, P.D. and Stamou, G. (1993). Non-homogeneous Markovian models in ecological modelling: a study of the zoobenthos dynamics in Thermaikos Gulf, Greece, *Ecol. Model.* 66, 197–215.

Taylor, G.J., S.I. McClean, and Millard, P. (2000). Stochastic model of geriatric patient bed occupancy behaviour, *J. R. Stat. Soc.* 163(1), 39–48.

Tsaklidis, G. (1994). The evolution of the attainable structures of a homogeneous Markov system with fixed size, *J. Appl. Probab.* 31, 348–361.

Vassiliou, P.-C.G. (1982). Asymptotic behaviour of Markov systems, *J. Appl. Probab.* 19, 851–857 .

Vassiliou P.-C.G. (1997). The evolution of the theory of non-homogeneous Markov systems, *Appl. Stochastic. Models Data Anal.* 13(3–4), 159–176.

Part V

Life Table Data, Survival Analysis, and Risk in Household Insurance

18

Comparing the Gompertz-Type Models with a First Passage Time Density Model

Christos H. Skiadas[1] and Charilaos Skiadas[2]

[1] Technical University of Crete, Greece
[2] Hanover College, Indiana, USA

Abstract: In this chapter we derive and analyse Gompertz-type probability density functions and compare these functions to a first passage time density function. The resulting Gompertz-type pdfs are mirror images of each other, each skewed in a specific direction, whereas the first passage-type model gives functions with both left and right skewness depending on parameter values. We apply these pdfs to the life table data for females in the United States, 2004, and to the medfly data provided in Carey et al. (1992). Our application shows that the mortality data in the two cases have opposite skewness. The results support that the underlying mortality mechanism is qualitatively different in the cases.

Keywords and phrases: Gompertz, dynamic model, probability density function, life table data, Carey data, Weibull

18.1 Introduction

There is an extensive bibliography concerning the famous Gompertz model and its applications to life table data. The questions raised by Robine and Ritchie (1993) in response to Carey et al. (1992) suggest comparing the medfly life span modelling data to the human life span data.

As Carey et al. (1992) state, the experiment was mainly designed to explain longevity. However, the majority of the data collected, for more than $1,200,000$ medflies, are not appropriate to support longevity studies, but instead are mainly to be used for explaining the mortality law for medflies. It is worth noting that the medfly data were easily verified as following the Gompertz probability density function. Later on, Weitz and Fraser (2001) suggested a first passage probability density function to model the medfly data. The proposed model type was first derived independently by Schrödinger (1915) and Smoluchowsky (1915), and in a more recent form by Siegert (1951), and it is known as the *inverse Gaussian distribution.* In recent years we can find various applications of this model to lifetime data analysis and reliability, and other fields.

C.H. Skiadas (ed.), *Advances in Data Analysis*, Statistics for Industry and Technology, DOI 10.1007/978-0-8176-4799-5_18,

A more general *first passage density function* was proposed by Janssen and Skiadas (1995) to model the human life table data. This model was applied to the data provided from the life table records of Belgium and France, and it gave a very good fitting. However, it was a model difficult to work with, as it contained many parameters. A simpler first exit time density function was proposed by Skiadas (2006) and Skiadas and Skiadas (2007), with applications to life table data from Greece. The special case of a quadratic health state function was discussed in Skiadas et al. (2007). The first exit (or *hitting*) time density has the form:

$$g_{\mathrm{DM}}(t) = c(kt) - ({}^{(3/2)}e)^{-\left(\left(\ell-(kt)^b\right)^2/2t\right)}, \tag{18.1}$$

where c, ℓ, and k are parameters, and b is a constant mainly related to the skewness of the probability density function. The (simpler) inverse Gaussian distribution model corresponds to the case where $b = 1$.

The model (18.1) was tested using the life table data from Greece (1992–2000), and it showed very good fitting. More important, the term

$$H(t) = \ell - (kt)^b, \tag{18.2}$$

called the *health state function* in Skiadas and Skiadas (2007), provides useful information on the mortality data applied. Namely, it describes the perceived average health state of an individual for a given age. Equation (18.2) indicates a health state that decays over time, slowly at first and faster as the age of individuals approaches its natural limits.

18.2 The Gompertz-type models

The Gompertz model was first proposed by Gompertz (1825), and a thorough analysis of it can be found in Winsor (1932). In differential equation form, the model has the equation

$$(\ln x)' = -b \ln x, \tag{18.3}$$

or equivalently

$$\dot{x} = -bx \ln x, \tag{18.4}$$

where x is a function of time t, and b is a positive parameter expressing the rate of growth of the system. Without loss of generality the function x can be assumed bounded ($0 < x \leq 1$, with $x = 1$ corresponding to the entire population), so that $\dot{x}$ is the probability density function of the growth process. Direct integration of (18.4) gives as solution the *Gompertz function*:

$$x = e^{\ln(x_0)e^{-bt}} \tag{18.5}$$

The probability density function of the Gompertz model is then given by

$$g(t) = \dot{x} = -b\ln(x_0)e^{-bt}e^{\ln(x_0)e^{-bt}}. \tag{18.6}$$

An interesting variant of the Gompertz function arises when we replace x by $1-x$ in the right side of the Gompertz differential equation, resulting in a mirror image of the Gompertz model:

$$\dot{x} = -b(1-x)\ln(1-x). \tag{18.7}$$

Direct integration gives as solution the *mirror Gompertz function*:

$$x = 1 - e^{\ln(1-x_0)e^{bt}}. \tag{18.8}$$

The probability density function of the mirror Gompertz model is then given by

$$g_{\mathrm{MGM}}(t) = \dot{x} = -b\ln(1-x_0)e^{bt}e^{\ln(x_0)e^{bt}}. \tag{18.9}$$

This model arises by considering the *relative decay* of the system, instead of the *relative growth.* It has skewness opposite that of the Gompertz model, as now the maximum growth rate is achieved when

$$x = 1 - \frac{1}{e}$$

instead of when $x = 1/e$. A comparison of the two models is given in Figure 18.1, with the mirror Gompertz model appearing in gray. Both models are referred to in the literature as "the Gompertz model", with different disciplines preferring one model over the other. The second variant is favoured in the actuarial sciences, as it is more intimately related to mortality.

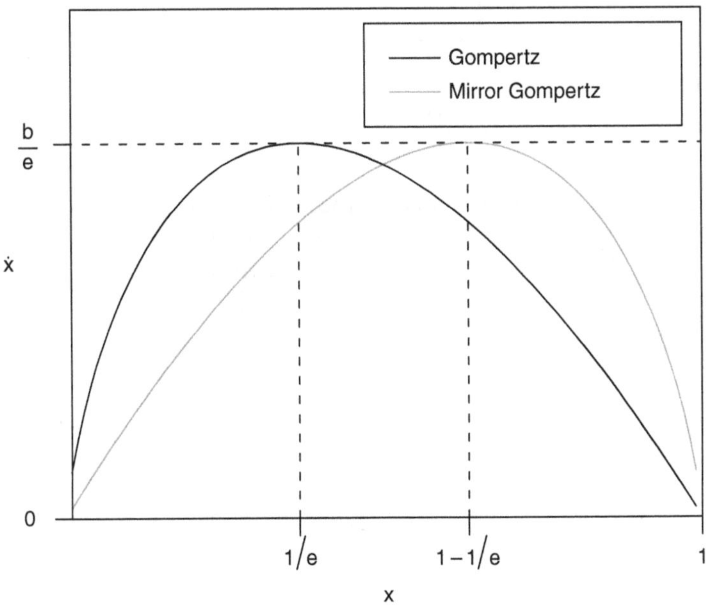

Figure 18.1. The two Gompertz-type models

18.3 Application to life table and the Carey medfly data

The Carey data are provided in his famous *Science* paper (Carey et al., 1992). Since then, several papers with further analyses and applications have appeared. The data used in this study are selected from a laboratory experiment where the life span of 1, 203, 646 medflies was measured.

Weitz and Fraser (2001) used the medfly data to test the inverse Gaussian distribution as a model resulting from the *first exit time theory*. The fitting of this model was quite good. In the present study we fit the more general model given by equation (18.1) to the data, and we test whether the exponent b diverges from unity ($b = 1$ being exactly the case studied in Weitz and Fraser, 2001). In the same study we test the argument of Robine and Ritchie (1993) regarding a comparative study of the medfly life span and the human life span. The United States 2004 life table data for females was used for the comparative study. Four models are tested. The equations used for the data fit are:

$$g_{\mathrm{DM}}(t) = c(kt)^{-3/2} e^{-(\ell-(kt)^b)^2/2t} \qquad \text{(dynamic model)}$$

$$g_{\mathrm{G}}(t) = ce^{-kt} e^{-\ell e^{-kt}} \qquad \text{(Gompertz model)}$$

$$g_{\mathrm{MG}}(t) = ce^{kt} e^{-\ell e^{kt}} \qquad \text{(mirror Gompertz)}$$

$$g_{\mathrm{W}}(t) = c(kt)^{\ell-1} e^{-(kt)^{\ell}}. \qquad \text{(Weibull)}$$

For the dynamic model, the parameter b was 1.4 for the Medfly data and $b = 5.76$ for the life table data for females 2004 in the United States.

Table 18.1. Fit comparison for USA 2004, females

	USA Data Females 2004 Fit			
Model	c	k	ℓ	MSE (10^{-4})
Dynamic Model	0.0724	0.01875	17.318	5.85
Mirror Gompertz	0.0000221	0.09776	0.000241	13.71
Weibull	0.09095	0.01167	8.64	20.17

The results are summarized in Table 18.1 for USA 2004 females. As can be seen from Table 18.1, the best fit is done by the dynamic model with parameter $b = 5.76$ and a mean squared error MSE = 5.85, followed by the mirror Gompertz with a MSE = 13.71, and finally the Weibull model with MSE = 20.17.

Table 18.2. Fit comparison for Carey medfly data

	Carey Medfly Data Fit			
Model	c	k	ℓ	MSE (10^{-4})
Dynamic Model	0.396	0.2085	8.193	11.56
Gompertz	1.31335	0.13715	9.61624	9.68
Weibull	0.1197	0.046596	2.656	16.91

For the Carey medfly data the fitting results are summarized in Table 18.2. The best model fit is now given by the Gompertz model, followed closely by the dynamic model with $b = 1.4$. The Weibull model performs much worse. Regarding the Weitz and Fraser (2001) application with $b = 1$, the resulting fitting error is MSE $= 13.59$, that is, higher than both the Gompertz and the dynamic models. The estimated parameters for the Weitz and Fraser application are $c = 1.32336$, $k = 0.47726$, and $\ell = 10.2200$ by following the method of nonlinear least squares estimation applied here. The Weibull model, in both cases presented here, did not give results as good as the Gompertz and the dynamic models.

Figure 18.2 illustrates the fit comparison between the medfly data and the USA 2004 females data by using the Gompertz and mirror Gompertz models, respectively (gray lines), as well as the proposed dynamic model (black line). The timescale was rescaled according to the method proposed by Robine and Ritchie (1993). It is clear that the medfly data are very well presented by the Gompertz model, whereas the human mortality data are well expressed by the mirror Gompertz.

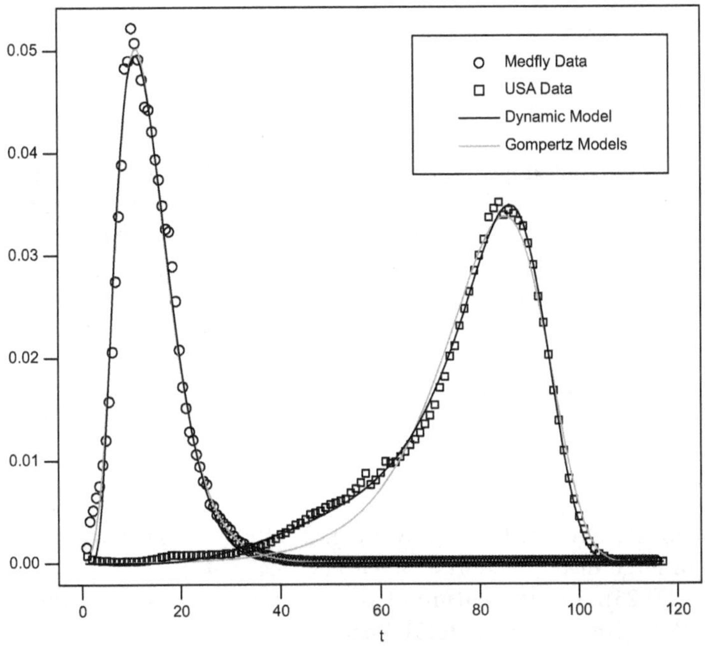

Figure 18.2. Gompertz, mirror Gompertz, and dynamic models applied to the medfly and USA 2004 female data

18.4 Remarks

According to a theory based on the *tangent approximation* (see Jennen, 1985; Jennen and Lerche, 1981) the hitting time distribution for the case of the Health State Process $H(t)$ presented above, is approximated by:

$$\begin{aligned} g(t) &= \frac{|H(t) - tH'(t)|}{t} p(t) \\ &= \frac{|H(t) - tH'(t)|}{\sqrt{2\pi\sigma^2 t^3}} \exp\left(-\frac{H(t)^2}{2\sigma^2 t}\right), \end{aligned} \tag{18.10}$$

where $p(t)$ is the transition probability density function. Based on this distribution, we propose that the life table data be modelled with the function:

$$g(t) = \frac{|\ell + (b-1)(kt)^b|}{\sqrt{2\pi\sigma^2 t^3}} \exp\left(-\frac{(\ell - (kt)^b)^2}{2\sigma^2 t}\right). \tag{18.11}$$

This function offers fits very similar to those of the model suggested here, and has better simulation properties, that will be examined in upcoming papers.

18.5 Conclusion

In this chapter we presented a comparative study including two Gompertz-type models and a dynamic model. The Weibull model was also tested in the applications. The application of the Gompertz and mirror Gompertz, and of the dynamic model, to explain the behavior of mortality data was very promising, both from a fitting point of view, but also from an explanatory point of view.

References

Carey, J., Liedo, P., Orozco, D., and Waupel, J. (1992). Slowing of mortality rates at older ages in large medfly cohorts. *Science*, 258:457–461.

Gompertz, B. (1825). On the nature of the function expressing of the law of human mortality. *Phil. Trans. R Soc.*, 36:513–585.

Janssen, J. and Skiadas, C. (1995). Dynamic modelling of life-table data. *Appl. Stoch. Models Data Anal.*, 11(1):35–49.

Jennen, C. (1985). Second-order approximation for Brownian first exit distributions. *Ann. Probab.*, 13:126–144.

Jennen, C. and Lerche, H. R. (1981). First exit densities of Brownian motion through one-sided moving boundaries. *Z. Wahrsch.uerw. Gebiete*, 55:133–148.

Robine, J. and Ritchie, K. (1993). Explaining fruit fly longevity. *Science*, 260:1665.

Schrödinger, E. (1915). Zür theorie der fall- und steigversuche an teilchenn mit brownsche bewegung. *Phys. Zeit.*, 16:289–295.

Siegert, A. (1951). On the first passage time probability problem. *Phys. Rev.*, 81: 617–623.

Skiadas, C. (2006). Stochastic modeling of Greek life table data. *Commun. Dependability Qual. Manag.*, 9(3):14–21.

Skiadas, C., Matalliotakis, G., and Skiadas, C. (2007). An extended quadratic health state function and the related density function for life table data. In Skiadas, C., editor, *Recent Advances in Stochastic Modeling and Data Analysis*, pp. 360–369. World Scientific, Singapore.

Skiadas, C. and Skiadas, C. (2007). A modeling approach to life table data. In Skiadas, C., editor, *Recent Advances in Stochastic Modeling and Data Analysis*, pp. 350–359. World Scientific, Singapore.

Smoluchowsky, M. (1915). Notiz über die berechning der brownschen molkularbewegung bei des ehrenhaft-millikanchen versuchsanordnung. *Phys. Zeit.*, 16: 318–321.

Weitz, J. and Fraser, H. (2001). Explaining mortality rate plateaus. *Proc. Natl. Acad. Sci.*, 98(26):15383–15386.

Winsor, C. (1932). The Gompertz curve as a growth curve. *Proc. Natl. Acad. Sci.*, 18(1):1–8.

19

A Comparison of Recent Procedures in Weibull Mixture Testing

Karl Mosler and Lars Haferkamp

Universität zu Köln, D-50923 Köln, Germany (e-mail: mosler@statistik.uni-koeln.de)

Abstract: The chapter considers recent approaches to testing homogeneity in a finite mixture model, the modified likelihood ratio test (MLRT) of Chen et al. 2001, and the D-tests of Charnigo and Sun (2004). They are adapted to Weibull mixtures with and without a Weibull-to-exponential transformation of the data. Critical quantiles are calculated by simulation. To cope with the dependency of quantiles on the unknown shape parameter, a corrected D-statistics is implemented and explored. First results are given on the power of these tests in comparison with that of the ADDS test by Mosler and Scheicher (2008).

Keywords and phrases: Mixture diagnosis, survival analysis, unobserved heterogeneity, overdispersion, goodness-of-fit

19.1 Introduction

A practically important problem is to decide whether for given data a Weibull mixture specification should be preferred over a nonmixed Weibull model, that is, whether the data contain unobserved parameter heterogeneity. Various procedures have been proposed in the literature for this specification problem, among them graphical devices (Jiang and Murthy, 1995) and statistical tests (Mosler and Scheicher, 2008). For a comparison of these tests in exponential mixtures, see Mosler and Haferkamp (2007).

In this chapter we consider three recent approaches to testing mixture homogeneity: the modified likelihood ratio test (MLRT) of Chen et al. 2001, the D-test of Charnigo and Sun (2004), and the ADDS-test. The ADDS test is due to Mosler and Seidel (2001) for exponential and Mosler and Scheicher (2008) for Weibull mixtures.

The subsequent discussion focuses on Weibull scale mixtures that have at most two components and common shape parameter, i.e., on densities

$$\begin{aligned} &f(x;\beta_1,\beta_2,\pi_1,\gamma) \\ &= \pi_1 \frac{\gamma}{\beta_1}\left(\frac{x}{\beta_1}\right)^{\gamma-1} e^{-(x/\beta_1)^\gamma} + (1-\pi_1)\frac{\gamma}{\beta_2}\left(\frac{x}{\beta_2}\right)^{\gamma-1} e^{-(x/\beta_2)^\gamma}, \end{aligned} \tag{19.1}$$

C.H. Skiadas (ed.), *Advances in Data Analysis*, Statistics for Industry and Technology, DOI 10.1007/978-0-8176-4799-5_19,

for $x \geq 0$ and parameters $\gamma, \beta_1, \beta_2 > 0$, $0 < \pi_1 < 1$. If $\gamma = 1$, an exponential mixture arises.

To test for mixture homogeneity, consider a random sample $X_1, \ldots, X_n$ from (19.1). The alternative hypothesis corresponds to $\beta_1 \neq \beta_2$, while the null is signified by $\beta_0 = \beta_1 = \beta_2$ and an arbitrary $\pi_1 \in]0, 1[$, say $\pi_1 = \frac{1}{2}$.

19.2 Three approaches for testing homogeneity

The MLRT of Chen et al. 2001 is a penalized likelihood ratio test. It is based on the usual log-likelihood plus a penalty term,

$$l^M(\beta_1, \beta_2, \pi_1, \gamma) = \sum_{i=1}^{n} \log f(\beta_1, \beta_2, \pi_1, \gamma) + C \log[4\pi_1(1 - \pi_1)] \,. \tag{19.2}$$

Here, $C > 0$ is a constant that weighs the penalty. Following previous authors (Charnigo and Sun, 2004) we use a fixed constant $C = \log 10$. In a first step, $l^M(\beta_1, \beta_2, \pi_1, \gamma)$ is maximized to obtain estimators $\hat{\beta}_1^M, \hat{\beta}_2^M, \hat{\pi}_1^M$, and $\hat{\gamma}^M$. Similarly, under H_0, $l^M(\beta_0, \beta_0, \frac{1}{2}, \gamma_0)$ is maximized to obtain estimators $\hat{\beta}_0^M, \hat{\gamma}_0^M$. In a second step, the test statistic

$$MLRT = 2\left[l^M(\hat{\beta}_1^M, \hat{\beta}_2^M, \hat{\pi}_1^M, \hat{\gamma}^M) - l^M(\hat{\beta}_0^M, \hat{\beta}_0^M, \frac{1}{2}, \hat{\gamma}_0^M)\right] \tag{19.3}$$

is evaluated. Asymptotically, under H_0 and some regularity conditions, $MLRT$ is distributed as the fifty–fifty mixture of a χ_1^2 variable and a constant at 0 (Chen et al., 2001). Note that in the special case of exponential mixtures, the MLRT statistic is (19.3) with both $\hat{\gamma}^M$ and $\hat{\gamma}_0^M$ substituted by the constant 1.

To solve the Weibull homogeneity problem, the MLRT can be used in two different ways, as described above and after a Wei2Exp transformation of the data. Either the four parameters under H_1 and the two under H_0 are estimated via penalized likelihood (19.2) and the statistic (19.3) is used as it stands, or, alternatively, γ_0 is estimated under H_0, and the estimate $\hat{\gamma}_0$ is used to transform the data, $X_i \mapsto X_i^{\hat{\gamma}}$. Then, from the transformed data, the parameters β_1, β_2, and π_1 are estimated via penalized likelihood (19.2) with $\gamma = 1$ and the MLRT statistic is evaluated.

The D-test, in its original form, measures the area between two densities, one fitted under H_0, and the other fitted under H_1. Firstly, the parameters of the null distribution and the alternative distribution are estimated by some consistent estimators $\hat{\beta}_0, \hat{\gamma}_0, \hat{\beta}_1, \hat{\beta}_2, \hat{\pi}_1$, and $\hat{\gamma}$. Then the D-statistic

$$D = \int_0^\infty \left[f(x; \hat{\beta}_1, \hat{\beta}_2, \hat{\pi}_1, \hat{\gamma}) - f(x; \hat{\beta}_0, \hat{\beta}_0, \frac{1}{2}, \hat{\gamma}_0)\right]^2 w(x)dx \tag{19.4}$$

$$= \sum_{i=0}^{2}\sum_{j=0}^{2} \hat{\pi}_i \, \hat{\pi}_j \frac{\hat{\gamma}_i \hat{\gamma}_j}{\hat{\beta}_i^{\hat{\gamma}_i} \hat{\beta}_j^{\hat{\gamma}_j}}$$

$$\times \int_0^\infty x^{\hat{\gamma}_i - 1} x^{\hat{\gamma}_j - 1} \exp\left(-\left(\frac{x}{\hat{\beta}_i}\right)^{\hat{\gamma}_i} - \left(\frac{x}{\hat{\beta}_j}\right)^{\hat{\gamma}_j}\right) w(x)dx \,, \tag{19.5}$$

where the notation $\hat{\pi}_0 = -1$ and $\hat{\pi}_2 = 1 - \hat{\pi}_1$ is used.

In the special case of an exponential mixture model, D is the same with the constant 1 in place of $\hat{\gamma}_0$ and $\hat{\gamma}_1$. In this case, as Charnigo and Sun (2004) show, D has an asymptotic null distribution which is equivariant to β_0. They provide tables of critical quantiles and report that, in the diagnosis of exponential scale mixtures, the simple D-test is slightly outperformed by the MLRT when n is small ($n \leq 100$). Charnigo and Sun (2004) therefore propose weighted forms of the D-test which put more weight to differences in the tails of the two densities: in place of the differential dx in the integral formula (19.5) they use $x\,dx$ or $x^2\,dx$. In the sequel, these weighted variants of the D-test are signified by 'w1D' and 'w2D', respectively.

Like the MLRT, the D-test can be employed either to the original data from a Weibull mixture model or to transformed data that have been subject to a Wei2Exp transformation with an estimated shape parameter $\hat{\gamma}_0$. In Section 19.3 we investigate the effect of estimating γ_0 on the critical regions of the D-tests and the MLRT.

The ADDS test combines a dispersion score (DS) test with a classical goodness-of-fit test. The DS statistic,

$$DS = \left(\frac{n(n-1)}{n+1}\right)^{1/2} \frac{1}{(\overline{X})^2} \left[S^2 - \frac{1}{2n}\sum_{i=1}^{n} T_i^2\right], \tag{19.6}$$

is combined with a goodness-of-fit statistic of Anderson–Darling type,

$$AD = \left(1 + \frac{0.6}{n}\right)\left(n - \frac{1}{n}\sum_{i=1}^{n}(2i-1)\left(\log\left(1 - e^{-T_{(i)}/\overline{X}}\right) + \frac{T_{(i)}}{\overline{X}}\right)\right), \tag{19.7}$$

where $\overline{X}$ and S^2 denote the sample mean and variance and $T_{(i)}$ is the ith order statistic. Reject H_0 if either DS or AD is too large.

Under H_0, both test statistics do not depend on the scale parameter β. Mosler and Seidel (2001) have demonstrated that the power of the ADDS test for exponential mixtures is always at least comparable to that of a bootstrap LRT, a moment LRT, and a DS test. On large classes of alternatives the tests are outperformed by the ADDS test.

19.3 Implementing MLRT and D-tests with Weibull alternatives

In order to avoid estimation of all Weibull parameters under H_0 and H_1, we first employ Wei2Exp forms of the MLRT and the D-test. That is, the data $x_1, \ldots, x_n$ are transformed to $x_1^{\hat{\gamma}}, \ldots, x_n^{\hat{\gamma}}$, and MLRT and D-tests for homogeneity in exponential mixtures are done with the transformed data. We simulated the quantiles of each test with estimated γ under H_0.

In our second approach we apply the D-tests and the MLRT directly to the data. Chen et al. 2001 and Charnigo and Sun (2004) report implementations of their tests for exponential mixtures.[1] However, when implementing the D-tests in a Weibull

[1] We thank these authors for kindly giving us their computer codes.

mixture model, severe difficulties arise. One difficulty is in choosing a good estimation method for the parameters of a Weibull mixture distribution. The other is to find an approximate functional dependency of the relevant quantiles of the D-test statistic and the parameters of the null hypothesis to calculate proper critical values.

The statistic (19.5) was calculated by numerical integration.[2] Three weightings of the D-test were investigated, $w_1(t) = t$, $w_2(t) = t^2$, and $w_g(t) = t^{\hat{\gamma}}$. In estimating the parameters under H_0 and H_1, we used the *Nelder–Mead simplex algorithm* (Olsson, 1979).[3] The Nelder–Mead simplex algorithm is included in *R* and works together with some methods of the *MASS* package (Venables and Ripley, 2002). We used a multiple initial value procedure to avoid being trapped at local maxima. The procedure was also applied with the MLRT to estimate the parameters of the penalized likelihood function.

After all we found from our simulations that the critical quantiles of the D-test depend heavily on shape parameter γ_0 under the null (see Figure 19.1). The same is partially true for the MLRT. The ADDS test, in comparison, shows no relevant dependency on γ_0; with increasing γ_0 it becomes only slightly more conservative. To cope with this observed dependency on γ_0 and with an additional combined dependency on scale β_0 and level α, we introduced corrected statistics as follows,

$$T^* = T \cdot \frac{h(\hat{\beta}_0)}{i(\hat{\gamma}_0, \alpha)} \, .$$

Here, $h(\beta_0)$ is a function that is specific to each test, and $i(\gamma_0, \alpha)$ is an interpolation function obtained from simulation (with $n = 1000$) of the $(1 - \alpha)$-quantiles of the statistic with different shape parameters γ_0. The value of $i(\gamma_0, \alpha)$ has been determined by linear interpolation of the simulated quantiles of the shape parameters $\gamma_0 \in \{1, 1.5, 2, 3, 5\}$. We chose $h(\beta)$ as a linear function:

$$h(\beta) = \begin{cases} \beta, & \text{D-test}, \\ 1 + 0.3(\beta - 1), & \text{w1D}, \\ 1, & \text{w2D}, \\ 1 + 0.3(\beta - 1), & \text{wgD}, \\ 1 + 0.2(\beta - 1), & \text{MLRT}. \end{cases}$$

This function has been determined from simulated 95% quantiles of the various tests.

With the corrected statistics, critical quantiles have been determined by simulation for $n = 100, 1000$ and $\alpha = 0.1, 0.05, 0.01$. The number of replications was always 5000.

---

[2] The integration was done in the interval $[0.5x_{\min}, 2x_{\max}]$ by using the QUADPACK *R*-routines (adaptive quadrature of functions) from Piessens et al. (1983). $x_{\min}$ and $x_{\max}$ denote smallest and largest observations.

[3] We also tried the methods of Kaylan and Harris (1981) and Albert and Baxter (1995). The Nelder–Mead algorithm proved to be the fastest one; its likelihood comes near to that of the method of Kaylan and Harris. The *PAEM* of Albert and Baxter yielded a worse likelihood.

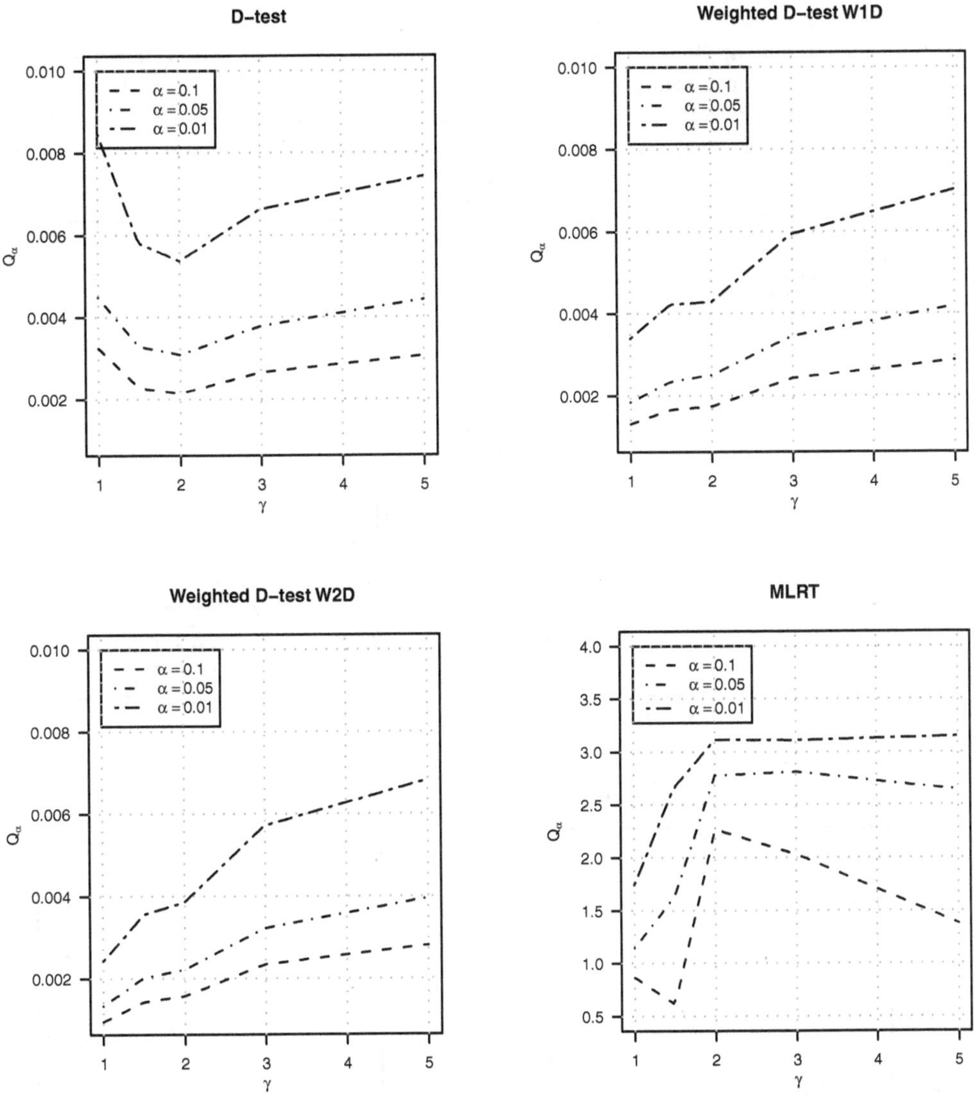

Figure 19.1. Dependency of critical quantile Q_α on the true shape parameter γ, for the D-test, the weighted D-tests (w1D, w2D), and the MLRT (without Wei2Exp transformation); $n = 1000$, $\alpha = 0.01, 0.05, 0.10$

19.4 Comparison of power

In a power simulation study we considered both forms of the MLRT and the D-tests, with and without employing a Wei2Exp transformation. In addition we used several weighted versions of the D-test. The power of these tests was evaluated under different alternatives and contrasted with that of the ADDS test.

Figure 19.2 exhibits the power of the nonweighted D-test, the w2D-test, and the MLRT, each being applied to Wei2Exp transformed data, and the ADDS test. The alternatives are lower contaminations, that is, mixtures of a Weibull distribution with another Weibull distribution having smaller scale. The main result of the simulation study is that under lower contamination the ADDS is the only one that develops

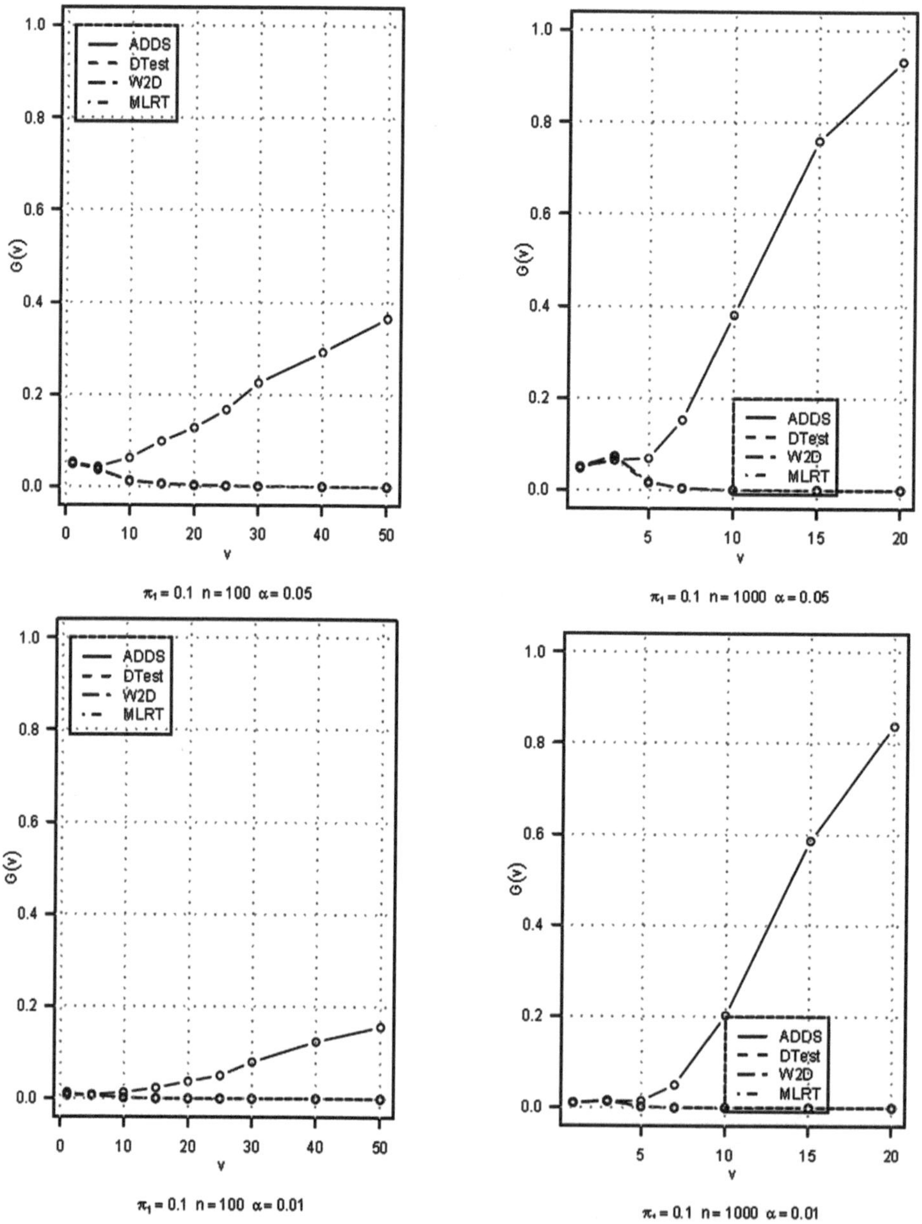

Figure 19.2. Power under lower contaminations: D-test, w2D-test (quadratically weighted D-test), and MLRT with Wei2Exp transformation, ADDS test. Comparison of power under the alternative $S(t) = 0.9\exp(-t^\gamma) + 0.1\exp(-(vt)^\gamma)$, depending on scale ratio $v \geq 1$

reasonable power, while the other three tests break completely down. It has been further shown that under upper contaminations, that is, mixtures with larger scaled Weibull distributions, the ADDS test is only slighty less powerful than the others.

A similar power comparison has been done for the MLRT and two D-tests that are applied to the nontransformed data. Here, it came out that under lower contaminations the nonweighted D-test outperforms the other three tests when the sample size is small ($n = 100$), while for larger samples ($n = 1000$) the four tests develop similar power. Under upper contaminations, the ADDS test proves to be better than the others, which develop poor power only. This holds true even for fifty–fifty mixtures.

The relatively poor performance of the MLRT and the D-tests may be attributed to parameter estimation on the alternative and, in addition, to the dependency of quantiles on parameters. The power of these tests will possibly improve if correction factors are determined in a less simple way. In particular, instead of employing the same linear function h for all α, one could employ some nonlinear interpolations depending on α. Also the number of replications ($R = 5000$) could be increased to obtain more precise results. However, it is rather obvious that the qualitative results of this chapter will not change.

19.5 Conclusion

In a nutshell: While the MLRT and the D-tests perform well in other models (Chen et al., 2001; Charnigo and Sun, 2004), their application appears to be not recommendable for homogeneity testing in a Weibull mixture model. Here, the ADDS test provides a reasonable diagnostic alternative.

References

J. R. G. Albert and L. A. Baxter. Applications of the EM algorithm to the analysis of life length data. *Applied Statistics*, 44:323–341, 1995.

R. Charnigo and J. Sun. Testing homogeneity in a mixture distribution via the l^2 distance between competing models. *Journal of the American Statistical Society*, 99:488–498, 2004.

H. Chen, J. Chen, and J. D. Kalbfleisch. A modified likelihood ratio test for homogeneity in finite mixture models. *Journal of the Royal Statistical Society*, B 63:19–29, 2001.

R. Jiang and D. N. P. Murthy. Modeling failure data by mixture of two Weibull distributions: A graphical approach. *IEEE Transactions on Reliability*, 41:241–247, 1995.

A. R. Kaylan and C. M. Harris. Efficient algorithms to derive maximum-likelihood estimates for finite exponential and Weibull mixtures. *Computers and Operations Research*, 8:97–104, 1981.

K. Mosler and L. Haferkamp. Size and power of recent tests for homogeneity in exponential mixtures. *Communications in Statistics–Simulation and Computation*, 36:493–504, 2007.

K. Mosler and C. Scheicher. Homogeneity testing in a Weibull mixture model. *Statistical Papers*, 49:315–332, 2008.

K. Mosler and W. Seidel. Testing for homogeneity in an exponential mixture model. *Australian and New Zealand Journal of Statistics*, 43:231–247, 2001.

D. M. Olsson. Estimation for mixtures of distributions by direct maximization of the likelihood function. *Journal of Quality Technology*, 11:153–159, 1979.

R. Piessens, E. deDoncker-Kapenga, C. Uberhuber, and D. Kahaner. *Quadpack: A Subroutine Package for Automatic Integration.* Springer Verlag, New York, 1983.

W. N. Venables and B. D. Ripley. *Modern Applied Statistics with S-PLUS.* Springer, New York, 4th edition, 2002.

20

Hierarchical Bayesian Modelling of Geographic Dependence of Risk in Household Insurance

László Márkus, N. Miklós Arató, and Vilmos Prokaj

Department of Probability Theory and Statistics, Eötvös Loránd University, Budapest, Hungary

Abstract: We analyse here the spatial dependence structure for the counts of certain type of claims occurring in household insurance in Hungary. We build up a Bayesian hierarchical model for the Gaussian Markov random field driven nonhomogeneous spatial Poisson process of claims counts. Model estimation is carried out by the MCMC technique, by applying sparse matrices, and a novel approach for updates by radial and angular components to obtain better mixing in over 3000 dimensions. We design a procedure that tailors the acceptance rate automatically during burn-in to a prescribed (optimal) value while updating the chain.

Keywords and phrases: Bayesian hierarchical models, disease-mapping, Gaussian Markov random fields, household insurance risk, Markov chain Monte Carlo (MCMC), spatial statistics

20.1 Introduction

It is a common practice in household, automobile, etc. insurance to charge the risk premium per unit exposure time according to the geographical area to which the contract belongs. It means the premium changes geographically even when all risk factors (other than location) are the same. However, companies apply various principles leading to very different spatial dependences in premium rating. More often than not these principles reflect common sense approaches, rather than exact risk estimations. In view of customer sensitivity to "unjustly" increased premiums it is highly desirable to estimate the spatial variation in risk and to price accordingly. Nevertheless, judging by the few available publications, the problem attained relatively little attention in the past.

One of the early works is due to Taylor (1989), who links two-dimensional splines on a plane to the map of the region, in order to assess spatial variability in Australian household contents insurance. Boskov and Verrall (1994) elaborate premium rating by postcode area, suggesting a Bayesian spatial model for the counts of claims. Their ideas are closely related to those of Besag et al. (1991), who allow for both spatially structured and unstructured heterogeneity in one model. One set of error components are

C.H. Skiadas (ed.), *Advances in Data Analysis*, Statistics for Industry and Technology, DOI 10.1007/978-0-8176-4799-5_20,

i.i.d. reflecting an unstructured pattern of heterogeneity, while the other set exhibits spatial correlation among neighbouring areas and remains uncorrelated otherwise. This construction is usually referred to as structured heterogeneity. The model is substantially refined in Brouhns et al. (2002), where geographically varying unobserved factors cause an unknown part of variation of the spatial parameters. The model identification mixes a frequentist approach to estimate the effect of all risk factors, other than location, with a Bayesian approach to evaluate the risk of each district.

Dimakos and Frigessi di Rattalma (2002) propose a fully Bayesian approach to nonlife premium rating, based on hierarchical models with latent variables for both claim counts and claim size. In this way they avoid the removal of the effect of all nonspatial risk factors by data preprocessing. Inference is based on the joint posterior distribution and is performed by Markov chain Monte Carlo (MCMC). They show that interaction among latent variables can improve predictions significantly.

Denuit and Lang (2004) incorporate spatial effects into a semiparametric additive model as a component of the nonlinear additive terms with an appropriate prior reflecting neighbourhood relationships. The functional parameters are approximated by penalised splines. This leads to a unified treatment of continuous covariates and spatially correlated random effects.

The spatial estimation of insurance risk is very much the same as risk estimation for disease mapping in spatial epidemiology. The risk-categorisation of the localities is equivalent to the colouring of a map and statistical techniques are well elaborated for that; see, e.g., Green and Richardson (2002) and references therein.

20.2 Data description, model building, and a tool for fit diagnosis

We have household insurance data at our disposal from the Hungarian division of a certain international insurance company, who initiated the present study. In order to illustrate our methods and findings, the company allowed us to use the data of a frequent and not expensive claim-type (we were required to avoid further specification), from 628,087 policies. We analyse the counts of claims occurring in the period of 16-Sept-1998 to 30-Sept-2005; that means 171,426 claims originating from 1,877,176 policy-years. The data record of the jth policy consists of the observed number of claims z_j, the exposure time τ_j (given as the number of days the policy was valid *within* the studied period), the location of the insured property (given as one of 3171 postcode areas of the country, hereinafter referred to as localities), the categorised population size of the localities, and the building-, wall-, and roof-types. The locations are classified into one of the ten categories according to population size. We have four categories of buildings, three categories of walls and six categories of roofs. In what follows we refer to these four categorical variables as risk factors. Our goal is to estimate the expected number of claims per unit exposure time (one year) for all 3171 localities. The naive estimation (observed counts/exposure) is obviously unreliable, especially in localities with no policy, or with just a few. So, the estimation must rely heavily on the spatial structure of the data, meaning the information available in the neighbouring locations.

At policy level we assume that the claim counts datum z_j of the jth individual policy is the observed realisation of the random variable Z_j, which is conditionally Poisson distributed, given the intensity. The Poisson intensity parameter λ_{Z_j} of Z_j depends on τ_j the exposure time (the time the policyholder spent in risk), which is known to us as data. Furthermore, the intensity depends on a general risk factor effect κ_j derived from the above-defined (nonspatial) risk factors, characterising the policy. Finally, the intensity parameter depends on the locality to which the contract belongs through the *spatial relative risk*, that we write in an exponential form as e^{ϑ_i}. It is the same for every policy belonging to location i. Suppose in addition that interdependence among claim counts is created solely through the intensity parameters; i.e., Z_js are conditionally independent Poisson variables, given the values of the intensities. Our final assumption is that the effects of the exposure time, risk factors, and the spatial relative risk are multiplicative on the intensity. To give an appropriate form for the intensities it is convenient to single out a joint scale parameter λ, and thus keep the mean of the exponents ϑ_i in the spatial relative risk at zero. So, we have for the intensity $\lambda_{Z_j} = \lambda \cdot \tau_j \cdot \kappa_j \cdot e^{\vartheta_i}$. where i stands for the location of policy j.

If we were to build a fully Bayesian model we should put priors on all possible nonempty risk classes created from the localities and the values of risk factors, and sample from a posterior of dimension well over 10^6. Clearly, the estimation of that high-dimensional model would not be viable. On the other hand, aggregating to regions would reduce the dimension of the problem, but the request of the company was to carry through an analysis on location level in order to avoid the pooling effect of the regions. So, we proceed to estimate *separately* the risk factor effect and the spatial relative risk.

For the first instance suppose λ and all e^{ϑ_i}s to be equal to 1. Then the κ_js are easily estimable by a generalised linear model. Regarding the estimated κ_js as if they were given data, we proceed further with the estimation of the spatial relative risk. Once we've done so, we renew the estimation of κ_js treating the estimated λ and e^{ϑ_i} as known and refit the generalised linear model. By iterating these steps better prediction of the claim counts can be achieved.

When aggregating the claims from policy to location level the conditional independence induces that the claim counts Y_i at the ith location are distributed as Poisson$(\lambda \cdot e^{\vartheta_i} \cdot \sum_j (\tau_j \cdot \kappa_j))$, where the summation in j goes over all policies belonging to location i. The quantity $\sum_j (\tau_j \cdot \kappa_j)$ can be interpreted as the modified exposure of location i and we denote it by t_i. So, in the first stage of model hierarchy we are dealing with a nonhomogeneous spatial Poisson process with intensities dependent on an overall scale, modified exposures, risk factors, and locations. As τ_js are given and the estimated κ_js are treated as data, the modified exposure is also regarded as a known quantity for every location. Hence, the two components λ and ϑ_i of the Poisson intensities remain to be estimated.

At the next hierarchical stage of model building we prescribe a structure on the logarithm of the spatial relative risks (log-srr. for short), that is, the exponents ϑ_i. These location-dependent components of the Poisson rates comply with a Gaussian Markov random field (GMRF) model. That means we have the vector $\boldsymbol{\Theta} = (\vartheta_i)$ of dimension equalling the number of localities (3171), that has a zero mean multidimensional normal distribution with spatially structured covariance matrix. The covariance between ϑ_i and ϑ_j depends on the *neighbourhood* relationship of the ith and jth localities.

In order to determine neighbourhoods for the localities we used the Hungarian Administrative Boundaries database as provided by the Hungarian Geodesic and Remote Sensing Institute. We choose localities to be neighbours when sharing a common piece of boundary, and call them adjacent to each other. Neighbourhood is then summarised either in the adjacency matrix $\mathbf{A}$, or the adjacency graph $\mathcal{A}$. The i, jth element of $\mathbf{A}$ is one or zero according to the ith and jth localities being neighbours or not, respectively. The ith and jth nodes of the adjacency graph $\mathcal{A}$ are connected by an edge when the ith and jth localities are adjacent to each other. The indirect kth-order neighbourhood (i is the second-order neighbour of j if there is at least one neighbour of i neighbouring j as well, and so on for higher order) is represented (with multiplicity) by the appropriate power of the $\mathbf{A}$ matrix. In the graph representation it corresponds to a path of length k leading from i to j. The i, jth element of $\mathbf{A}^k$, the kth power of the adjacency matrix, counts the paths of length k from i to j.

To proceed with the model building we suppose for the log-srr.-s $\boldsymbol{\Theta}$ a conditional autoregression (CAR) model, as described, e.g., in Clayton and Kaldor (1987), Cressie (1993) or Banerjee et al. (2004, Chapter 3). Remark here that other GMRF specifications as in Besag et al. (1991), or Arató et al. (2006) would also be possible, but the analysis of model choice is subject to further study, so we do not elaborate on it here. The CAR model in a slightly modified form as we use it prescribes the correlation matrix $\boldsymbol{\Sigma}$ of the Gaussian vector $\boldsymbol{\Theta}$ as $\boldsymbol{\Sigma} = \tau^{-2} \cdot \mathbf{D}(\mathbf{I} - \varrho\mathbf{A})^{-1}\mathbf{D}$, with $\mathbf{I}$ denoting the unit matrix as usual, and $\mathbf{D}$ is an arbitrarily chosen diagonal matrix, with all-positive diagonal elements. The assumption $\varrho < \varrho_{\max}$, with $\varrho_{\max}$ being reciprocal to the maximal eigenvalue of $\mathbf{A}$, guarantees that the covariance defined this way is valid; that is, the matrix $(\mathbf{I} - \varrho\mathbf{A})^{-1}$ is positive definite and remains so when multiplied from both sides with the same positive diagonal matrix. We restrict ϱ to be positive, as it is not particularly meaningful to suppose a negative spatial association for adjacent regions. A heuristic explanation of the chosen covariance structure decomposes the covariance of the log-srr.-s between two localities according to the degree of neighbourhood. That means only part of the covariance is due to the immediate neighbourhood relation, a smaller part inherited from neighbours of the neighbour, and then an even smaller part from their neighbours, and so on. These "parts" decrease exponentially with base ϱ, and summed up, give the actual covariance between the ith and jth location. Multiplying $\mathbf{A}^k$ by ϱ^k, one gets the part of the covariance stemming from the kth degree neighbourhood relations, as $\mathbf{A}^k$ counts those ones. Summing up for all ks we get the full covariance in the form of a power series of $\varrho \cdot \mathbf{A}$ equalling $(\mathbf{I} - \varrho\mathbf{A})^{-1}$. The diagonal $\mathbf{D}$ matrix serves to assign different uncertainties (i.e., variance) to the log-srr at different locations. In terms of the adjacency graph $\mathbf{D}$ assigns weights (of uncertainty) to the vertices. This may help to deal with overdispersion in the first-level Poisson model. A short exposure time creates more uncertain information on the parameters of a locality than a longer one. It may be worth while to reflect it in the setup by weighting the variances of θ_i according to the modified exposure through the $\mathbf{D}$ matrix, further justifying its introduction. As the lsrr. θ_i is on the logarithmic scale, it is natural to chose $d_{i,i}$ to be inversely proportional to the logarithm of the modified exposure time. While doing so, we have to take care of the localities with zero exposure by perturbing the zeros randomly with V, a very small positive variable, setting $d_{i,i}^{-1} = \log(t_i + V \cdot \chi_{\{t_i=0\}})$. It is not meaningful, however, to estimate $d_{i,i}$, as it reflects prior knowledge.

We use the probability map (Cressie, 1993, formula (6.2.1) Chapter 6.) for diagnosing model fit, which, in our case, is created from the estimates of the Poisson intensity

$\widehat{\lambda}_i$ at location i, equalling the expected number of claims there. Hypothesising the estimated intensity $\widehat{\lambda}_i$ to be the true one, we compute the probability of deviation of the observed values y_i from the expected one. To be more specific, when the expected is less than the observed (i.e., $\widehat{\lambda}_i \leq y_i$), compute the probability of exceedance of the observed sample $\mathrm{P}(Y_i \geq y_i)$, whereas when the expected is greater than the observed ($\widehat{\lambda}_i > y_i$) compute the probability of shortfall to the sample $\mathrm{P}(Y_i \leq y_i)$, under the hypothesis $Y_i \sim \mathrm{Poisson}(\widehat{\lambda}_i)$. This procedure is analogous to the computation of the p-value in hypothesis testing. Finally, after categorising, we display these probabilities for every location on a map. Small values in the probability map mean that the observed value is unlikely at the given estimated intensity, that questions the goodness of the estimation. High probabilities, on the other hand, do not necessarily mean a good estimate; it may reflect overfitting, when the estimator "learns" the sample too well. To recognise overfitting determine the distribution of the values in the probability map. When, e.g., expected is greater than observed, $\widehat{\lambda}_i \geq y_i$, the computed $\mathrm{P}(Y_i \leq y_i)$ equals $F(y_i)$, where $F(.)$ is the probability distribution function of Y_i, i.e., the $\mathrm{Poisson}(\widehat{\lambda}_i)$ distribution. Had Y_i a continuous distribution, $F(Y_i)$ would be uniform on [0,1]. The effect of the discrete (Poisson) sample can be eliminated by perturbing the values with an appropriately tailored uniform random variable.

20.3 Model estimation, implementation of the MCMC algorithm

We estimate the model parameters ϱ, τ, and $\mathbf{\Theta}$ by Markov chain Monte Carlo simulation. The challenge in the implementation of the MCMC algorithm is the very high dimension. We have to reach convergence and mixing in a 3173-dimensional state space. All the papers mentioned in the introduction addressed significantly lower dimensional problems and their publicly available programs did not seem to work in our case.

The base ϱ of the exponential decay in correlation is a crucial hyperparameter of the third level of hierarchy, to be estimated. In line with the Bayesian setup we put a prior on ϱ and in order to fulfil the requirement $0 < \varrho < \varrho_{\max}$ suppose $\varrho/\varrho_{\max}$ is distributed according to a beta(p, q) law. Choosing $p = 1, q = 1$ results in a vague prior.

In Bayesian parameter estimation of Gaussian vectors a gamma-prior is usual for τ^2 the precision parameter of the covariance. This choice is often favoured because it results in a gamma-posterior, allowing for Gibbs sampling. However, the precision parameter τ^2 is strongly related to the radius $||\mathbf{L}^T\mathbf{\Theta}||^2 = \langle \Sigma^{-1}\mathbf{\Theta}, \mathbf{\Theta}\rangle$ of the transformed Gaussian vector $\mathbf{L}^T\mathbf{\Theta} = \widetilde{\mathbf{\Theta}}$, where $\mathbf{L}^T$ comes from the Cholesky decomposition of the inverse of the covariance matrix: $\mathbf{L}\mathbf{L}^T = \tau^2 \cdot \mathbf{D}^{-1}(\mathbf{I} - \varrho\mathbf{A})\mathbf{D}^{-1}$. In the very high (3171) dimension the density of the length of $\widetilde{\mathbf{\Theta}}$ has a very narrow bandwidth, centered at $\sqrt{d-1}/\tau^{-1}$, resulting in a very narrow environment around the Gaussian ellipsoid of $\mathbf{\Theta}$, where the values of $\mathbf{\Theta}$ are likely. It is a simple geometrical consideration, that if we add to a vector *in high dimension* another vector of fixed length with uniform random angle, then with high probability it will *increase* the length of the original vector and decrease it with only a very small probability (unlike in two dimensions, where each of these probabilities hardly differs from 1/2). This means, random walk

Metropolis updates will lengthen the vector, whereas the precision in the likelihood tries to keep its length in the narrow band. If we were to avoid frequent rejection, most of the updates have to be of very small length in order to remain within the mentioned narrow environment of likely values. To allow for longer updates in order to achieve better mixing and faster convergence we suggest updating the length r of the Gaussian vector, $\widetilde{\boldsymbol{\Theta}}$ separately from the direction of $\boldsymbol{\Theta}$ represented by the unit vector U. Instead of parametrising our model with the triplet of $(\tau, \varrho, \boldsymbol{\Theta})$, we reparametrise it by four parameters (τ, ϱ, r, U), sample from their posterior via the MCMC, and obtain samples of the original triplet by backtransforming.

The probability distribution of the squared length r^2 of $\widetilde{\boldsymbol{\Theta}}$ is $\Gamma((d/2), \delta)$, where d is the dimension, $d = 3171$, and $\delta = \tau^2/2$. We choose this Gamma distribution as the prior for length r, given τ, but (similarly to Green and Richardson, 2002) the distribution of the precision τ^2 can be integrated out from the conditional one of r, and that is how we use it in the acceptance – rejection step. The update of r goes by a geometric normal random walk.

The unit vector $U = \widetilde{\boldsymbol{\Theta}}/r$ representing the direction of $\widetilde{\boldsymbol{\Theta}}$ has a uniform distribution on the unit sphere. To obtain U_n, the nth update of U we first generate the zero mean Gaussian random vector T_n of dimension d with i.i.d components: $T_n \sim \mathcal{N}_d(0, \sigma^2\mathbf{I})$. Then we compute U_n as the unit vector in the direction of $\widetilde{\boldsymbol{\Theta}}_{n-1} + T_n$, where $\widetilde{\boldsymbol{\Theta}}_{n-1}$ is the actual value of $\widetilde{\boldsymbol{\Theta}}$. The spread parameter σ^2 plays an important role in tailoring the acceptance rate, as we describe below.

Next we backtransform the updated $\widetilde{\boldsymbol{\Theta}}$ to obtain the new value for $\boldsymbol{\Theta}$ in the actual iteration step. That means we have to compute $(\mathbf{L}^T)^{-1}\widetilde{\boldsymbol{\Theta}}$, with the Cholesky decomposition $\mathbf{L}\mathbf{L}^T$ of the updated inverse covariance matrix. These are computationally very demanding operations, so efficient computing is essential at this point. As one locality has on average six or seven neighbours, the adjacency matrix is a sparse matrix, so we can invoke sparse matrix operations to accelerate significantly the computations; that is why we utilise the Cholesky decomposition (cf., e.g., Knorr-Held and Rue, 2002 and references therein). With this we compute $\boldsymbol{\Theta}$ as $\boldsymbol{\Theta} = r(\mathbf{L}^T)^{-1}U$. By introducing r into the computation, we simplify the computation of the quadratic form $\boldsymbol{\Theta}\boldsymbol{\Sigma}^{-1}\boldsymbol{\Theta}$ in the likelihood. On the other hand, whenever ϱ changes we have to recompute the Cholesky decomposition.

For the joint scale parameter λ of the Poisson intensities there is a way to derive an estimation without MCMC sampling. For given ϱ, τ, and $\boldsymbol{\Theta}$ it is straightforward to maximise the log-posterior in λ. The maximum is attained at

$$\widehat{\lambda}_{ML} = \frac{\left(\sum_{i=1}^N y_i\right)}{\sum_{i=1}^N e^{\vartheta_i} t_i}$$

providing for the conditional maximum likelihood estimator of λ, given the rest of the parameters. Though the use of this conditional maximum likelihood estimator is more natural in the frequentist setup, it can also be regarded as the limit of Bayesian estimates in the following sense. Put a usual gamma-prior on λ with some α shape and β rate. When β is negligible compared to the sum $\sum_{i-1}^N e^{\vartheta_i} \cdot t_i$, then the

$$\lambda_0 = \frac{\lambda}{\sum_{i=1}^N e^{\vartheta_i} \cdot t_i}$$

variable is conditionally independent of the rest of the parameters, and when β tends to 0, then the joint distribution of ϱ and $\mathbf{\Theta}$ given λ tends to the joint distribution of ϱ and $\mathbf{\Theta}$ given $\widehat{\lambda}_{ML}$. After a new proposal is obtained for $\mathbf{\Theta}$ (or, to be more precise, either for its length, or for its direction as we proposed above) the conditional max-likelihood estimator $\widehat{\lambda}_{ML}$ of λ is plugged in to compute the updated log-posterior.

We carried out about 2.8 million full iteration steps in a three-hour run (meaning ca. 280 iterations per second) on a PC, with a program written as a C/C++ addin to R. The first 0.8 million was regarded as burn-in, and we applied a thinning by 200 steps.

The MCMC algorithm renews the sampled values of ϱ, r, U by random walk updates. The spread, the variance of the Gaussian steps in the walk, allows for automatic control of the acceptance rate in the following way. At the nth step during burn-in only, when a proposal is accepted, we increase, whereas when rejected, decrease the logarithm of the spread by constant times $1/n$. As is usual in stochastic approximation we choose the pair of accept – reject constants so that the average change in the spread is zero if the acceptance rate is equal to the prescribed rate we are bound to achieve. As the 24% acceptance rate is optimal, we set this as the target value. The deviance of the acceptance rates from the 24%, the chain produced *after* the 800,000 iteration burn-in period, was below 1% in the application, indeed.

Storing all or a substantial part (of order 10,000 at least) of the 2.8 million simulated values for the 3171 coordinates θ_i, is not viable, because it requires a lot of memory and hampers the computation. So we could not determine quantiles or histograms for the majority of the parameters (a few, of course, can always be singled out).

To present a detailed diagnostic analysis of the convergence and mixing of the proposed MCMC procedure exceeds by far the frame work of the present chapter. Every 200th iteration is selected to display in Figure 20.1 a 10,000-step-long trace plot of the log-posterior, indicating that the chain reached convergence. The autocorrelation function (ACF) of the trace of the first coordinate θ_1 of $\mathbf{\Theta}$ is also presented there, showing that the 200th values are practically uncorrelated. Observing that the same is true for traces of the other coordinates, and parameters $\varrho, \tau, \widehat{\lambda}_{ML}$ (not shown), we conclude that the mixing of the chain is quite acceptable.

Another topic is model choice and the goodness of fit that can only be addressed very briefly in the present chapter, and that we intend to return to and come up with a detailed analysis in a follow-up paper. Here in Figure 20.2 we only present the residuals (first graph), that is, the difference of the observed claims and the estimated Poisson intensities for every locality, and the p-values of the probability map. The distributions of

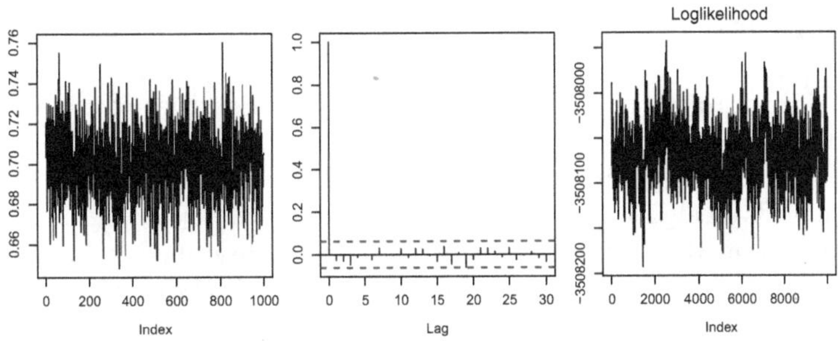

Figure 20.1. Simple diagnostic plots for MCMC convergence and mixing

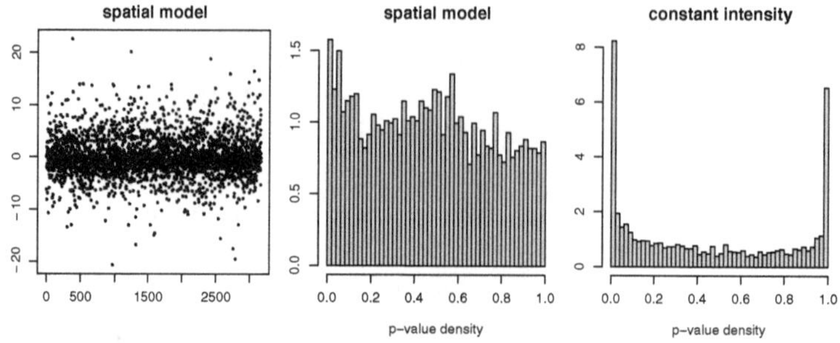

Figure 20.2. Residuals of predicted claims from the spatial model: comparison of p-values for the spatial and the constant-intensity model

the p-values are compared for the homogeneous spatial Poisson process (localities have all equal intensities) and our suggested spatial model, in terms of density histograms (second and third graphs). Clearly, the distribution deviates from the theoretical in the homogeneous case, and follows it for our model. Finally we compare on two maps the naive estimate of the intensities (left) with the estimate from our suggested model (right) in Figure 20.3. Our model decreases the range of intensities substantially, and smooths them into more homogeneous areas, showing a clearer spatial pattern, very much as we expected.

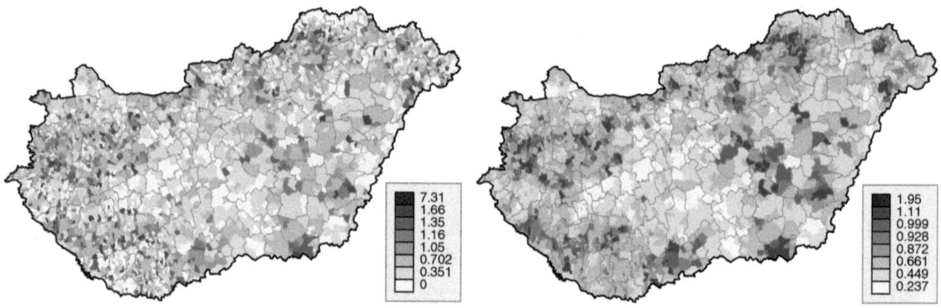

Figure 20.3. Maps of naive (*left*) and spatial (*right*) estimations of intensities

---

20.4 Conclusion

We presented a locationwise estimate of the Poisson intensities of claim counts for a certain type of claims in household insurance, and addressed the territorial dependence of risk through the lsrr. parameter of the model. We fitted our very high-dimensional model via an MCMC algorithm where we proposed a novel updating for the GMRF of the lsrr.-s. We invented an automatism for obtaining a prescribed (optimal) acceptance rate while updating the chain. Detailed convergence and mixing diagnostic and comparison with different MCMC implementations can be subjects for further discussion.

The comparison of different GMRF models for the lsrr. such as, e.g., the BYM one, is our immediate goal and we intend to present our findings in a follow-up paper.

Acknowledgements. This research was partially supported by the Hungarian National Research Fund OTKA, grant No.: T047086.

References

Arató, N.M., Dryden, I.L., Taylor, C.C., 2006. Hierarchical Bayesian modelling of spatial age-dependent mortality, *Computational Statistics and Data Analysis*, **51**(2), 1347–1363.

Banerjee, S., Carlin, B.P., Gelfand, A.E., 2004. *Hierarchical Modeling and Analysis for Spatial Data.* Chapman and Hall/CRC Press, Boca Raton, FL.

Besag, J., York, J., Mollië, A., 1991. Bayesian image restoration with two applications in spatial statistics (with discussion), *Annals of the Institute of Statistical Mathematics*, **43**, 1–59.

Boskov, M., Verrall, R.J., 1994. Premium rating by geographic area using spatial models, *ASTIN Bulletin*, **24**, 131–143.

Brouhns, N., Denuit, M., Masuy, B., Verrall, R., 2002. Ratemaking by geographical area: A case study using the Boskov and Verrall model, *Publications of the Institut de statistique, Louvain-la-Neuve*, Discussion paper **0202**, 1–26.

Clayton, D.G., Kaldor, J., 1987. Empirical Bayes estimates of age-standardized relative risks for use in disease mapping, *Biometrics* **43**, 671–681.

Cressie, N.A.C., 1993. *Statistics for Spatial Data.* John Wiley & Sons, New York.

Denuit, M., Lang, S. 2004. Non-life rate-making with Bayesian GAMs, *Insurance: Mathematics and Economics*, **35**, 627–647.

Dimakos, X.K., Frigessi di Rattalma, A., 2002. Bayesian premium rating with latent structure, *Scandinavian Actuarial Journal*, **2002** 162–184.

Green, P.J., Richardson, S., 2002. Hidden Markov models and disease mapping, *Journal of the American Statistical Association*, **97**(460), 1055–1070.

Knorr-Held, L., Rue, H., 2002. On block updating in Markov random field models for disease mapping, *Scandinavian Journal of Statistics*, **29**, 597–614.

Taylor, C.G., 1989. Use of spline functions for premium rating by geographic area, *ASTIN Bulletin*, **19**(1), 89–122.

Part VI

Neural Networks and Self-Organizing Maps

21

The FCN Framework: Development and Applications

Yiannis S. Boutalis[1], Theodoros L. Kottas[1], and Manolis A. Christodoulou[2]

[1] Department of Electrical and Computer Engineering,Democritus University of Thrace, 67100 Xanthi, Greece (e-mail: `ybout@ee.duth.edu, tkottas@ee.duth.gr`)
[2] Department of Electronic and Computer Engineering,Technical University of Crete, 73100 Chania, Greece (e-mail: `manolis@ece.tuc.gr`)

Abstract: The *Fuzzy Cognitive Network* (FCN) framework is a proposition for the operational extension of fuzzy cognitive maps to support the close interaction with the system they describe and consequently become appropriate for adaptive decision making and control applications. They constitute a methodology for data, knowledge, and experience representation based on the exploitation of theories such as fuzzy logic and neurocomputing. This chapter presents the main theoretical results related to the FCN development based on theorems specifying the conditions for the uniqueness of solutions for the FCN concept values. Moreover, case application studies are given, each one demonstrating different aspects of the design and operation of the framework.

Keywords and phrases: Fuzzy cognitive networks, fuzzy cognitive maps, uniqueness of solutions, contraction mapping theorem

21.1 Introduction

Fuzzy Cognitive Maps (FCM) are inference networks using cyclic directed graphs that represent the causal relationships between concepts (Kosko, 1997; Kosko, 1986a). They use a symbolic representation for the description and modeling of the system. In order to illustrate different aspects in the behavior of the system, a fuzzy cognitive map consists of nodes where each one represents a system characteristic feature. The node interactions represent system dynamics. An FCM integrates the accumulated experience and knowledge on the system operation, as a result of the method by which it is constructed, i.e., by using human experts who know the operation of the system and its behavior. Different methodologies to develop FCM and extract knowledge from experts have been proposed in (Stylios and Groumpos, 1999, 2004).

Fuzzy cognitive maps have already been used to model behavioral systems in many different scientific areas. For example, in political science (Axelrod, 1976), fuzzy cognitive maps were used to represent social scientific knowledge and describe decision-making methods (Kottas et al., 2004, Zhang et al., 1989, Georgopoulos et al.,

C.H. Skiadas (ed.), *Advances in Data Analysis*, Statistics for Industry and Technology, DOI 10.1007/978-0-8176-4799-5_21,

2003). Kosko enhanced the power of cognitive maps considering fuzzy values for their nodes and fuzzy degrees of interrelationships between nodes (Kosko, 1986a, 1997). He also proposed the differential Hebian rule (Kosko, 1986b) to estimate the FCM weights expressing the fuzzy interrelationships between nodes based on acquired data. After this pioneering work, fuzzy cognitive maps attracted the attention of scientists in many fields and they have been used in a variety of different scientific problems. Fuzzy cognitive maps have been used for planning and making decisions in the field of international relations and political developments and to model the behavior and reactions of virtual worlds. FCMs have been proposed as a generic system for decision analysis (Zhang et al., 1989, 1992) and as coordinators of distributed cooperative agents.

An extension of FCM called the Dynamic Cognitive Network (DCN) appears in Miao et al. (2001), where the concepts are also allowed to receive values from multivalued sets and the weights of the interconnection arcs are replaced by transfer functions to account for causal dynamics. DCNs are also used for decision support tasks. The fuzzy causal network (Liu and Zang, 2003; Zhang et al., 2006) is another extension of traditional FCM, which is also used for decision support based on the principle of causal discovery in the presence of uncertainty and incomplete information. Neutrosophic cognitive maps (Smarandache, 2001; Kandasamy and Smarandache, 2003) are generalisations of FCMs and their unique feature is the ability to handle indeterminacy in relations between two concepts. Recently, Fuzzy Cognitive Networks (FCN) (Kottas et al., 2007a) were presented as a complete computational and storage framework to facilitate the use of FCM in cooperation with the physical system they describe.

Regarding FCM weight estimation and updating, recent publications (Huerga, 2002; Papageorgiou and Groumpos, 2004; Papageorgiou et al., 2004; Aguilar, 2002) extend the initially proposed differential Hebbian rule (Kosko, 1986b) to achieve better weight estimation. Another group of methods for training FCM structure involves genetic algorithms and other exhaustive search techniques (Koulouriotis et al., 2001; Papageorgiou et al., 2005; Khan et al., 2004; Stach et al., 2005), where the training is based on a collection of particular values of input–output historical examples and on the definition of the appropriate fitness function to incorporate design restrictions.

Traditionally, FCMs are used for decision support or diagnosis tasks without immediate interaction with the system they describe. Once the graph is constructed and the arc weights are estimated based on either experts' opinion gathering or acquired data, the FCM is left to operate alone and produce its results without interaction with the physical system. In recent years, the use of FCM to control physical processes has been proposed in Stylios and Groumpos (1999, 2004), Stylios et al. (2006), Kottas et al. (2005), and Kottas et al. (2007a), where the FCM is used either as a direct controller of the physical process or as an expert supervisor of a traditional controller. The operation of the FCM in close cooperation with the real system it describes remains still an open issue. Such an on-line interaction with the real system might require continuous changes in the weight interconnections, which reflect the experts' knowledge about the node interdependence. This knowledge might not be entirely correct or, perhaps, the system has undergone changes during its operation. Moreover, as shown in this chapter, different operation conditions might require different weight assignments. The motivation for supporting such kind of operation comes from the success of FCMs in their traditional fields of applications and from the fact that traditional control problems from the field of electrical and mechanical engineering are successfully tackled using conventional fuzzy systems and fuzzy controllers. As shown in this chapter, the fuzzy cognitive network approach can serve as a reliable approach for these problems too.

One issue that needs more theoretical investigation is the conditions under which the concept values of FCM reach an equilibrium point and if this point is unique. According to Kosko (1997), starting from an initial state, simple FCMs will follow a path that ends in a fixed point or limit cycle, while more complex ones may end in an aperiodic or "chaotic" attractor. These fixed points and attractors could represent *meta rules* of the form, "If input then attractor or fixed point." The relation of the existence of these attractors or fixed points to the weight interconnections of the FCM has not been fully investigated. This is, however, of paramount importance if one wants to use FCMs in reliable adaptive system identification and control schemes.

In this chapter, we study first the existence of the above fixed points by using an appropriately defined contraction mapping theorem. It is proved that when the weight interconnections fulfill certain conditions, related to the size of the FCM, the concept values will converge to a unique solution regardless of their initial states. Otherwise the existence or the uniqueness of equilibria cannot be assured. In view of these results meta rules of the form, "If weights then fixed point," are more appropriate to represent the behavior of an FCM. Fuzzy cognitive networks can work on the basis of such meta rules. The FCN framework is an attempt to operationally extend FCMs to support the close interaction with the system they describe and consequently become appropriate for control applications and adaptive decision making (Kottas et al., 2007a, 2005; Boutalis et al., 2005). The framework consists of (a) the representation level (the cognitive graph), (b) the updating mechanism that receives feedback from the real system and (c) the storage of the acquired knowledge throughout the operation. This way, a fuzzy cognitive graph representation obtains dynamic features aiming at the control of real systems and presents a modeling and control alternative, when a precise mathematical model of the system is not available. To distinguish the proposed operational framework from traditional FCMs we call it *fuzzy cognitive network*.

In FCNs the nodes are labeled as control (or input), reference, output, and simple operation nodes. In the proposed operational framework, the FCN reaches its equilibrium point using direct feedback from the node values of the real system and the limitations imposed by the reference nodes. The interconnection weights are adjusted on-line during this operation by using an extended delta rule, which exploits system feedback and provides smooth and fast convergence, preventing at the same time weight values from being saturated. Moreover, the updating procedure is further enhanced and accelerated by using information from previous equilibrium points of the system operation. This is achieved by dynamically building a database, which, for each encountered operational situation, assigns a fuzzy if-then rule connecting the involved weight and node values. The range of the node and weight variables is dynamically partitioned to define appropriate membership functions. This way, the weight updating using system feedback gradually starts from values that are closer to the desired ones and therefore the procedure is significantly sped-up.

The chapter is organized as follows. Section 21.2 describes the representation and mathematical formulation of fuzzy cognitive maps. Section 21.3 presents theoretical results, where the proof of the existence solution of the concept values of a fuzzy cognitive map is given. FCNs are invoked in Section 21.4 where the main components of the framework, its updating, and storage mechanism are presented.

Selected application case studies are presented in Sections 21.5 and 21.6 demonstrating the various aspects of the framework. Finally Section 21.7 concludes the chapter.

21.2 Fuzzy cognitive maps

Fuzzy cognitive maps is a modeling methodology for complex systems, originated from the combination of fuzzy logic and neural networks. The graphical illustration of an FCM is a signed fuzzy graph with feedback, consisting of nodes and weighted interconnections. The nodes of the graph are related to concepts that are used to describe main behavioral characteristics of the system. Nodes are connected by signed and weighted arcs representing the causal relationships that exist among concepts. Graphical representation illustrates which concept influences other concepts, showing the interconnections between them. This simple illustration permits thoughts and suggestions in reconstructing FCM, such as the adding or deleting of an interconnection or a concept. In conclusion, an FCM is a fuzzy-graph structure, which allows systematic causal propagation, in particular forward and backward chaining.

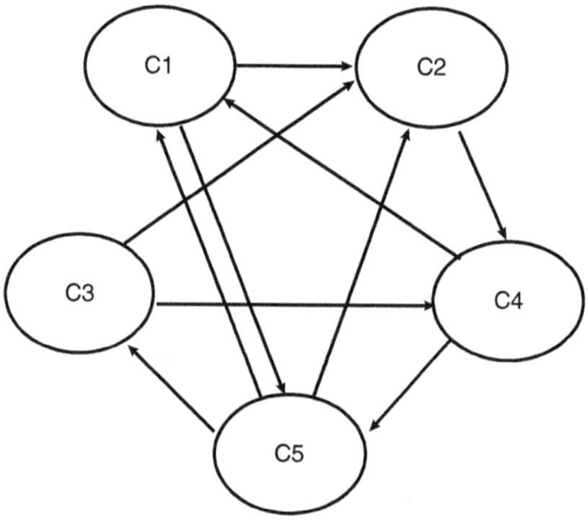

Figure 21.1. An FCM with five nodes

21.2.1 Fuzzy cognitive map representation

A graphical representation of FCMs is depicted in Figure 21.1. Each concept represents a characteristic of the system; in general it represents events, actions, goals, values, and trends of the system. Each concept is characterized by a number A_i that represents its value and it results from the transformation of the real value of the system variable represented by this concept, in the interval [0,1]. All concept values of the form Vector A which is expressed as

$$A = \left[A_1 \; A_2 \; \cdots \; A_n \right]^T$$

with n being the number of the nodes (in Figure 21.1 $n = 5$). Causality between concepts allows degrees of causality and not the usual binary values, so the weights of the interconnections can range in the interval $[-1, 1]$.

The existing knowledge of the behavior of the system is stored in the structure of nodes and interconnections of the map Each node-concept represents one of the key factors of the system. Relationships between concepts have three possible types; either express positive causality between two concepts ($W_{ij} > 0$) or negative causality ($W_{ij} < 0$) or no relationship ($W_{ij} = 0$). The value of W_{ij} indicates how strongly concept C_i influences concept C_j. The sign of W_{ij} indicates whether the relationship between concepts C_i and C_j is direct or inverse. The direction of causality indicates whether concept C_i causes concept C_j, or vice versa. These parameters have to be considered when a value is assigned to weight W_{ij}. For the FCM of Figure 21.1 matrix W is equal to:

$$W = \begin{bmatrix} 0 & 0 & 0 & W_{41} & W_{51} \\ W_{12} & 0 & W_{32} & 0 & W_{52} \\ 0 & 0 & 0 & 0 & W_{53} \\ 0 & W_{24} & W_{34} & 0 & 0 \\ W_{15} & 0 & 0 & W_{45} & 0 \end{bmatrix}.$$

The equation that calculates the values of concepts of fuzzy cognitive map, according to Stylios et al. (2006) is equal to:

$$A_i(k) = f\left(\sum_{\substack{j=1 \\ j \neq i}}^{n} W_{ij}^T A_j(k-1) + A_i(k-1)\right) \tag{21.1}$$

where $A_i(k)$ is the value of concept C_i at discrete time k, $A_i(k-1)$ the value of concept C_i at discrete time $k-1$, $A_j(k-1)$ the value of concept C_j at discrete time $k-1$, and W_{ij} is the weight of the interconnection from concept C_j to concept C_i. f is a sigmoid function used in the fuzzy cognitive map, which squashes the result in the interval [0,1] and is expressed as

$$f = \frac{1}{1 + e^{-x}}.$$

Equation (21.1) can also be written as

$$A(k) = f(W^{ext} \cdot A(k-1)) \tag{21.2}$$

where W^{ext} is such that:

$$W_{ij}^{ext} \underset{i \neq j}{=} W_{ji},\ W_{ij}^{ext} \underset{i=j}{=} d_{ii},$$

where d_{ii} is a variable that takes on values in the interval $[0, 1]$, depending upon the existence of "strong" or "weak" self-feedback to node i. Note that the case d_{ii} close to 0 is generic, while the d_{ii} close to 1 is an exception. See among other examples, the virtual undersea world example in Kosko (1997, p. 513) where only two out of 24 nodes are using self-feedback. For the FCM of Figure 21.1 matrix W^{ext} is equal to:

$$W^{ext} = \begin{bmatrix} d_{11} & 0 & 0 & W_{41} & W_{51} \\ W_{12} & d_{22} & W_{32} & 0 & W_{52} \\ 0 & 0 & d_{33} & 0 & W_{53} \\ 0 & W_{24} & W_{34} & d_{44} & 0 \\ W_{15} & 0 & 0 & W_{45} & d_{55} \end{bmatrix}.$$

From now on, in this chapter the matrix W^{ext} is just called W. Equation (21.2) can be rewritten as

$$A(k) = f(W \cdot A(k-1)). \tag{21.3}$$

In the next section we derive conditions that determine the existence of a unique solution of (21.3).

21.3 Existence and uniqueness of solutions in fuzzy cognitive maps

In this section we check the existence of solutions in equation (21.3). We know that the allowable values of the elements of FCM vectors A lie in the closed interval $[0, 1]$. This is a subset of $\Re$ and is a complete metric space with the usual l_2 metric. We define the regions where the FCM has a unique solution, which does not depend on the initial condition since it is the unique equilibrium point.

21.3.1 The contraction mapping principle

We now introduce the contraction mapping theorem (Rudin, 1964).

Definition 1. *Let X be a metric space, with metric d. If φ maps X into X and there is a number $c < 1$ such that*

$$d(\varphi(x), \varphi(y)) \leq cd(x, y) \tag{21.4}$$

for all $x, y \in X$, then φ is said to be a contraction of X into X.

Theorem 1. *Rudin* (1964) *If X is a complete metric space, and if φ is a contraction of X into X, then there exists one and only one $x \in X$ such that $\varphi(x) = x$.*

In other words, φ has a unique fixed point. The uniqueness is a triviality, for if $\varphi(x) = x$ and $\varphi(y) = y$, then (21.4) gives $d(x, y) \leq cd(x, y)$, which can only happen when $d(x, y) = 0$.

Equation (21.3) can be written as

$$A(k) = G(A(k-1)) \tag{21.5}$$

where $G(A(k-1))$ is equal to $f(W \cdot A(k-1))$.

In FCMs $A \in [0, 1]^n$ and it is also clear according to (21.3) that $G(A(k-1)) \in [0, 1]^n$. If the following inequality is true,

$$d(G(A), G(A')) \leq cd(A, A')$$

then G has a unique fixed point A such that:

$$G(A) = A.$$

Before presenting the main theorem we need to explore the role of f as a contraction function.

Theorem 2. *The scalar sigmoid function* f, *(*$f = 1/(1+e^{-x})$*) is a contraction of metric space* X *into* X, *where* $X = [a,b], a \leq 0, b \geq 1$ *according to Definition* 1, *where:*

$$d(f(x), f(y)) \leq cd(x,y). \tag{21.6}$$

Proof. Here f is the sigmoid function, x,$y \in X$, X is as defined above, and c is a real number such that $0 \leq c < 1$.

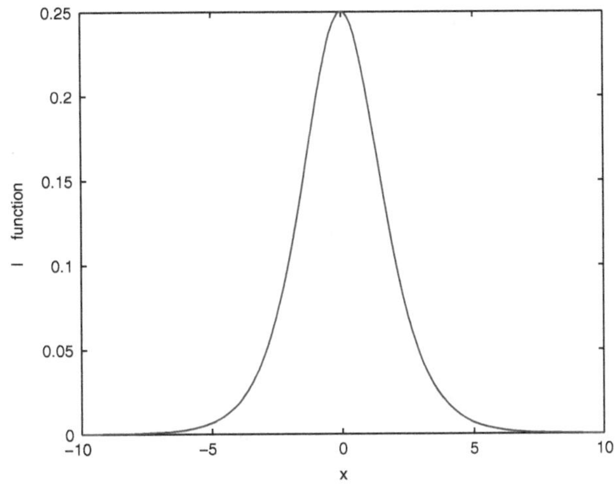

Figure 21.2. Inclination of sigmoid function f

The inclination l of a sigmoid function f is equal to:

$$l = \frac{\partial f}{\partial x} = \frac{e^{-x}}{(1+e^{-x})^2} = \frac{1}{e^x}\left(\frac{1}{1+e^{-x}}\right)^2 = \frac{1}{e^x} f^2 \tag{21.7}$$

for $x \in X$. Equation (21.7) is plotted in Figure 21.2. According to Figure 21.2 one can see that the inclination l of $f(x)$ in the bounded set X is always smaller than 1/4, as follows,

$$\frac{1}{4} \geq l \tag{21.8}$$

and l also equals

$$\frac{d\left(f(x), f(y)\right)}{d\left(x,y\right)} = l. \tag{21.9}$$

From (21.8) and (21.9) we get:

$$\frac{d\left(f(x), f(y)\right)}{d\left(x,y\right)} = l < 1. \tag{21.10}$$

Thus there is always a number c for which $0 \leq c < 1$, such that (21.10) is:

$$\frac{d\left(f(x), f(y)\right)}{d\left(x, y\right)} < c < 1. \tag{21.11}$$

Theorem 3. *There is one and only one solution for any concept value A_i of any FCM, if:*

$$\left(\sum_{i=1}^{n} \|w_i\|^2\right)^{1/2} < 4 \tag{21.12}$$

where w_i is the ith row of matrix W and $\|w_i\|$ is the l_2 norm of w_i.

Proof. Let X be the complete metric space $[0,1]^n$ and $G : X \to X$ be a map such that:

$$d(G(A), G(A')) \leq cd(A, A') \tag{21.13}$$

for some $0 \leq c < 1$.

Vector G is equal to:

$$G = \begin{bmatrix} f(w_1 \cdot A) \\ f(w_2 \cdot A) \\ f(w_3 \cdot A) \\ \vdots \\ f(w_n \cdot A) \end{bmatrix} \tag{21.14}$$

where n is the number of concepts of the FCM, f is the sigmoid function defined above, w_i is the ith row for matrix W of the FCM, where $i = 1, 2, ..., n$, and by $\cdot$ we denote the inner product between two equidimensional vectors which both belong to $\Re^n$.

Assume A and A' are two different concept values for the FCM. Then we want to prove the following inequality,

$$\|G(A) - G(A')\| \leq c \, \|A - A'\| \, . \tag{21.15}$$

But $\|G(A) - G(A')\|$ according to (21.14) equals

$$\|G(A) - G(A')\| = \left(\sum_{i=1}^{n} (f(w_i \cdot A) - f(w_i \cdot A'))^2\right)^{1/2}.$$

According to Theorem 2 for the scalar argument of $f(.)$ which is $w_i \cdot A$ in the bounded and closed interval $[-a, a]$ with a being a finite number it is true that:

$$|f(w_i \cdot A) - f(w_i \cdot A')| \leq c_i' \, |(w_i \cdot A) - (w_i \cdot A')|$$

for every $i = 1, 2, ..., n$. Thus

$$|f(w_i \cdot A) - f(w_i \cdot A')| \leq c' \, |(w_i \cdot A) - (w_i \cdot A')|$$

where $c' = \max(c_1', c_2', ..., c_n')$.

By using the Cauchy–Schwartz inequality we get:

$$c'|(w_i \cdot A) - (w_i \cdot A')| = c'|w_i \cdot (A - A')| \leq c'\|w_i\|\|A - A'\|.$$

Subsequently, we get:

$$\|G(A) - G(A')\| = \left(\sum_{i=1}^{n} (f(w_i \cdot A) - f(w_i \cdot A'))^2\right)^{1/2} \leq \left(\sum_{i=1}^{n} (c' \|w_i\| \|A - A'\|)^2\right)^{1/2}.$$

Finally:

$$\|G(A) - G(A')\| \leq c' \|A - A'\| \left(\sum_{i=1}^{n} \|w_i\|^2\right)^{1/2}.$$

A necessary condition for the above to be a contraction is:

$$c' \left(\sum_{i=1}^{n} \|w_i\|^2\right)^{1/2} < 1 \tag{21.16}$$

From equation (21.8) we have that:

$$c' \leq 1/4$$

so that the condition of equation (21.16) now becomes:

$$\left(\sum_{i=1}^{n} \|w_i\|^2\right)^{1/2} < 4. \tag{21.17}$$

21.3.2 Exploring the results

FCM with two concepts

Suppose that we have an FCM with two nodes. The weight matrix W_2 of this FCM is:

$$W_2 = \begin{bmatrix} d_{11} & w_{21} \\ w_{12} & d_{22} \end{bmatrix}.$$

According to Theorem 3 in order that an FCM with two nodes has a unique concept solution inequality (21.12) must be true. In this case (21.12) is written as

$$d_{11} + w_{21}^2 + w_{12}^2 + d_{22} < 16.$$

Since $|w_{21}| \leq 1$, $|w_{12}| \leq 1$ and d_{ii} can at most both take the value of 1, one can easily see that the above inequality is **always true** and particularly:

$$1 + w_{21}^2 + w_{12}^2 + 1 \leq 4 < 16.$$

FCM with three concepts

Suppose that we have an FCM with three nodes. The weight matrix W_3 of this FCM is:

$$W_3 = \begin{bmatrix} d_{11} & w_{21} & w_{31} \\ w_{12} & d_{22} & w_{32} \\ w_{13} & w_{23} & d_{33} \end{bmatrix}.$$

Taking into account that the magnitude of every weight value of W_3 is less than one equation (21.12) is now written:

$$d_{11} + w_{21}^2 + w_{31}^2 + w_{12}^2 + d_{22} + w_{32}^2 + w_{13}^2 + w_{23}^2 + d_{33} \leq 9 < 16$$

where it is obvious that, for an FCM with three concepts, the condition for the uniqueness is **always true.**

FCM with four concepts

Suppose that we have an FCM with four nodes. The weight matrix W_4 of this FCM is:

$$W_4 = \begin{bmatrix} d_{11} & w_{21} & w_{31} & w_{41} \\ w_{12} & d_{22} & w_{32} & w_{42} \\ w_{13} & w_{23} & d_{33} & w_{43} \\ w_{14} & w_{24} & w_{34} & d_{44} \end{bmatrix}.$$

The square root of the sum of the square l_2 norm of each row of matrix W_4 is equal to:

$$\sqrt{\sum_{i=1}^{4} \|w_i\|^2} = \sqrt{\|w_1\|^2 + \|w_2\|^2 + \|w_3\|^2 + \|w_4\|^2}. \tag{21.18}$$

The l_2 norm of each row is equal to:

$$\|w_i\| = \sqrt{\sum_{j=1}^{4} w_{ij}^2},$$

where i denotes the ith row of matrix W_4 and j denotes the column index. Equation (21.18) is now:

$$\sqrt{\sum_{i=1}^{4} \|w_i\|^2} = \sqrt{\|w_1\|^2 + \|w_2\|^2 + \|w_3\|^2 + \|w_4\|^2}$$

$$\Rightarrow \sqrt{\sum_{i=1}^{4} \|w_i\|^2} = \sqrt{\sqrt{\sum_{j=1}^{4} w_{j1}^2}^{\,2} + \sqrt{\sum_{j=1}^{4} w_{j2}^2}^{\,2} + \sqrt{\sum_{j=1}^{4} w_{j3}^2}^{\,2} + \sqrt{\sum_{j=1}^{4} w_{j4}^2}^{\,2}}$$

$$\Rightarrow \sqrt{\sum_{i=1}^{4} \|w_i\|^2} = \sqrt{\sum_{j=1}^{4} w_{j1}^2 + \sum_{j=1}^{4} w_{j2}^2 + \sum_{j=1}^{4} w_{j3}^2 + \sum_{j=1}^{4} w_{j4}^2} = \sqrt{\sum_{j=1}^{4} \left(\sum_{i=1}^{4} w_{ji}^2 \right)}$$

$$\Rightarrow \sqrt{\sum_{i=1}^{4} \|w_i\|^2} = \sqrt{\sum_{j=1}^{4} (d_{jj}) + \sum_{j=1}^{4} \left(\sum_{i=1, i \neq j}^{4} w_{ji}^2 \right)}$$

Since for the nondiagonal elements $|w_{ji}| < 1$, then : $w_{ji}^2 < 1$.

Finally the above equation concludes:

$$\sqrt{\sum_{j=1}^{4}(d_{jj}) + \sum_{j=1}^{4}\left(\sum_{i=1,i\neq j}^{4} w_{ji}^2\right)} \leq \sqrt{4+12} = 4$$

Subsequently, we get:

$$\sqrt{\sum_{i=1}^{4}\|w_i\|^2} \leq 4.$$

According to Theorem 3 in order that only one solution exists for the concepts of an FCM the following inequality must be true,

$$\sqrt{\sum_{i=1}^{4}\|w_i\|^2} < 4.$$

We finally get the next conclusion: since generically most of the d_{ii} will be zero, "an FCM with four concepts has a unique solution generically."

FCM with more than four concepts

Suppose that we have an FCM with more than four nodes. The weight matrix W_n of the FCM is:

$$W_n = \begin{bmatrix} d_{11} & w_{21} & w_{31} & \dots & w_{n1} \\ w_{12} & d_{22} & w_{32} & \dots & w_{n2} \\ w_{13} & w_{23} & d_{33} & \dots & w_{n3} \\ \dots & \dots & \dots & \dots & \dots \\ w_{1n} & w_{2n} & w_{3n} & \dots & d_{nn} \end{bmatrix}$$

where $n > 4$. The square root of the sum of the square l_2 norm of each row of matrix W_n is given by:

$$\sqrt{\sum_{i=1}^{n}\|w_i\|^2} = \sqrt{\|w_1\|^2 + \|w_2\|^2 + \cdots + \|w_n\|^2} = \sqrt{\sum_{j=1}^{n}\left(\sum_{i=1}^{n} w_{ji}^2\right)}$$

$$\Rightarrow \sqrt{\sum_{i=1}^{n}\|w_i\|^2} = \sqrt{\sum_{j=1}^{n}(d_{jj}) + \sum_{j=1}^{n}\left(\sum_{i=1,i\neq j}^{n} w_{ji}^2\right)} \leq \sqrt{\sum_{j=1}^{n}(d_{jj})} + \sqrt{\sum_{j=1}^{n}\left(\sum_{i=1,i\neq j}^{n} w_{ji}^2\right)}.$$

Finally we conclude that for an FCM with $n > 4$ concepts Theorem 3 is true when:

$$\sqrt{\sum_{j=1}^{n}\left(\sum_{i=1,i\neq j}^{n} w_{ji}^2\right)} \leq 4 - \sqrt{\sum_{j=1}^{n}(d_{jj})}. \tag{21.19}$$

Therefore, when $n > 4$ the condition for the uniqueness of solution of (21.3) depends on the number of diagonal d_{ii} elements of the FCM that are nonzero and the size of the

FCM. However, equation (21.19) provides us with an upper bound for the weights of the FCM. When the weights are within this bound the solution of (21.3) is unique and therefore the FCM will converge to one value regardless of its initial concept values. This in turn gives rise to a meta rules representation of the FCM having the form "**If weights then fixed point**". This representation is employed by fuzzy cognitive networks, which are presented in Section 21.4.

21.3.3 FCM with input nodes

So far we have not considered the existence of input nodes. This kind of node is also called "*steady*" node in Kottas et al. (2007a) in the sense that they influence but are not influenced by the other nodes of the FCM. We now show that the results obtained in the previous section are still valid. For the FCM of Figure 21.3 C_1 is such an input (or steady) node. Its weight matrix W is equal to:

$$W = \begin{bmatrix} 0 & 0 & 0 & 0 \\ w_{12} & d_{22} & 0 & w_{42} \\ w_{13} & w_{23} & d_{33} & w_{43} \\ 0 & w_{24} & w_{34} & d_{44} \end{bmatrix}$$

while vector A containing the node values is:

$$A = \left[U_1 \; A_2 \; A_3 \; A_4.\right]^T.$$

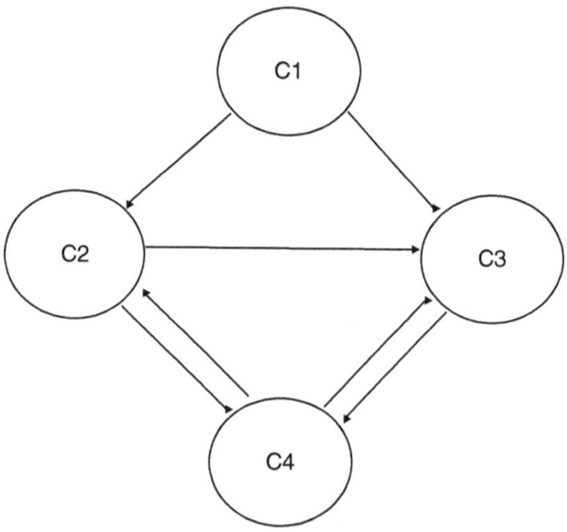

Figure 21.3. FCM with one input node

For the FCM of Figure 21.3, matrix G in equation (21.14) assumes now the following form,

$$G = \begin{bmatrix} U_1 \\ f(w_2 \cdot A) \\ f(w_3 \cdot A) \\ f(w_4 \cdot A) \end{bmatrix}$$

where U_1 is the input to FCM nodes. In a more general form matrix G can be written as

$$G = \begin{bmatrix} U_1 \\ U_2 \\ \vdots \\ U_m \\ f(w_{m+1} \cdot A) \\ f(w_{m+2} \cdot A) \\ f(w_{m+3} \cdot A) \\ \vdots \\ f(w_{m+n} \cdot A) \end{bmatrix} \tag{21.20}$$

corresponding to vector $A = \begin{bmatrix} U_1 \ U_2 \cdots U_m \ A_{m+1} \ A_{m+2} \cdots A_{m+n} \end{bmatrix}^T$, where m is the number of inputs and n is the number of the other concept nodes in FCM. Under this definition equation (21.5) assumes the same form:

$$A(k) = G(A(k-1)). \tag{21.21}$$

The next theorem proves that for matrix G and vector A defined above the results of Theorem 3 are still valid.

Theorem 4. *For an FCM with input nodes, with its concept values driven by* (21.21), *where A and G are described in* (21.20), *there is one and only one solution for any concept value A_i if equation* (21.12) *is fullfiled; that is,*

$$\left(\sum_{i=1}^{n} \|w_{m+i}\|^2 \right)^{1/2} < 4$$

where w_{m+i} is the $(m+i)$th row of matrix W and $\|w_{m+i}\|$ is the l_2 norm of w_{m+i}.

Proof. Assume A and A' are two different concept values for the FCM having one or more inputs. Then we want to prove again inequality (21.15); that is,

$$\|G(A) - G(A')\| \leq c \|A - A'\| .$$

But since input node values are not influenced by the other nodes of the FCM $\|G(A) - G(A')\|$ according to (21.20) is equal to:

$$\|G(A) - G(A')\| = \left(\sum_{i=1}^{m} (U_i - U_i)^2 + \sum_{i=1}^{n} (f(w_{m+i} \cdot A) - f(w_{m+i} \cdot A'))^2 \right)^{1/2}$$

where m is the number of inputs and n is the number of the other nodes in the FCM. The above equation is equivalent to the following,

$$\|G(A) - G(A')\| = \left(0 + \sum_{i=1}^{n} (f(w_{m+i} \cdot A) - f(w_{m+i} \cdot A'))^2 \right)^{1/2} .$$

Therefore, $\|G(A) - G(A')\|$ assumes quite the same form with that appearing in Theorem 3 leading to the same condition; that is,

$$\left(\sum_{i=1}^{n} \|w_{m+i}\|^2\right)^{1/2} < 4.$$

21.4 The fuzzy cognitive network approach

As shown in Section 21.3 the concept values of the FCM with a specified matrix W have a unique solution as far as (21.12) and consequently (21.19) are fulfilled. The perspective of transforming FCMs into a modeling and control alternative requires first to update its weight matrix W so that the FCM can capture different mappings of the real system, and second to store these different kinds of mappings. The fuzzy cognitive network (Kottas et al., 2007a) has been proposed as an operational extension framework of FCM, which updates its weights and reaches new equilibrium points based on the continuous interaction with the system it describes. Moreover, for each equilibrium point a fuzzy rule based storage mechanism of the form "If weights then fixed point" is provided, which facilitates and speeds up its operation. The components of FCN are briefly presented below.

21.4.1 Close interaction with the real system

The operation of the FCN in close cooperation with the real system it describes might require continuous changes in the weight interconnections, depending on feedback received from the real system. Figure 21.4 presents the interactive operation of the FCN with the physical system it describes. The weight updating procedure is described below.

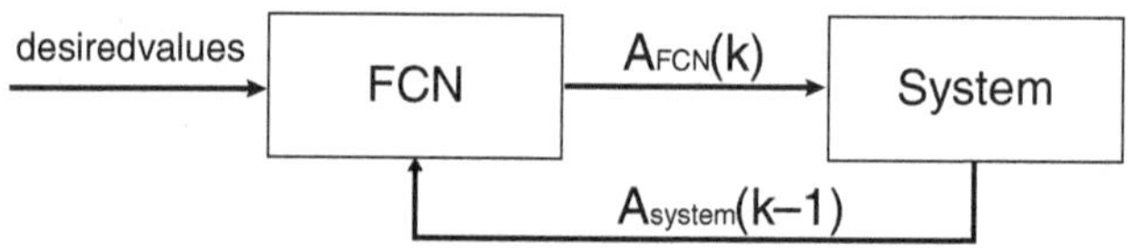

Figure 21.4. Interactive operation of the FCN with the physical system

21.4.2 Weight updating procedure

The updating method takes into account feedback node values from the real system. Using the updated weights the FCN reaches a new equilibrium point. In this approach the updating is made based on the conventional delta rule, which is described by the following equations,

$$p_j = A_j^{system}(k) - G_i(A^{FCN}(k-1)) \tag{21.22}$$

$$W_{ij}(k) = W_{ij}(k-1) + R_{ij}(ap_j(1-p_j))A_i^{FCN}(k) \tag{21.23}$$

where p_j is the error, a is the learning rate (usually set at $a = 0.1$), G_i is the ith element of matrix G, and R_{ij} is a calibration variable which prevents the FCN node and weight values from being driven in their saturation point. R_{ij} can be computed by the following formula (Kottas et al., 2005)

$$R_{ij} = \eta \frac{\sum_{i=1}^{i=n} |W_{ij}|}{|W_{ij}|} \quad \text{if } W_{ij} \neq 0 \quad \text{and} \quad R_{ij} = 0 \quad \text{if } W_{ij} = 0$$

where constant value η is used to drive values R_{ij} in the range [0, 1]. In most practical situations $\eta = 0.1$. The vector A^{FCN} in equation (21.22) refers to the response of the FCN, after it receives feedback from the system. In the case where the control objective is that one or more nodes reach a desired value then for these nodes equation (21.22) is rewritten as

$$p_j = A_j^{desired}(k) - G_i(A_i^{FCN}(k-1)). \tag{21.24}$$

After the weights updating, equation (21.1) will give new equilibrium concept values to the FCN. If the weights are chosen such that they meet the condition derived in Section 21.3, these values will be unique. The calculated node values will be applied to the real system, which in turn provides feedback to the FCN to be used by the new updating cycle according to Figure 21.4. Error p_j appearing in equations (21.22) and (21.23) is actually estimated for each one of the nodes j of the FCN, regardless of its label. Equation (21.24) is used for calculating the error of desired value nodes, while equation (21.22) is used for the errors of all other node values. When the real node values coming as a feedback from the system are fed to the FCN, this may present nonzero error in all of its nodes. The error becomes zero only when the weights are updated so that the node values of the FCN match exactly the values of the corresponding physical quantities.

21.4.3 Storing knowledge from previous operating conditions

The procedure described in the previous subsection modifies FCN's knowledge about the system by continuously modifying the weight interconnections and consequently the node values. During the repetitive updating operation the procedure uses feedback from the system variables. This means that in each iteration all the intermediate weight and node values, some of which are control values, are fed to the real system and its response is used to give the new updating direction. It is obvious that this procedure continuously annoys the physical system, something that in many cases is undesirable. In the sequence we propose a methodology that alleviates this annoyance and further speeds up the updating procedure. This is done by storing the previous acquired operational situations in a fuzzy if-then rule database, which associates in a fuzzy manner the various weights with the corresponding equilibrium node values. This storage mechanism actually creates meta rules of the form "If weights then equilibrium point" in accordance with the results of Section 21.3. The procedure is explained as follows.

Suppose for example that the FCM of Figure 21.1 has a unique equilibrium point

$$A = \left[A1 \; A2 \; A3 \; A4 \; A5 \right]^T$$

which is connected with the weight matrix W:

$$W = \begin{bmatrix} d_{11} & 0 & 0 & a_{41} & a_{51} \\ a_{12} & d_{22} & a_{32} & 0 & a_{52} \\ 0 & 0 & d_{33} & 0 & a_{53} \\ 0 & a_{24} & a_{34} & d_{44} & 0 \\ a_{15} & 0 & 0 & a_{45} & d_{55} \end{bmatrix}.$$

In order that A is a unique solution of equation (21.1) weight matrix W has to be such that inequality (21.12) is fulfilled. For weight matrix W inequality (21.12) takes the form:

$$a_{41}^2 + a_{51}^2 + a_{12}^2 + a_{32}^2 + a_{52}^2 + a_{53}^2 + a_{24}^2 + a_{34}^2 + a_{15}^2 + a_{54}^2 < 16 - \sqrt{\sum_{j=1}^{5} (d_{jj})}$$

where $n = 5$ is the number of concepts of the FCN.

Suppose also that the FCN in another operation point is related to the following weight matrix W, which also fulfills (21.12),

$$W = \begin{bmatrix} d_{11} & 0 & 0 & b_{41} & b_{51} \\ b_{12} & d_{22} & b_{32} & 0 & b_{52} \\ 0 & 0 & d_{33} & 0 & b_{53} \\ 0 & b_{24} & b_{34} & d_{44} & 0 \\ b_{15} & 0 & 0 & b_{45} & d_{55} \end{bmatrix}$$

with the unique equilibrium point being:

$$A = \left[B1\ B2\ B3\ B4\ B5 \right]^T .$$

Inequality (21.12) for the weight matrix W has now the form:

$$b_{41}^2 + b_{51}^2 + b_{12}^2 + b_{32}^2 + b_{52}^2 + b_{53}^2 + b_{24}^2 + b_{34}^2 + b_{15}^2 + b_{54}^2 < 16 - \sqrt{\sum_{j=1}^{5} (d_{jj})}.$$

The fuzzy rule database, which is obtained using the information from the two previous equilibrium points, is depicted in Figures 21.5 and 21.6 and is resolved as follows.

There are two rules for the description of the above two different equilibrium situations:

Rule 1

if node C1 is mf1 *and* node C2 is mf1 *and* node C3 is mf1 *and* node C4 is mf1 *and* node C5 is mf1

then w_{12} is mf1 *and* w_{15} is mf1 *and* w_{24} is mf1 *and* w_{32} is mf1 *and* w_{34} is mf1 *and* w_{41} is mf1 *and* w_{45} is mf1 *and* w_{51} is mf1 *and* w_{52} is mf1 *and* w_{53} is mf1

Rule 2

if node C1 is mf2 *and* node C2 is mf2 *and* node C3 is mf2 *and* node C4 is mf2 *and* node C5 is mf2

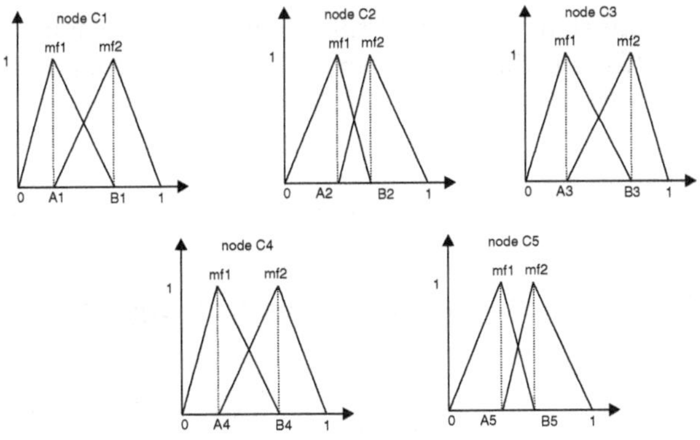

Figure 21.5. Left-hand side (if-part)

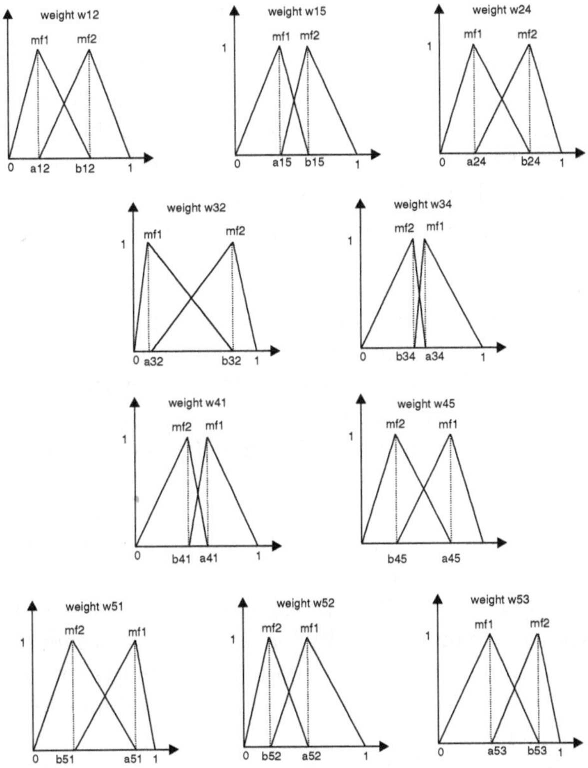

Figure 21.6. Right-hand side (then-part)

then w_{12} is mf2 *and* w_{15} is mf2 *and* w_{24} is mf2 *and* w_{32} is mf2 *and* w_{34} is mf2 *and* w_{41} is mf2 *and* w_{45} is mf2 *and* w_{51} is mf2 *and* w_{52} is mf2 *and* w_{53} is mf2

The number and shape of the fuzzy membership functions of the variables of both sides of the rules are gradually modified as new desired equilibrium points appear to the system during its operation. To add a new triangular membership function in the fuzzy description of a variable, the new value of the variable must differ from one already encountered value more than a specified threshold. The threshold comes usually as a compromise between the maximum number of allowable rules and the detail in fuzzy representation of each variable.

21.5 Controlling a wastewater anaerobic digestion unit (Kottas et al., 2006)

Rapid industrialization has resulted in the generation of a large quantity of effluents with high organic contents. These effluents, if treated properly, can contribute in environmental protection and energy recovery (Forster and Wase, 1987; Miyamoto, 1997). Sewage sludge, food industry wastes, and wastewater are attractive substrates for the production of biogas (Marchaim, 1992). In recent years, considerable attention has been paid toward the development of reactors for the anaerobic treatment of wastes leading to the conversion of organic molecules into biogas (Nemerow and Dasgupta, 1991). Within the most popular designs are the upflow anaerobic sludge bed (UASB) and the expanded granular sludge bed (EGSB) reactors. Successful operation of both reactors relies on the occurrence of granular sludge with excellent settling properties and high activity (Skiadas et al., 2003). Furthermore, the growth of anaerobic microorganisms depends on numerous factors, including residence time, temperature, redox potential, pH, and nutrient composition (Aivasidis and Diamantis, 2005). Monitoring and control of the anaerobic process is essential for stable operation, prevention of process failure, and utilization of maximum reactor capacity. In this section we present a method for controlling such an anaerobic process using FCN. The experiments were carried over an experimental pilot plant reactor.

The pilot plant is a unit of anaerobic sludge bed reactor. The system was equipped with pH and temperature control. The reactor was used for methanization of preacidified food industry wastewater and operated continuously at an increasing volumetric organic loading rate by increasing wastewater flowrate. The experimental unit is presented in Figure 21.7. The original wastewater (diluted peach pulp) was acidified using a continuous stirred tank reactor (CSTR) at hydraulic residence time (HRT) equal to 6–8 h. The pilot plant was monitored daily for biogas and methane production, pH, and temperature. Additionally, at steady-state conditions samples were obtained from the influent and effluent of the reactor and analyzed for total and soluble COD, ethanol, and acetic, propionic, and butyric acid.

The graph shown in Figure 21.8 represents a fuzzy cognitive network for the anaerobic digestion process. This graph was produced based on the experience gained from the operation of the experimental unit. The graph has 11 nodes, where nodes $C1$, $C2$, $C3$, and $C11$ stand for wastewater flowrate, reactor pH, reactor temperature, and water's reactor reflow, respectively. These nodes are steady value (input) nodes and at the same time control nodes, since a change of their values affects the values of the output nodes. Nodes $C4$, $C5$, $C6$, $C7$, $C8$, $C9$, and $C10$ are output nodes representing HRT, soluble COD inflow, soluble COD outflow, flow of CO_2, flow of the rest gases of the gas flow, gas flow, and flow of the CH_4, respectively. The volume of the sludge bed is not considered to be a control node in this graph because it is kept constant

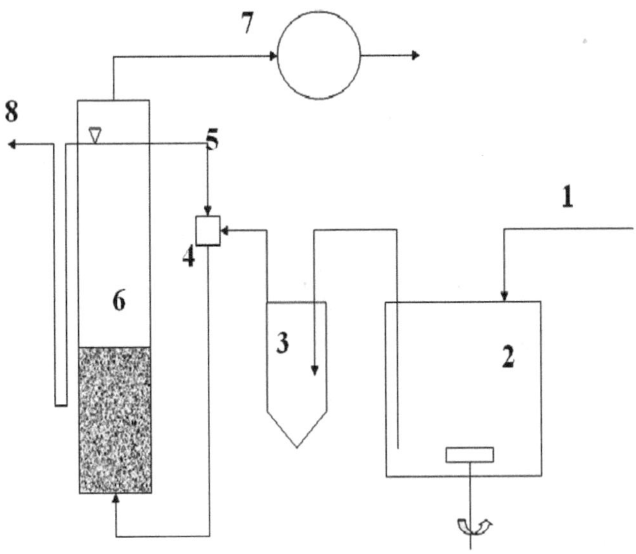

Figure 21.7. Schematic representation of the pilot plant used for anaerobic wastewater treatment: (1) raw wastewater, (2) acidification tank, (3) sedimentation tank, (4) pH conditioning tank, (5) recycle stream, (6) UASB reactor, (7) biogas measurement and analysis, (8) treated effluent

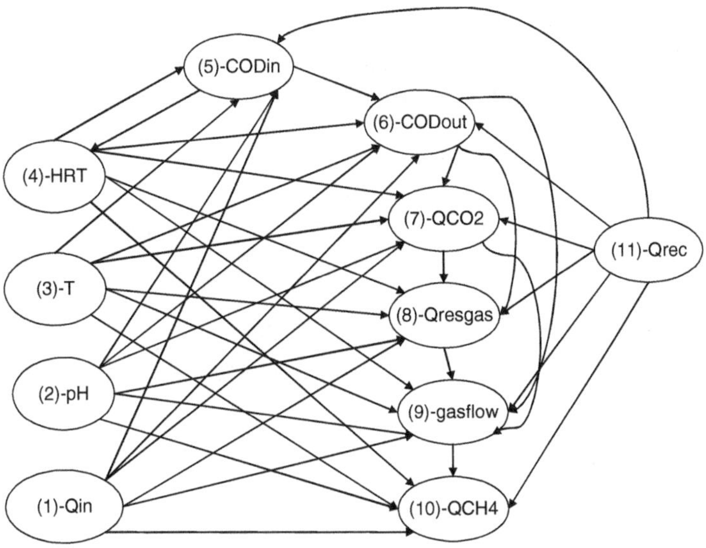

Figure 21.8. The FCN designed for the control of the anaerobic digestion process

by regularly subtracting sludge. With this graph representation W and A assume the following form.

$$W = \begin{bmatrix} 0 & 0 & 0 & 0 & 0 & 0 & 0 & 0 & 0 & 0 & 0 \\ 0 & 0 & 0 & 0 & 0 & 0 & 0 & 0 & 0 & 0 & 0 \\ 0 & 0 & 0 & 0 & 0 & 0 & 0 & 0 & 0 & 0 & 0 \\ 0 & 0 & 0 & 1 & W_{5,4} & 0 & 0 & 0 & 0 & 0 & 0 \\ W_{1,5} & W_{25} & W_{3,5} & W_{4,5} & 1 & 0 & 0 & 0 & 0 & 0 & W_{11,5} \\ W_{1,6} & W_{2,6} & W_{3,6} & W_{4,6} & W_{5,6} & 1 & 0 & 0 & 0 & 0 & W_{11,6} \\ W_{1,7} & W_{2,7} & W_{3,7} & W_{4,7} & 0 & W_{6,7} & 1 & 0 & 0 & 0 & W_{11,7} \\ W_{1,8} & W_{2,8} & W_{3,8} & W_{4,8} & 0 & W_{6,8} & 0 & 1 & 0 & 0 & W_{11,8} \\ W_{1,9} & W_{2,9} & W_{3,9} & W_{4,9} & 0 & W_{6,9} & W_{7,9} & W_{8,9} & 1 & 0 & W_{11,9} \\ W_{1,10} & W_{2,10} & W_{3,10} & W_{4,10} & 0 & 0 & 0 & 0 & W_{9,10} & 1 & W_{11,10} \\ 0 & 0 & 0 & 0 & 0 & 0 & 0 & 0 & 0 & 0 & 0 \end{bmatrix}$$

and $A = \left[U_1\ U_2\ U_3\ A_4\ A_5\ A_6\ A_7\ A_8\ A_9\ A_{10}\ \ U_{11} \right].$

21.5.1 Control of the process using the FCN

Once the FCN has been trained using experimental data, it is capable of adjusting the values of the control nodes in order to drive the real system to a new desired equilibrium point. The control mechanism is described below.

Suppose that the system is in a specific equilibrium point with node values given from the next D vector:

$$D = \left[D_1\ D_2\ D_3\ D_4\ D_5\ D_6\ D_7\ D_8\ D_9\ D_{10}\ \ D_{11} \right].$$

In that specific point the designer-engineer demands to move the real system to a new equilibrium point, which is different from the one that the system already has. Once the desired values of the node or nodes have been determined, the control system must decide the values of the control nodes ($C1$, $C2$, $C3$, and $C11$) in order to drive the real system to the desired equilibrium point. Suppose, for example, that the specifications require the desired value of node C_{10} to be E_{10}. The error p_{10} for the value of node C_{10} is:

$$p_{10} = E_{10} - G_{10}(A^{system}) = E_{10} - \frac{1}{1 + e^{-\left(\sum\limits_{i=1, i\neq 10}^{n} A_i^{system} W_{ij} + A_{10}^{system}\right)}}.$$

By taking the partial derivative of the above equation in respect to the control node values the following delta rule is derived, which determines the required change of the control nodes' values:

$$\begin{aligned} A_1(k) &= A_1(k-1) + p_{10}(1 - p_{10})W_{1,10} \\ A_2(k) &= A_2(k-1) + p_{10}(1 - p_{10})W_{2,10} \\ A_3(k) &= A_3(k-1) + p_{10}(1 - p_{10})W_{3,10} \\ A_{11}(k) &= A_{11}(k-1) + p_{10}(1 - p_{10})W_{11,10}. \end{aligned}$$

In a more general form the above equations can be rewritten as follows,

$$p_j = A_j^{desired}(k) - G_i(A^{system}(k-1)) \tag{21.25}$$

$$A_i^{control}(k) = A_i^{control}(k-1) + p_j(1-p_j)W_{ij} \tag{21.26}$$

where the "control" superscript is used to indicate that the updating is performed only to the control nodes.

By using the fuzzy rule database of the already trained FCN we calculate the values of the interconnections related to the new control nodes' values. We repetitively apply equations (21.25) and (21.26) in order to minimize the error p_{10}. Once the error p_{10} reaches zero (actually becomes sufficiently small), the FCN control mechanism sends the new control nodes' values to the physical system. When the real system is triggered by the control values it returns feedback from the measurable nodes' values. In the case where the feedback value of node C_{10} is not the desired one, this means that the FCN is facing an operational condition not encountered during its training stage. It can enrich its knowledge by using the mechanism described in the previous section. First, equations (21.22) and (21.23) are repetitively executed in order to adjust the FCN weights, which in turn reflect the new operational knowledge for the FCN. Next, the fuzzy rule database is updated according to the procedure described in the previous section so that it incorporates the new acquired knowledge. A pictorial representation of the above procedure is given in Figure 21.9.

The method presented here presents some advantages in comparison to traditional control methods. The most important innovation of the FCN control mechanism is that it does not require any mathematical knowledge for the description of the real system. Actually FCNs combine the knowledge of experts and the operational knowledge (data) derived from the operation of the system. The experts' knowledge is mostly used to construct the cognitive graph and probably to give initial weight sets. Experimental data from the real system are used to enrich the knowledge of the FCN on the sys-

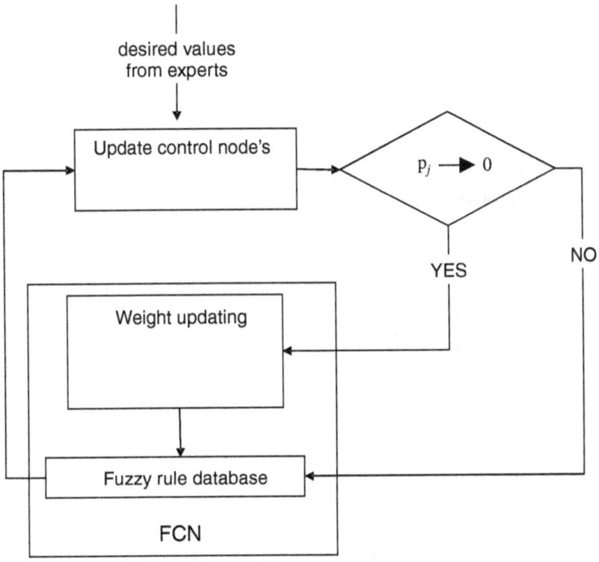

Figure 21.9. Control structure in order to achieve the desired equilibrium point defined from the experts

tem's operating conditions. Moreover, during its set points control actions, the FCN adapts its knowledge. Therefore the proposed control mechanism is an adaptive control scheme.

21.5.2 Results

To test the proposed methodology off-line we used data obtained from the operation of the experimental anaerobic digestion unit of the laboratory of wastewater management and treatment technologies of Democritus University of Thrace (DUTH). For various representative values of the control nodes, the resulting steady-state values of the other nodes were measured. The 11 node values obtained this way form a vector representing a real operational condition of the system and can be used to train or test the FCN. A number of 142 such data vectors were experimentally obtained. There 98 of these data vectors remaining selected to initially train the FCN using the procedure described above. The were 44 data vectors were used to test the generalization ability of the trained FCN.

Figure 21.10 shows 28 of the 44 test data values of the three control nodes Q_{in}, T, and pH. These values were selected randomly. However, they are arranged and displayed in respect to the time (in days) they were measured. Figure 21.11 shows the production of CH_4 (node 10) when the above control node values are imposed on the real system and on the FCN alone, respectively. The dotted spots represent the measured CH_4 values, while the solid line connects the values of CH_4, which are estimated by the FCN. It can be observed that the estimation error produced by the off-line training procedure is relatively small having a mean value of 7.4. However, Figure 21.11 clearly demonstrates that the FCN can provide relatively accurate results even for operational conditions about which it has not been taught.

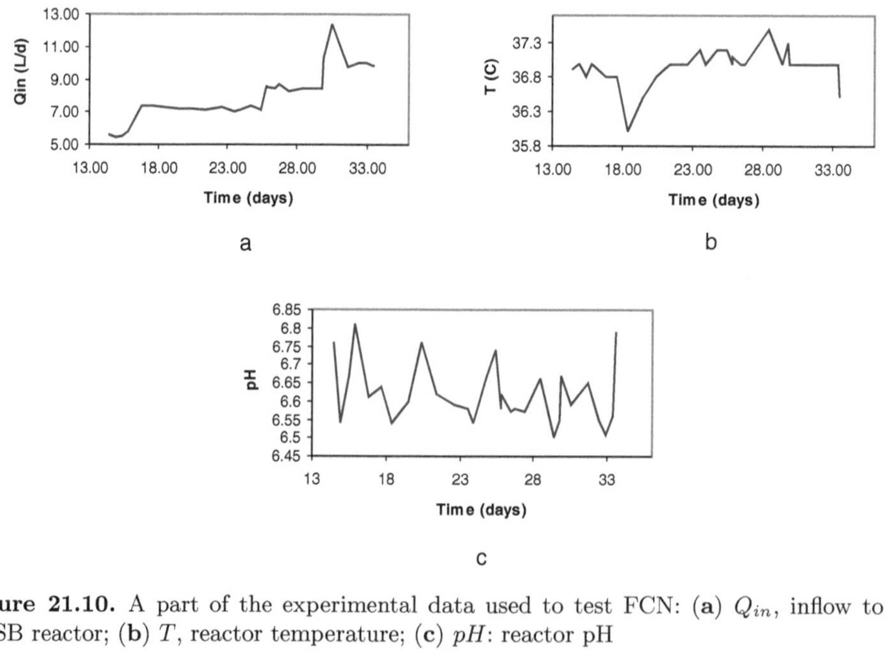

Figure 21.10. A part of the experimental data used to test FCN: (**a**) Q_{in}, inflow to the UASB reactor; (**b**) T, reactor temperature; (**c**) pH: reactor pH

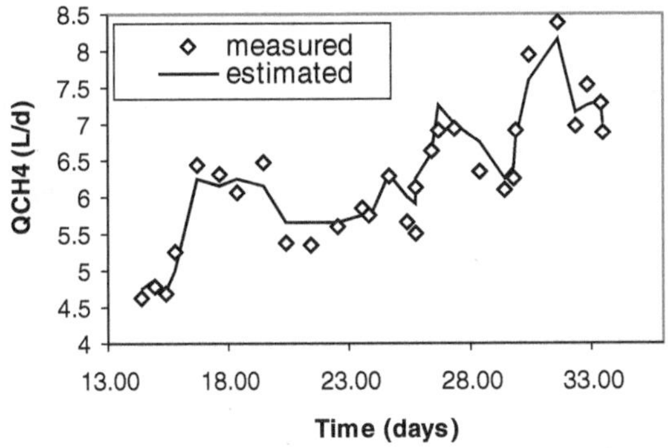

Figure 21.11. A comparison between estimated and measured QCH4 values for the experimental anaerobic digestion process

The above procedure tests only the approximation and generalization ability of the FCN. To test the proposed control mechanism introduced in the previous subsection we perform the following experiment. Suppose we want to regulate Q_{in}, T, and pH in order to change the value of QCH_4. Let us select the time instant 33,52d, where according to our measurements QCH_4 is equal to value 6,87158. At that specific time the values of control nodes are: Node 1(Q_{in}): 9,85, node 2(pH): 6,79, node 3(T): 36,5 and node 11(Q_{rec}): 58. Suppose now, that we require raising QCH_4 to 8,072414 (this value is one of the already measured experimental values, not used for training) in a relatively small time interval. The latter implies that control node 11 cannot actually affect the process and therefore we should rely only on changes in the other three control nodes. The control mechanism must regulate the values of control nodes (Q_{in}, T, and pH) in order to reach the desired value of QCH_4. By applying equation (21.25) with the desired value of QCH_4, we calculate p_{10} which is equal to 1,200834. For node 11 (Q_{rec}) there is no need to change its value because we assumed that practically it is kept constant. Applying p_{10} to equation (21.26) (with $i = 1, 2, 3$ and $i \neq 11$), the FCN regulates, according to the procedure described in the previous section, the values of nodes 1, 2, and 3 to the values: Node 1(Q_{in}): 11,79, node 2(pH): 6,60, node 3(T): 36,7, and node 11(Q_{rec}): 58. It has to be noted here that the experimental data for nodes 1, 2, 3, and 11 associated with $QCH_4 = 8{,}072414$ are: Node 1(Q_{in}): 11,7835, node 2(pH): 6,5912, node 3(T): 36,727, and node 11(Q_{rec}): 58. Therefore, the control procedure provides realistic estimations for the desired values of the control nodes. Various control case studies have been applied. In the sequel we refer to one of these in order to describe in detail the methodology followed.

The control paradigm is aiming at driving the produced gas flow from the initial value 6.22 l/d to 5.34 l/d by regulating Q_{in} without regulating pH. Figure 21.12 shows characteristic graphs. Due to the slow process variations the FCN does not respond immediately to the measured values unless a certain time period has elapsed since its last control action. This period was selected to be 400 sec, that is, almost 7 min.

Looking at Figure 21.12 we observe that the control actions can be divided into three phases. Phase A–B (400 sec): although Q_{in} was changed from 4.22 l/d to 2.87 l/d gas production was changed only slightly, while pH was kept constant. This new situation drove FCN to further changing Q_{in} from 2.87 l/d to 3.87 l/d.

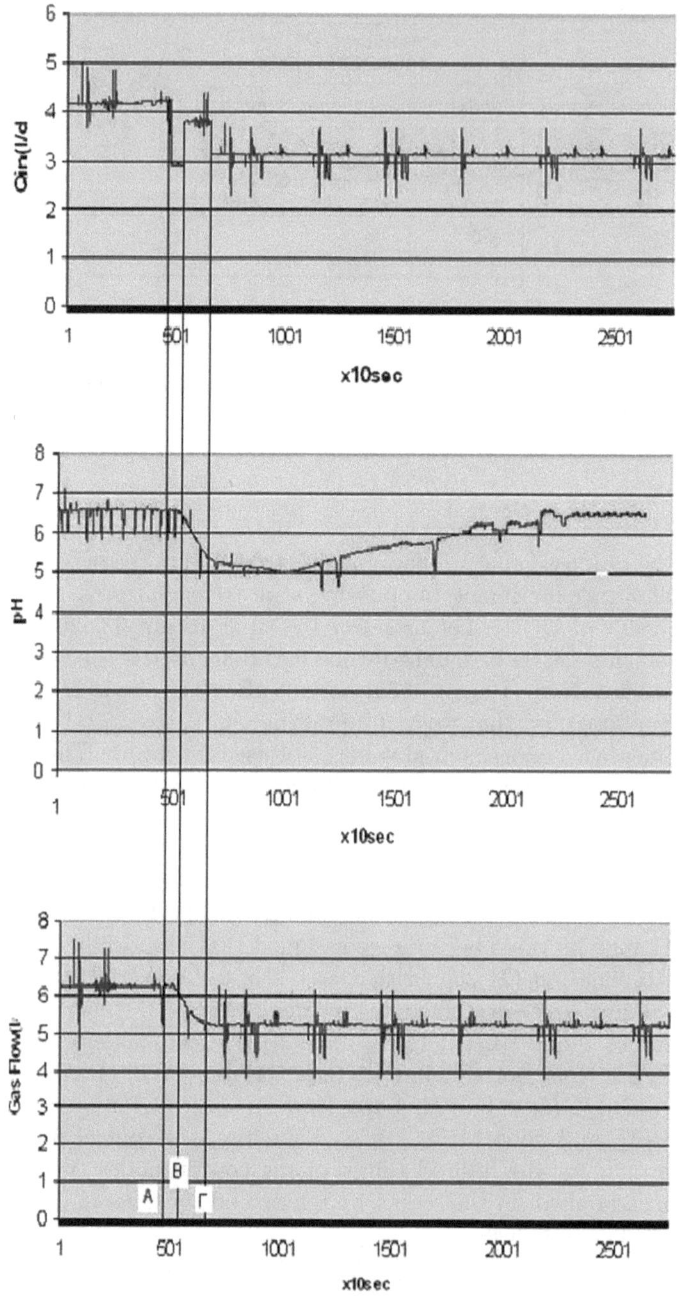

Figure 21.12. Characteristic graphs of a control experiment

Phase B–C (800 sec): after 800 sec of operation with the new value $Q_{in} = 3.87$ l/d the gas flow reaches the desired value. However, since the pH value is not one of the values that the FCN knows from its initial training it proceeds to again change Q_{in} from 3.87l/d to 2.24l/d. Finally, after 3800 sec without further changes in Q_{in} the experimental unit reaches the desired equilibrium point and the FCN stores the new acquired knowledge. In this experiment, in order to avoid producing frequent useless control commands, apart from the use of the 400 sec time interval between consecutive control actions, the following strategy was found to give the best results.

- If the FCN observes a rapid change in one of the process characteristic quantities (slope of change grater than 50 percent) it considers the new operation situation as a new equilibrium situation.
- If the observed slope is less than 50 percent then the FCN postpones the application of any control action for the next 400 sec. After this time elapses then it checks again for the existence of a new equilibrium situation.

21.5.3 Discussion

This example presents a highly nonlinear system, where the initial mathematical knowledge about it is practically nonexistent or not used at all. The particular structure of the network gave us the opportunity to determine an alternative procedure for determining the values of the control nodes. Since the control nodes are steady nodes, and therefore they cannot be influenced by other nodes of the FCN, their values, appropriate for driving the real system in a desired operation condition are not determined indirectly by the weight updating of the entire FCN, but rather by using a procedure similar to weight updating applied now to control node value updating (equation (21.26)). This is a useful extension of the proposed updating mechanism to cover the applications where the control nodes are steady value nodes.

21.6 The FCN approach in tracking the maximum power point in PV arrays (Kottas et al., 2007b)

The studies on the photovoltaic systems are increasing extensively because they can be considered as a large, secure, essentially inexhaustible and broadly available resource as a future energy supply. However, the output power induced in the photovoltaic modules is influenced by the intensity of solar cell radiation and temperature of the solar cells. Therefore, to maximize the efficiency of the renewable energy system, it is necessary to track the maximum power point of the PV array. A PV array is by nature a nonlinear power source, which under constant uniform irradiance has a current–voltage (I–V) characteristic like that shown in Figure 21.13.

There is a unique point on the curve, called the maximum power point (MPP), at which the array operates with maximum efficiency and produces maximum output power. It is well known that the MPP of a PV power generation system depends on array temperature, solar insolation, shading conditions, and PV cells ageing, so it is necessary to constantly track the MPP of the solar array. A switch-mode power converter, called a maximum power point tracker (MPPT), can be used to maintain the PV array's operating point at the MPP. The MPPT does this by controlling the PV array's voltage or current independently of those of the load. If properly controlled by an MPPT algorithm, the MPPT can locate and track the MPP of the PV array.

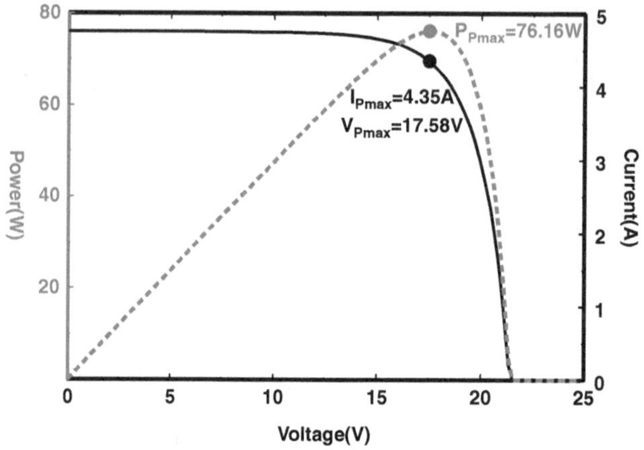

Figure 21.13. PV array I–V and P–V characteristics

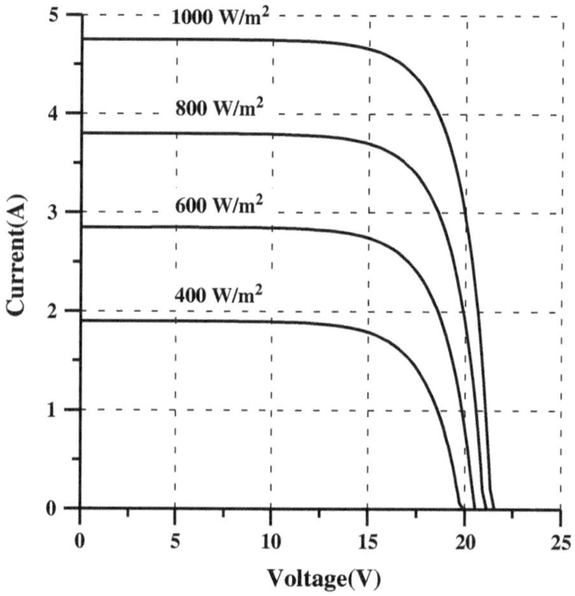

Figure 21.14. PV array I–V characteristics at various insolation levels

However, the location of the MPP in the I–V plane is not known a priori. It must be located, either through model calculations or by a search algorithm. Figure 21.14 shows a family of PV I–V curves under increasing irradiance, but at constant temperature. Needless to say there is a change in the array voltage at which the MPP occurs. For years, research has focused on various MPP control algorithms to draw the maximum power of the solar array (Hiyama et al., 1995; Ro and Rahman, 1998; Miyatake et al., 2002; Bahgat, 2005; Salameh and Taylor, 1990; Koutroulis et al., 2001; Masoum et al., 2002; Won, 1994; Simoes and Pranceschetti, 1999).

FCNs can also be used for MPP quite efficiently (Kottas et al., 2007b; Karlis et al., 2007). The FCN can be designed so that its nodes represent essential operational(voltage, current, insolation, temperature) and control (current) variables of the PV system. The node interconnection weights are determined using data which are constructed so that they cover the operation of a PV system under a wide range of different climatic conditions. Once the FCN is trained it can be mounted on any PV system. The performance of the method using climatic data for a specific PV system of the market, such as changing insolation and temperature and seasonal variations is very satisfactory.

The graph shown in Figure 21.15 is the FCN representation of the components of a photovoltaic system, which are involved in its operation and can determine its maximum power point performance. The graph has six nodes, which are related to the following physical quantities of the photovoltaic system.

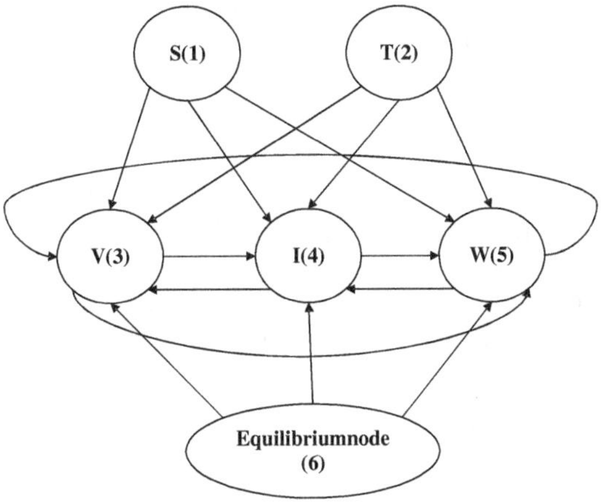

Figure 21.15. An FCN designed for the photovoltaic project

Node C1 represents the irradiation with range in the interval [0 1]. Zero is the minimum point of the irradiation (usually $0\,\mathrm{mW/cm^2}$) and one is the maximum point, corresponding to $100\,\mathrm{mW/cm^2}$.

Node C2 represents the temperature which also must be in the interval [0 1]. Zero is the minimum point of the temperature (usually $-30\,°\mathrm{C}$) and one is the maximum point, usually $70\,°\mathrm{C}$.

Node C3 represents the optimum voltage of the photovoltaic system for the climatological data obtained at the specific point of time, which also must be in the interval [0 1]. Zero is the minimum point of the voltage (usually 0 Volt) and one is the maximum point V_{max}.

Node C4 represents the optimum current of the photovoltaic system for the climatological data obtained at the specific point of time, which also must be in the interval [0 1]. Zero is the minimum point of the current (usually 0 Ampere) and one is the maximum point I_{max}.

Node C5 expresses the optimum output power of the photovoltaic system for the climatological data obtained at the specific point of time, which also must be in the interval [0 1]. Zero is the minimum point of the power (usually 0 Watt) and one is the maximum point W_{max}, where W_{max} is a characteristic given from PV operational data under T_{min} and S_{max}.

Node C6 is an artificial design node, the value of which is used to regulate the equilibrium point in the nodes C3, C4, and C5. The value of C6 is steady and equals 1. The weights W_{63}, W_{64}, and W_{65} are originally set to zero and are allowed to change only when one or more weights affecting nodes 3, 4, and 5 exceed the value of absolute 1. For example, the value of weight W_{63} is allowed to be updated when the weights that affect node C3 (W_{13}, W_{43}, and W_{53}) are going to take values larger than the absolute value 1. In this situation weight W_{63} is activated and its value is no longer set to zero. By using equilibrium node C6 and the weights connecting this node with nodes C3, C4, and C5, we manage to regulate the values of nodes C3, C4, and C5 by always keeping values of the graph weights below absolute value 1.

In this configuration nodes C1, C2, and C6 are steady value nodes and nodes C3, C4, and C5 could be control nodes, but only node C4 is chosen to be the control node. Its value is used to regulate the current of the system. The regulation of the current of the system means that a different power is now the output power of the photovoltaic. Control nodes are the nodes the values of which will be used by the real system as control actions. Node C4 is used to calculate the optimum current needed to regulate the output power of the photovoltaic in the maximum point.

21.6.1 Simulation of the PV system

Using the equivalent circuit of a solar cell (Figure 21.16) the nonlinear I–V characteristics of a solar array are extracted, neglecting the series resistance (Hua and Shen, 1998):

$$I_0 = I_{ph} - I_{rs}(e^{(qV_0)/(kTA)} - 1) - \frac{V_0}{R_{sh}} \tag{21.27}$$

where I_0 is the PV array output current (A); V_0 is the PV array output voltage (V); q is the charge of an electron; k is Boltzmann's constant in J/K; A is the p–n junction ideality factor; T is the cell temperature (K); and I_{rs} is the cell reverse saturation current. The factor A in equation (21.27) determines the cell deviation from the ideal p–n junction characteristics.

The photocurrent I_{ph} depends on the solar radiation and the cell temperature as stated in the following equation,

$$I_{ph} = (I_{scr} + k_i(T - T_r))\frac{S}{100} \tag{21.28}$$

where $I_s cr$ is the PV array short circuit current at reference temperature and radiation, k_i is the short circuit current temperature coefficient (A/K), and S is the solar radiation

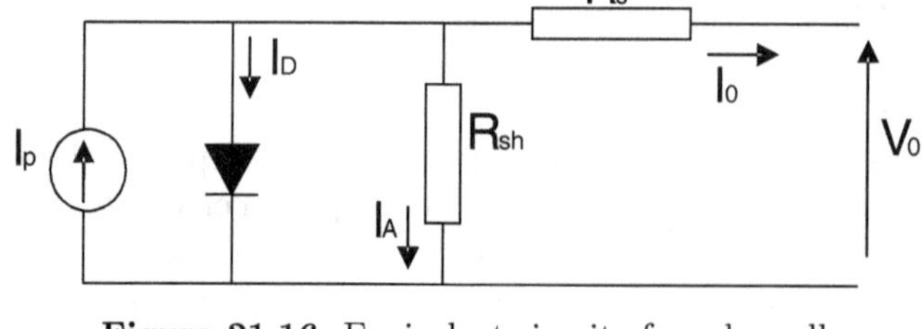

Figure 21.16. Equivalent circuit of a solar cell

(mW/cm^2). The reverse saturation current I_{rs} varies with temperature according to the following equation,

$$I_{rs} = I_{rr} \left(\frac{T}{T_r}\right)^3 e^{(1.115/k'A)((1/T_r)-(1/T))} \tag{21.29}$$

where T_r is the cell reference temperature, I_{rr} is the reverse saturation current at T_r, and k' is the Boltzmann's constant in eV/K and the bandgap energy of the semiconductor used in the cell is equal to 1.115.

Finally, the next equation was used in the computer simulations to obtain the open circuit voltage of the PV array:

$$V_{oc} = \frac{AkT}{q} \ln\left(\frac{I_{ph} + I_{rs}}{I_{rs}}\right). \tag{21.30}$$

From equations (21.28), (21.29), and (21.30) we get:

$$I_{rr} = \frac{(I_{scr} + k_i(T - T_r))\frac{s}{100}}{e^{\frac{V_{oc}q}{AkT}} - 1} \left[\left(\frac{T_r}{T}\right)^3 e^{-(1.115/k'A)((1/T_r)-(1/T))}\right] \tag{21.31}$$

and from equation (21.27):

$$R_{sh} = \frac{V_{oc}}{-I_{rs}(e^{(qV_{oc})/(kTA)} - 1)}. \tag{21.32}$$

The required data for identifying the maximum operating point at any insolation level and temperature are the following.

(i) k_i
(ii) Open circuit voltage V_{oc} (for initial conditions $T_r = 25°\mathrm{C}$, $S = 100\,\mathrm{mW/cm^2}$)
(iii) Short circuit current I_{scr} (for initial conditions $T_r = 25°\mathrm{C}$, $S = 100\,\mathrm{mW/cm^2}$)
(iv) Maximum power voltage V_{mpp} (for initial conditions $T_r = 25°\mathrm{C}$, $S = 100\,\mathrm{mW/cm^2}$)
(v) Maximum power current I_{mpp} (for initial conditions $T_r = 25°\mathrm{C}$, $S = 100\,\mathrm{mW/cm^2}$)

all given by the PV array manufacturer

21.6.2 Control of the PV system using FCN

The FCN is first trained off-line by appropriately constructed meteorological data and using ideal node values based on the equations of Section 21.6.1. The off-line training is performed in an incremental manner. This means that for each training data vector which contains PV value variables corresponding to different operation conditions, the FCN weights are updated to comply with the data vector. Moreover, this new acquired knowledge is stored in a fuzzy rule database.

Once the FCN is trained off-line it can be connected to the PV system according to Figure 21.17. The FCN receives feedback from the PV system. The FCN weights appropriate for these feedback values are extracted by using the fuzzy rule database. Using these weights the FCN reaches a new equilibrium point using equation (21.21). The control node value of the FCN is then used to regulate the PV system in order

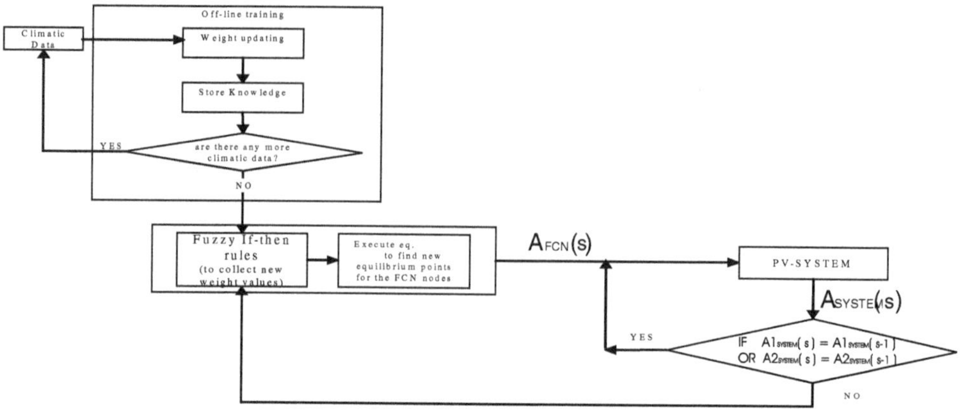

Figure 21.17. Simplified flowchart of the control process of the PV array using FCN

to give maximum power for the current conditions. The control method was tested for one year (climatological data from the year 2002 of the area of Xanthi, Greece, were given to the off-line training of the algorithm of Figure 21.17, while the PV array used for the simulated tests was the BP270L PV array). In Figure 21.18a climatological data derived from the day 14/01/2002 were tested in the algorithm of Figure 21.17, in order to test the trained FCN. The climatological data of the specified day were not given for off-line training. One can observe that the FCN trained by climatological data can offer highly efficient control laws in order to track the maximum power point of a PV array as can be seen in Figure 21.18b, where the estimated MPP from the FCN presents only 3.09% with respect to the ideal MPP values. This specific day was one

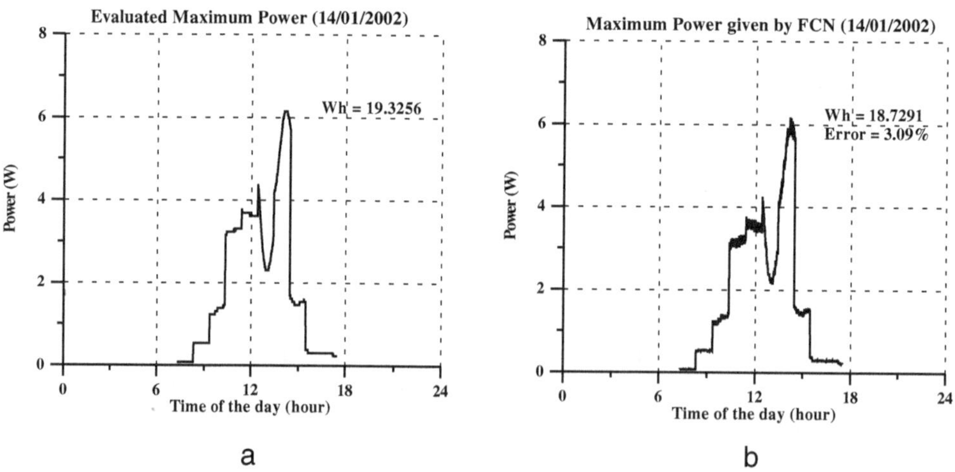

Figure 21.18. Comparison between (**a**) evaluated and (**b**) achieved using FCN MPP of the PV array for the least sunny day of the year 2002

of the less sunny days of the year, while the day at 09/07/2002, was one of the sunniest days of the year, but not the one with maximum power production, because the power produced depends not only on solar irradiation but also on temperature. Looking at Figures 21.18a and 21.18b, where the estimated MPP from the FCN presents a 2.24% error, one can conclude that the FCN MPP method can better estimate the MPP when irradiation and temperature are at high levels and the power tracked from the PV array is at high levels too. Finally, the average annual error of the FCN MPP method was estimated to be 2.01% when the method was tested for all days of year 2002.

21.6.3 Discussion

In this application the FCN was trained off-line to optimally control a PV array so that it reaches its MPP of operation. After the FCN is trained, its adaptive ability is unplugged and it operates based only on its already acquired and stored knowledge. An adaptive version of the same application is presented in Kottas et al. (2007b), where the FCN is combined with a conventional fuzzy MPP controller increasing the method further. The particularity appearing in this application is related to the structure of the cognitive graph itself and especially with the introduction of an equilibrium node (node 6). It was initially observed that if the FCM were designed based only on the other five nodes and on their interdependencies, there are operating conditions where the FCN weights cannot be kept under their saturation values. This implies that the cognitive graph, based only on the five nodes, is not sufficient for representing this particular system. Therefore, using artificial node(s) to extend the graph, the representation is more realistic. This observation demonstrates the need for devising new techniques, which will extend the initial structure of the cognitive graph, based on some measures of its inefficiency.

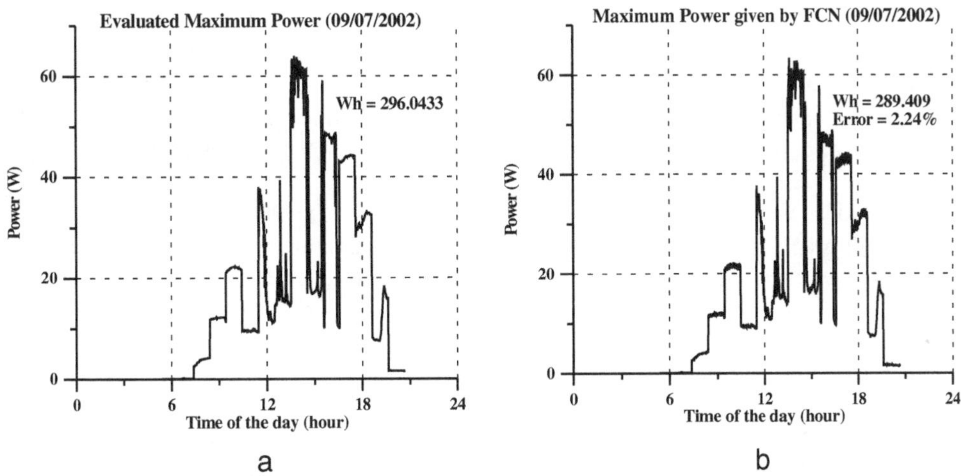

Figure 21.19. Comparison between (**a**) evaluated and (**b**) achieved using FCN MPP of the PV array for the sunniest day of the year 2002

21.7 Conclusions

The fuzzy cognitive network (FCN) framework is a proposition for the operational extension of fuzzy cognitive maps to support the close interaction with the system they describe and consequently become appropriate for adaptive decision making and control applications. Aspects related to its development and its application were presented in this chapter. Basic theoretical results based on theorems specifying the conditions for the uniqueness of solutions for the FCN concept values were initially given. These results give rise to a special form of knowledge representation through FCN, expressed with the help of a special form of meta rules. Two selected successful applications of FCN were presented. The first one is concerned with the control of a wastewater treatment unit and the second with the regulation of a PV system to reach its MPP operation. In the discussion subsection of these applications some specific aspects of the use of FCN in applications are shown.

References

Aguilar, J. (2002). Adaptive random fuzzy cognitive maps. *BERAMIA 2002, Lecture Notes in Artificial Intelligence 2527, F. J. Garijio, J. C. Riquelme and M. Toro, eds, Springer-Verlag Berlin Heidelberg*, pages 402–410.

Aivasidis, A. and Diamantis, V. (2005). Biochemical reaction engineering and process development in anaerobic wastewater treatment. *Advances in Biochemical Engineering/ Biotechnology*, volume 92, pages 49–76.

Axelrod, R. (1976). Structure of Decision, the Cognitive Maps of Political Elites. *Princeton University Press, Princeton, NJ.*

Bahgat, A. (2005). Maximum power point tracking controller for PV systems using neural networks. *Renewable Energy*, volume 30, pages 1257–1268.

Boutalis, S. Y., Kottas, L. T., Mertzios B., and Christodoulou, A. M. (2005). A fuzzy rule based approach for storing the knowledge acquired from dynamical FCMs. *5th International Conference on Technology and Automation*, pages 119–124.

Forster, C. and Wase, D. (1987). Environmental Biotechnology. *Ellis Horwood Limited, England.*

Georgopoulos, V. C., Malandraki, G. A., and Stylios, C. D. (2003). A Fuzzy Cognitive Map approach to differential diagnosis of specific language impairment. *Artificial Intelligence in Medicine*, volume 29, number 3, pages 261–278.

Hiyama, T., Kouzuma, S., and Imakubo, T. (1995). Identification of optimal operating point of PV modules using neural network for real time maximum power tracking control. *IEEE Transactions on Energy Conversion*, volume 10, number 2, pages 360–367.

Hua, C. and Shen, C. (1998). Study of maximum power tracking techniques and control of DC/DC converters for photovoltaic power system. *29th Annual IEEE Power Electronics Specialists Conference.*

Huerga, A. (2002). A balanced differential learning algorithm in fuzzy cognitive maps. *Proceedings of the Sixteenth International Workshop on Qualitative Reasoning.*

Kandasamy, V. and Smarandache, F. (2003). Fuzzy cognitive maps and neutrosophic cognitive maps. *ProQuest Information and Learning (University of Microfilm International).*

Karlis, A., Kottas, T., and Boutalis, Y. (2007). A novel maximum power point tracking method for PV systems using fuzzy cognitive networks (FCN). *Electric Power Systems Research*, volume 77, number 3–4, pages 315–327.

Khan, M., Khor, S., and Chong, A. (2004). Fuzzy cognitive maps with genetic algorithm for goal-oriented decision support. *International Journal Uncertainty, Fuzziness and Knowledge-based Systems*, volume 12, pages 31–42.

Kosko, B. (1986a). Fuzzy cognitive maps. *International Journal of Man-Machine Studies*, volume 24, pages 65–75.

Kosko, B. (1986b). Differential Hebbian learning. *Proceedings American Institute of Physics; Neural Networks for Computing*, pages 277–282.

Kosko, B. (1997). Fuzzy Engineering. *Prentice-Hall, Englewood Cliffs, NU.*

Kottas, L. T., Boutalis, S. Y., and Christodoulou, A. M. (2005). A new method for weight updating in Fuzzy cognitive Maps using system Feedback. *2nd International Conference on Informatics in Control, Automation and Robotics*, pages 202–209.

Kottas, L. T. Boutalis, S. Y. and Christodoulou, A. M. (2007a). Fuzzy cognitive networks: A general framework. *Intelligent Decision Technologies*, volume 1, number 4, pages 183–196.

Kottas, T. L., Boutalis, Y. S., Devedzic, and G., Mertzios, B. G. (2004). A new method for reaching equilibrium points in fuzzy cognitive maps. *Proceedings of 2nd International IEEE Conference of Intelligent Systems*, pages 53–60.

Kottas, T., Boutalis, Y., Diamantis, V., Kosmidou, O., and Aivasidis, A. (2006). A fuzzy cognitive network based control scheme for an anaerobic digestion process. *14th Mediterranean Conference on Control and Applications*, poster session.

Kottas, L. T., Boutalis, S. Y., and Karlis, A. (2007b). A new maximum power point tracker for PV arrays using fuzzy controller in close cooperation with fuzzy cognitive networks. *IEEE Transactions on Energy Conversion*, volume 21, number 3, pages 793–803.

Koulouriotis, D., Diakoulakis, I., and Emiris, D. (2001). Learning fuzzy cognitive maps using evolution strategies: A novel schema for modeling a simulating high-level behavior. *Proceedings of IEEE Congress on Evolutionary Computation*, volume 1, pages 364–371.

Koutroulis, E., Kalaitzakis, K., and Voulgaris, N. (2001). Development of a microcontroller-based, photovoltaic maximum power point tracking control system. *IEEE Transactions on Power Electronics*, volume 16, number 1, pages 46–54.

Liu, Z. Q. and Zhang, J. Y. (2003). Interrogating the structure of fuzzy cognitive maps. *Soft Computing*, volume 7, number 3, pages 148–153.

Marchaim, U. (1992). Biogas Processes for Sustainable Development. *FAO Agricultural Services Bulletin 95, Food and Agriculture Organization of the United Nations.*

Masoum, M., Dehbonei, H., and Fuchs, E. (2002). Theoretical and experimental analyses of photovoltaic systems with voltage- and current-based maximum power-point tracking. *IEEE Transactions on Energy Conversion*, volume 17, number 4, pages 514–522.

Miao, Y., Liu, Z., Siew, C., and Miao, C. (2001). Dynamical cogntive Network-an extension of fuzzy cognitive map. *IEEE Transactions on Fuzzy Systems*, volume 9, number 5, pages 760–770.

Miyamoto, K. (1997). Renewable Biological Systems for Alternative Energy Production. *FAO Agricultural Services Bulletin 128, Food and Agriculture Organization of the United Nations.*

Miyatake, M., Kouno, T., and Nakano, M. (2002). A simple maximum power tracking control employing fibonacci search algorithm for power conditioners of photovoltaic generators. *10th International Power Electronics and Motion Control Conference (EPE-PEMC 2002) Cavtat and Dubrovnik.*

Nemerow, N. and Dasgupta, A. (1991). Industrial and Hazardous Waste Treatment. *Van Nostrand Reinhold, New York.*

Papageorgiou, E. and Groumpos, P. (2004). A weight adaptation method for fuzzy cognitive maps to a process control problem. *Lecture Notes in Computer Science 3037 (Vol. II), M. Budak et al. (Intern. Conference on Computational Science, ICCS 2004, Krakow, Poland, 69 June), Springer Verlag*, pages 515–522.

Papageorgiou, E., Parsopoulos, K., Stylios, C., Groumpos, P., and Vrahatis, M. (2005). Fuzzy cognitive maps learning using particle swarm optimization. *International Journal of Intelligent Information Systems*, volume 25, number 1, pages 95–121.

Papageorgiou, E., Stylios, C., and Groumpos, P. (2004). Active Hebbian learning algorithm to train fuzzy cognitive maps. *International Journal of Approximate Reasoning*, volume 37, number 3, pages 219–247.

Ro, K. and Rahman, S. (1998). Two-loop controller for maximizing performance of a grid-connected photovoltaic-fuel cell hybrid power plant. *IEEE Transactions on Energy Conversion*, volume 13, number 3, pages 276–281.

Rudin, W. (1964). Principles of Mathematical Analysis. *McGraw-Hill Inc.*, pages 220–221.

Salameh, Z. and Taylor, D. (1990). Step-up maximum power point tracker for photovoltaic arrays. *Solar Energy*, volume 44, pages 57–61.

Simoes, M. and Franceschetti, N. (1999). Fuzzy optimization based control of a solar array. *Electric Power Applications, IEE Proceedings*, volume 146, number 5, pages 552–558.

Skiadas, I., Gavala, H., Schmidt, J., and Ahring, B. (2003). Anaerobic granular sludge and biofilm reactors. *Advances in Biochemical Engineering/Biotechnology*, volume 82, pages 35–67.

Smarandache, F. (2001). An introduction to neutrosophy, neutrosophic logic, neutrosophic set, and neutrosophic probability and statistics. *Proceedings of the First International Conference on Neutrosophy, Neutrosophic Logic, Neutrosophic Set, Neutrosophic Probability and Statistics University of New Mexico–Gallup*, volume 1, pages 5–22.

Stach, W., Kurgan, L., Pedrycz, W., and Reformat, M. (2005). Genetic learning of fuzzy cognitive maps. *Fuzzy Sets and Systems*, volume 153, number 3, pages 371–401.

Stylios, C. and Groumpos, P. (1999). A soft computing approach for modelling the supervisor of manufacturing systems. *Journal of Intelligent and Robotics Systems*, volume 26, number 34, pages 389–403.

Stylios, C. and Groumpos, P. (2004). Fuzzy cognitive maps in modeling supervisory control systems. *Journal of Intelligent and Fuzzy Systems*, volume 8, pages 83–98.

Stylios, C. and Groumpos, P., and Georgopoulos, V. (2006). A fuzzy cognitive maps approach to process control systems. *Journal of Intelligent and Robotics Systems*, volume 26, number 3, pages 389–403.

Won, C. (1994). A new maximum power point tracker of photovoltaic arrays using fuzzy controller. *25th Annual IEEE Power Electronics Specialists Conference*, volume 1, number 20–25, pages 396–403.

Zhang, W., Chen, S. and Bezdek, J. (1989). Pool2: A generic system for cognitive map development and decision analysis. *Proceedings of 2nd International IEEE Conference of Intelligent Systems*, volume 19, number 1, pages 31–39.

Zhang, W., Chen, S., Wang, W., and King, R. (1992). A cognitive map based approach to the coordination of distributed cooperative agents. *IEEE Transactions on Systems, Man, and Cybernetics*, volume 22, number 1, pages 103–114.

Zhang, J. Y., Liu, Z., and Zhou, S. (2006). Dynamic domination in fuzzy causal networks. *IEEE Transactions on Fuzzy Systems*, volume 14, number 1, pages 42–57.

22

On the Use of Self-Organising Maps to Analyse Spectral Data

Véronique Cariou[1] and Dominique Bertrand[2]

[1] Sensometrics and Chemometrics Laboratory, INRA – ENITIAA, rue de la Géraudière - BP 82 225, 44 322 Nantes Cedex 3 - France
[2] UR1268, Biopolymères Interactions Assemblages, INRA, F-44300 Nantes, France

Abstract: Self-organizing maps (SOM) have been widely used in different data analysis fields for both their clustering and visualisation properties. However, dealing with spectral data, artificial neural networks (ANN) have generally been applied within a supervised context rather than unsupervised one. In this chapter, we present how the use of self-organizing maps may help end-users to visualise spectral data. While representing the Kohonen map, external characteristics associated with spectra can be projected on the map. Dealing with high-dimensional data, a dimension reduction is proposed to provide synthetic representation of the map to the most relevant variables.

Keywords and phrases: Spectral data, self-organising maps, visualisation

22.1 Introduction

In many analytical applications, samples are characterised by digitised signals such as spectra or chromatograms. A collection of samples is described by a matrix $\boldsymbol{X}$ in which the rows represent the samples and the columns the measures. When data come from digitised curves, the variables are logically ordered and possibly labelled by the wavelength values or the time. The number of variables may be very large and often greater than the number of samples. Moreover, two contiguous variables in the signal share almost the same information and $\boldsymbol{X}$ presents a high degree of colinearity.

Applications of ANN methods on spectral data have mainly focused on a supervised approach. In most studies, qualitative analysis (unsupervised approach) was reduced to principal component analysis or a classical cluster analysis (see Alonso-Salces et al., 2005; Zhang et al., 2006). To our knowledge, the first implementations of SOM on spectral data were due to Caceres-Alonso and Garcia-Tejedor (1995) and Vandeerstraeten et al. (1998). In the last study, the authors showed how the Kohonen map enabled us to determine starch clusters based on their infrared spectra. More recently, a clustering of strawberry varieties was performed with the SOM method (see de Boishebert et al., 2006). Input data consisted in chemical signatures belonging to different varieties. It was confirmed that the map topology guaranteed the varieties' properties: signatures

C.H. Skiadas (ed.), *Advances in Data Analysis*, Statistics for Industry and Technology, DOI 10.1007/978-0-8176-4799-5_22,

corresponding to the same variety were kept in contiguous cells and different varieties were never mixed in the same unit. In Rossi et al. (2004), application of SOM on spectral data was discussed. The main contribution lays in a functional preprocessing to analyse such high-dimensional data. Finally, another study investigated the polyphenolic profiles of cider apple cultivars (see Alonso-Salces et al., 2005) using a first clustering analysis with SOM.

This study fits in the scope of SOM applications to spectral data since it offers both clustering and visualisation tools which may help end-users to analyse data curves. Here, we focus on different ways to visualise SOM results given spectral data:

- Representation of the codebooks, e.g., the patterns, onto the map
- Projection of external characteristics associated with the spectra
- Dimension reduction of spectra given external characteristics

The chapter is organized as follows. In Section 22.2, we outline the main steps of SOM and we introduce visualisation tools to interpret clustering results. In Section 22.3, we present the two different datasets: apple and mixture. We discuss the main results obtained using the SOM method. Finally, Section 22.4 presents further investigations in our work.

22.2 Self-organising map clustering and visualisation tools

Let us denote $\boldsymbol{X}$ the spectral dataset consisting of n rows (spectra) and p dimensions (wavelengths). The Kohonen algorithm is a particular clustering method which preserves the topology of a grid map (see Kohonen, 1995). The SOM consists of a regular, generally two-dimensional (2-D), grid of map neurons. Each neuron i is represented by a codebook $m_i = (m_{i1}, \ldots, m_{ij}, \ldots, m_{ip})$ where m_{ij} is the codebook value computed for the variable x_j. Each codebook corresponds to the prototype vector of individuals belonging to the neuron i. The neurons are linked one to another according to the map topology. During the SOM steps, this topology is taken into account through a neighbourhood function. First, the codebook's initialisation is performed through a principal component analysis applied to the spectral data instead of a random initialisation. Then, SOM is trained iteratively. At each training step, a sample vector $x(t)$ is randomly chosen from the input dataset. Then Euclidean distances between $x(t)$ and all the codebooks are computed.

The winner neuron, noted i^*, is defined as the closest of $x(t)$ given the Euclidean distance. To preserve the topology structure, all codebooks are next updated according to the proximity with the winning one:

$$m_i(t+1) = m_i(t) + \eta_{i^*}(i,t)(x(t) - m_i(t)) \tag{22.1}$$

where t is the time and η is the gain term based on a Gaussian function centered on neuron i^* which decreases during training. A standard neighbourhood function is the following,

$$\eta_{i^*}(i,t) = e^{-d(i^*,i)}/2r^2(t) \tag{22.2}$$

where d is the distance on the SOM map between the winner i^* and the neuron i and $r(t)$ denotes the neuron radius, decreasing during training. At the end of the SOM process, the partition of individuals is built up. Each individual is affected to the neuron which minimises the distance between this individual and its codebook.

Visualisation of the map. After SOM processing, different visualisation tools can be investigated in order to help end-users to interpret the results:

- Representation of the distance-matrix (called U-matrix) giving the codebooks' distances between contiguous neurons.
- Projection of the output map grid onto the principal components space. It provides a visualisation of the grid distortion after SOM process.
- Representation of the codebooks' map. Since spectral data correspond to curves, a convenient way to present clustering results is to visualise the prototype curve of each cluster, e.g., the codebooks.
- Representation of external characteristics associated with the spectra which are not used in the clustering process. For example, dealing with the apple dataset (see (22.3)), end-users may study the correspondence between the output map grid – based on spectra – and the development stage of apples. This is performed by labelling each neuron with the category mostly shared by the individuals belonging to it. Dealing with the mixture dataset, information on the composition of the powdered materials is provided for each sample. This information can be projected onto the map grid: each neuron value is equal to the mean computed on the set of individuals belonging to it.
- Visualisation of the map onto the different variables. Dealing with high-dimensional data, this leads to the representation of thousands of component planes. To bypass this problem, we propose a dimension reduction based on the codebooks' data matrix. In order to include external characteristics, we look for a small number of variables which provide the best correlation with them. This is performed through a partial least squares (PLS) regression (see Stahle and Wold, 1987 and Wold, 1995) performed on the codebooks' data matrix.

22.3 Illustrative examples

The two examples (namely mixture and apple) deal with near infrared (NIR) spectral data. Each observation is characterised by a set of variables representing the intensity of absorbed light as a function of the wavelength of the light.

The mixture collection has been built up for testing the linearity of a NIR spectrometer. Mixtures of powdered materials (wheat, pea, soya bean, maize) have been prepared according to a factorial design by varying the proportion of the different ingredients. The spectra of the mixtures have been acquired with a NIR instrument between 400 and 2498 at 2 nanometre intervals. The final data matrix had dimensions 354 rows (mixtures) and 1050 columns (wavelengths). A preliminary inspection of the data reveals that spectra mainly differ according to their height, e.g., their global intensity. In order to reduce this effect, usually due to instrumental condition variations, a standard normal deviate (SNV) correction (see Barnes et al., 1989) was applied. Original and corrected spectra are shown in Figure 22.1.

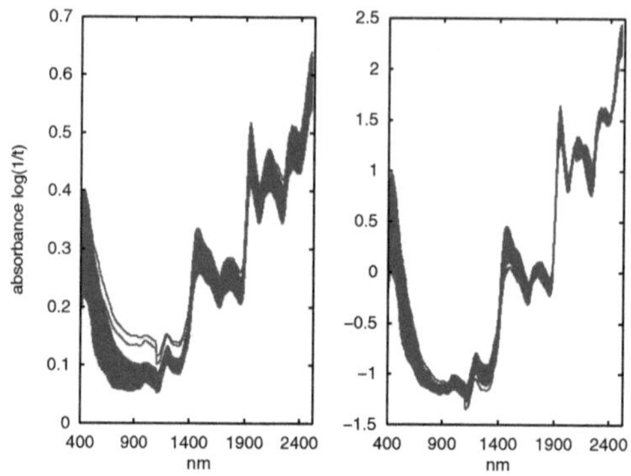

Figure 22.1. Initial mixture spectra and transformed ones using SNV

To apply the SOM algorithm, the Matlab package, available at *http://www.cis.hut.fi/projects/somtoolbox/*, has been used. A 3 × 7 SOM map is built up on the mixture corrected dataset. Figure 22.2 shows the distortion of the map on the first plane of the PCA while Figure 22.3 presents the codebooks' map grid. As with many spectral data, the mixture collection is mostly one-dimensional due to high colinearity. This is confirmed in Figure 22.2. After SOM training, there is a small map distortion on the first PCA plane. Moreover, studying the representation of the neurons' codebooks in Figure 22.3, one can observe the main curves' gradient on the first direction of the map.

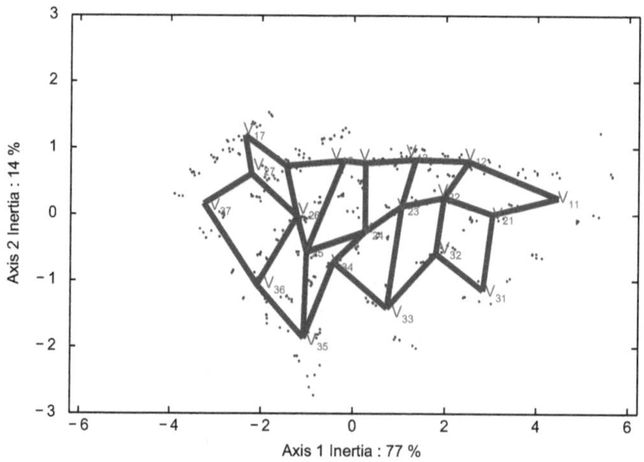

Figure 22.2. SOM map distortion on the first plane of the PCA

In a second stage, we are interested in studying the distribution of an external characteristic, namely the powdered materials, on the organised map. In Figure 22.4, the homogeneity of the map according to the proportions of wheat, maize, pea, and soya bean is revealed. In the center of the map, clusters correspond to complete mixtures.

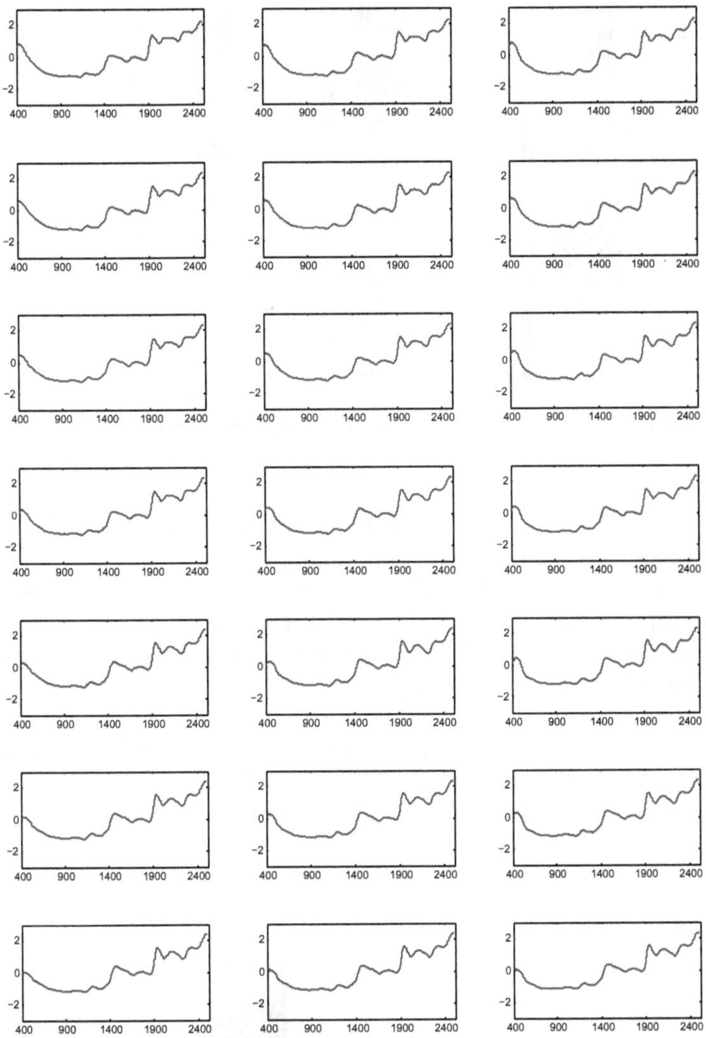

Figure 22.3. Representation of the neurons' codebooks

On the bottom-left of the map, clusters are characterised by high wheat proportion. On the right side, they are characterised by a higher maize proportion. Similarly, on the bottom-right (resp., top-right), a greater pea (resp., soya bean) proportion is observed.

In Figure 22.4, global information on ingredients' proportions is provided. In order to identify a small number of relevant spectral variables based on the correlation between their component planes and the ingredients' proportions, we propose to apply a PLS regression. The $\boldsymbol{Y}$ matrix is defined as the component planes of the wheat, maize, pea, and soya bean ingredients while the $\boldsymbol{X}$ matrix is the spectral component planes. As far as we deal with a map organisation, each component plane is first vectorised. Then, the SOM map is represented on the first PLS component planes. The first three

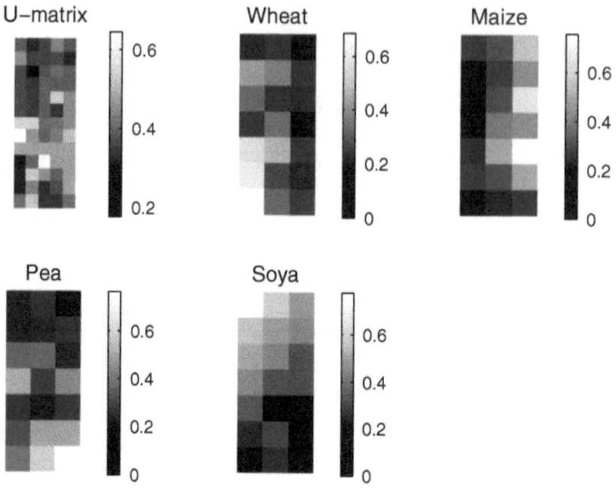

Figure 22.4. Map representation onto the composition external characteristics

components explain 85% of the inertia of $\boldsymbol{Y}$. Figure 22.5 shows the organization of the map on these three component planes.

The apple collection is related to the measure of the maturity of apples according to the time of ripening and the variety (cultivar) of the fruits. During ripening, it is supposed that the fruits may have a continuous evolution which can be detected from the spectral analysis. The fruits have been collected and analysed at six different dates which correspond to different development stages. This collection includes 1066 observations recorded at 550 wavelengths from 1100 to 2200 at 2-nanometre intervals. This dataset has been described in Guillermin et al. (2001). A SNV correction and second-order derivative are first applied to the data collection and a 2×6 map is built up.

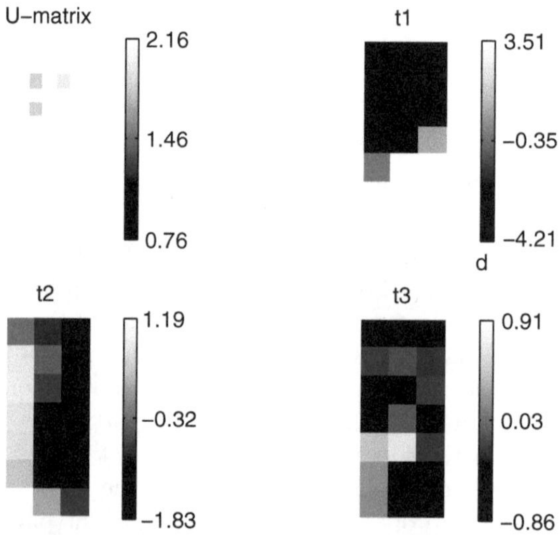

Figure 22.5. Map representation on the PLS components' planes

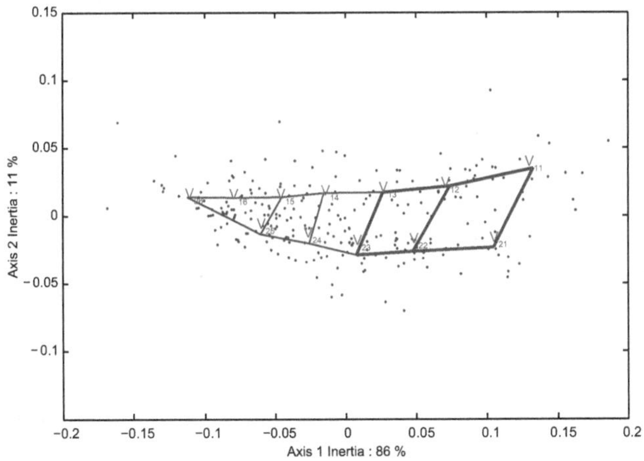

Figure 22.6. SOM map distortion on the first plane of the PCA

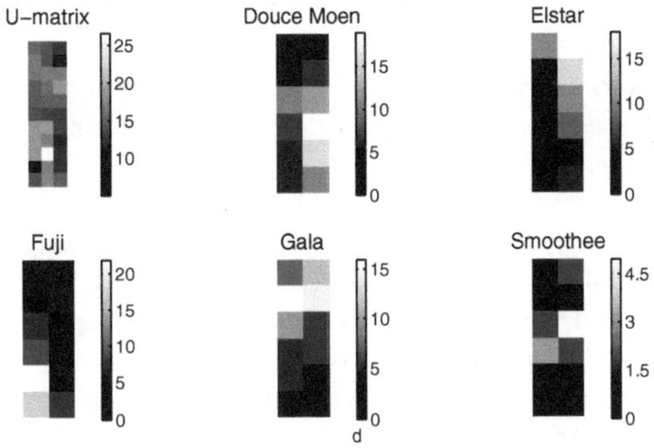

Figure 22.7. Distribution of the varieties onto a 12-unit map

As with mixture, the apple collection is highly unidimensional (see Figure 22.6). To visualise the apple variety, the composition of each neuron is presented on the map (see Figure 22.7). This map reveals the varieties' (Elstar (E), Gala (G), Fuji (F), Smoothee (S), and Douce Moen (D)) properties: signatures corresponding to the same variety were kept in contiguous cells.

22.4 Conclusion

In this chapter, we have presented how to visualise a large number of spectral data through SOM. Further works can be investigated:

- Testing other spectra corrections and mapping distances which would provide a better organisation of the map, especially with high-dimensional spectral data

- Spectral dimension reduction based on the SOM map in order to visualise a small set of component planes corresponding to a combination of initial variables (for example, intervals of wavelengths)
- Fuzzy membership of individuals to neurons in order to take into account the map topology

References

Alonso-Salces, R.M., Herrero, C., Barranco, A., Berrueta, L.A. , Gallo, B., and Vicente, F. (2005). Classification of apple fruits according to their maturity state by the pattern recognition analysis of their polyphenolic compositions. *Food Chemistry*, 93:113–123.

Barnes, R.J., Dhanoa, M.S., and Lister, S.J. (1989). Standard normal variate transformation and de-trending of Near-Infrared diffuse reflectance spectra. *Applied Spectroscopy*, 45:772–777.

Caceres-Alonso, P. and Garcia-Tejedor, A. (1995). Non-supervised neural categorisation of near infrared spectra. Application to pure compounds. *Journal of Near Infrared Spectroscopy*, 3:97–110.

de Boishebert, V., Giraudel, J-L., and Montury, M. (2006). Characterization of strawberry varieties by SPME-GC-MS and Kohonen self-organizing map. *Chemometrics and Intelligent Laboratory Systems*, 80, 1:13–23.

Guillermin, P., Bertrand, D., and Laurens, F. (2001). Application of near infrared spectroscopy to the non-destructive assessment of the development stages of apples. *In 6th International Symposium on Fruit, Nut, Vegetable Production Engineering*, pages 591–596, Potsdam, Germany.

Kohonen, T. (1995). *Self-Organizing Maps.* Springer, Berlin.

Rossi, F., Conan-Guez, B., and El Golli, A. (2004). Clustering functional data with the SOM algorithm. *In Proceedings of XIIth European Symposium on Artificial Neural Networks (ESANN 2004)*, pages 305–312, Bruges, Belgium, April.

Stahle, L. and Wold, S. (1987). Partial least squares analysis with cross-validation for the two-class problem: A Monte Carlo study. *Journal of Chemometrics*, 1: 185–196.

Vandeerstraeten, F., Wojciechowski, C., Dupuy, N., and Huvenne, J-P. (1998). Recognition of starch origin and modifications by chemometrics spectral data processing. *Analysis Magazine*, 26, 8:57–62.

Wold, S. (1995). PLS for multivariate linear modelling. *In QSAR: Chemometric Methods in Molecular Design, Methods and Principles in Medicinal Chemistry*, Weinheim, Germany.

Zhang, G., Ni, Y., Churchill, J., and Kokot, S. (2006). Authentification of vegetable oils on the basis of their physico-chemical properties with the aid of chemometrics. *Talanta*, 70:293–300.

23

Neuro-Fuzzy Versus Traditional Models for Forecasting Wind Energy Production

George Atsalakis, Dimitris Nezis, and Constantinos Zopounidis

Department of Production Engineering and Management, Technical University of Crete, Chania, Crete, Greece

Abstract: It is a well-known fact that the process of forecasting wind energy production is very popular with many researchers who are involved with RES (renewable energy sources). This chapter presents a wind energy production forecasting method, which was carried out with the use of an adaptive neural network with a fuzzy inference system (ANFIS). The model is tested with two different inputs: lagged values of the average speed of wind and the maximum speed of wind. The value of one-step-ahead energy production represents the output of the model. ANFIS uses a combination of the least-squares method and the backpropagation gradient descent method to estimate the optimal parameters of the model. The model is applied to a plant on the island of Evia, Greece. The results are compared with those of the autoregressive (AR) model and those of the autoregressive moving average (ARMA) model. The superiority of ANFIS is revealed.

Keywords and phrases: M11, ANFIS, wind energy production forecasting, renewable energy forecasting, soft computing forecasting

23.1 Introduction

In the very near future, wind energy will be welcomed by society, industry, and politics as a clean, practical, economical, and environmentally friendly alternative. After the 1973 oil crisis, renewable energy (RE) started to appear on the agenda and, hence, wind energy gained significant interest. As a result of extensive studies on this topic, wind energy has recently been applied in various industries (for instance, industries which produce electrical energy through the contribution of wind-turbines; Sahin, 2004).

This chapter specifically looks into wind energy forecasting with neuro-fuzzy techniques, which have been applied in many fields such as model identification of linear and nonlinear systems. In addition, wind energy forecasting is examined and is compared with different approaches in terms of performance, along a time series that is considered difficult to predict.

The wind energy and speed change are not continual throughout the entire year. For this reason, during the planning, design, operation, and maintenance of wind farms,

C.H. Skiadas (ed.), *Advances in Data Analysis*, Statistics for Industry and Technology, DOI 10.1007/978-0-8176-4799-5_23,

the sudden velocity variations are significant. If someone wants to calculate the velocity of wind for a specific region, a very good way is to use the known wind velocity maps, which are very useful forecasting tools for the meteorologists. Wind velocity maps provide a common basis for regional assessments and interpretations without regional prediction. The difficulty in predicting this meteorological parameter arises from the fact that it is a result of the complex interactions among large-scale forcing mechanisms such as pressure and temperature differences, the rotation of the earth, and local characteristics of the surface (Nielsen et al., 1999).

Furthermore, the installed wind energy capacity in Europe today is 20 GW, while the projections for 2010 according to the Kyoto protocol and the EC directives are up to 40–60 GW (Giebel, 2000). The continual development of wind energy creates the need to generate better forecasting tools for short-term forecasting of wind production in subsequent hours for the following 2–7 days. End-users (independent power producers, electric companies, transmission system operators, etc.) recognize the contribution of wind prediction for a secure and economic operation of the power system. More specifically, in a liberalised electricity market, prediction tools enhance the position of wind energy compared to other forms of dispatchable generation. Thus it is widely accepted that wind energy has come of age. It is remarkable to notice that as the electric companies become more skilled in this branch (wind energy forecasting), there are many benefits for themselves as well as for the customers. In some areas, wind energy delivers 20% or more of the electricity demand (Navarra in northern Spain, the Jutland part of Denmark, or the German land of Schleswig-Holstein). The minimum load can, at certain times, be covered solely from wind energy. It is hoped that many states of Europe will be connected to a great extent, in order to provide energy to one another. The lower variability of wind energy on the European scale has another benefit. Since wind energy is strong especially during winter, in the period with the highest demand, it can replace fossil fuel power plants without affecting the loss-of-load probability. The extent to which this is possible is called the capacity credit of wind energy (Giebel, 1996).

23.2 Related research

Recently more modern techniques that have come from the field of soft computing have been applied in energy forecasting. Artificial neural networks (a common engineering approach) were previously employed by Kalogirou (2000) for wind speed prediction. Other modern short-term techniques are the following: (1) Feedforward neural networks: simpler network types than the feedforward are the linear networks (LN). These networks have no hidden layer and the activation function of the output layer is linear (Hush and Horne, 1993). The weights and biases of this network can be trained using the Widrow–Hoff rule, which is a variation of the least mean squares algorithm. These networks produce linear approximations. (2) Radial basis function neural network: the parameters of the radial basis function networks can be determined in three steps: (a) using clustering algorithms detecting unit centres, (b) using the method of nearest neighbour founding widths, and (c) using weights found in the third layer by minimizing the sum squared error between the output and the actual data (Rnaweera et al., 1995).

(3) Elman recurrent network: the type of network used here was introduced by Elman and introduces additional input neurons called context (Connor et al., 1994).

Atsalakis and Ucenic (2006a,b), and Atsalakis et al. (2005) proposed a one-step-ahead neuro-fuzzy energy forecasting system. A genetically involved neural network proposed by Atsalakis is to forecast next-day wind energy (Atsalakis, 2007).

The paper of Cadenas and Rivera (2008) presents the short-term wind speed forecasting in the region of La Venta, Oaxaca, Mexico, by applying the technique of an artificial neural network (ANN) to the hourly time series representative of the site. The data were collected by the Comision Federal de Electricidad (CFE) through a network of measurement stations located in the place of interest. Diverse configurations of ANN were generated and compared through error measures, guaranteeing the performance and accuracy of the chosen models. The developed model for short-term wind speed forecasting indicated a very good accuracy to be used by the Electric Utility Control Centre in Oaxaca for the energy supply.

In the paper of Riahy and Abedi (2008), a new method, based on linear prediction, is proposed for wind speed forecasting. The method utilizes the 'linear prediction' method in conjunction with 'filtering' of the wind speed waveform. For verification purposes, the proposed method is compared with real wind speed data based on experimental results. The results show the effectiveness of the linear prediction method.

A novel approach for the simultaneous modelling and forecasting of wind signal components is presented by Goh et al. (2006). This is achieved in the complex domain by using novel neural network algorithms and architectures. First, a signal nonlinearity and component-dependent analysis are performed, which suggest the use of modular complex-valued recurrent neural networks (RNNs). This RNN-based modelling rests upon a combination of nonlinearity, complexity, and internal memory and allows for the multiple-steps-ahead forecasting of the wind signal in its complex form (speed and direction).

Another interesting paper on wind forecasting has been written by More and Deo (2003). Their work employs the technique of neural networks in order to forecast wind speeds at two coastal locations in India on a daily, weekly, as well as monthly basis. Both feedforward as well as recurrent networks are used. They are trained based on past data in an autoregressive manner using backpropagation and cascade correlation algorithms. A generally satisfactory forecasting as reflected in its higher correlation and lower deviations with actual observations is noted. The neural network forecasting is also found to be more accurate than traditional statistical time series analysis.

A useful study using Kalman filtering as a postprocessing method in numerical predictions of wind speed was presented by Louka et al. (2008). Two limited-area atmospheric models have been employed, with different options/capabilities of horizontal resolution, to provide wind speed forecasts. The application of the Kalman filter to these data leads to the elimination of any possible systematic errors, even in the lower resolution cases, contributing further to the significant reduction of the required CPU time. The results obtained showed a remarkable improvement in the model forecasting skill.

Apart from these forecasting techniques, there are some very interesting forecasting tools, which are proposed by some universities or scientific communities. This model, was developed at the University of Oldenburg (Beyer et al., 1999) and was named Previento (Focken et al., 2001). The Deutschlandmodell or the Lokalmodell (LM) of the German Weather Service (DWD) was used as the NWP model (Numerical Weather

Prediction (NWP)) and is the simulation of atmospheric processes on a computer with the aim of predicting the future development of the atmosphere based on knowledge of the actual state.

Monnich focused on the parameters that influence the accuracy of the results of a short–term forecasting method. After many trials and tests he formed the model to assess atmospheric stability. It was deemed better than some other models, as it receives as contributing factors, roughness, rotation of the earth, and others (Monnich, 2000). The use of MOS (model output statistics) was deemed very useful. However, since the NWP model changed frequently, the use of a recursive technique was recommended. Sometimes, it is observed that the theoretical and actual calculations of the power curve are inversely related because the wind turbines are repaired in regular time margins. It has been proven that the largest influence on the error came from the NWP model itself.

Martí Perez (2002) has developed a family of tools, which were called LocalPred and RegioPred. He involves adaptive optimisation of the NWP input, time series modelling, mesoscale modelling with MM5, and power curve modelling.

A new approach is described by Jorgensen et al. (2002). He integrated the power prediction module within the NWP itself and called it HIRPOM (HIRlam POwer prediction Model). Moreover, he attempted to attain better results than the initial model and for this reason, changed the parameter of horizontal model analysis, but did not manage to improve the overall effectiveness of the model. However, peak wind speeds were closer to the measured values for the high-resolution forecasts. For the higher resolution forecasts, the best model layers were ones closer to the ground than in the coarser models. For the errors, the author points out that phase errors (the timing of the frontal system) has a much larger influence on the error scores (and eventual payments) than level errors. For the same reason, Jorgensen carried out some interesting experiments, whereby he measured 25 bad forecasted days throughout the duration of 15 months for the Danish TSO Eltra and he proved that the data and results of the NWP posed a problem as they negatively influenced the forecasting results of his model.

Landberg (1994) developed a short-term prediction model based on physical reasoning similar to the methodology developed for the European Wind Atlas. It is the perfect example for the model chain in the introduction. He found that for the MOS to converge, about four months' worth of data were needed. If the wind from one of the upper NWP levels is used, the procedure is as follows: from the geostrophic wind and the local roughness, the friction velocity u^* is calculated by using the geostrophic drag law. This is then used in the logarithmic height profile, together with the local roughness. If the wind is already the 10 m wind, then the logarithmic profile can be used directly.

The Institute for Informatics and Mathematical Modelling (IMM) of the Technical University of Denmark developed a popular model, which is called The Wind Power Prediction Tool (WPPT) (Morales and Sipreolico, 2002). Initially, they used adaptive recursive least squares estimation with exponential forgetting in a multistep setup to predict from 0.5 up to 36 hours ahead. However, due to the lack of quality in the results for the higher prediction horizons, the forecasts were only used operationally up to 12 hours ahead. WPPT is a modelling system for predicting the total wind power production in a larger region based on a combination of online measurements of power production from selected wind farms, power measurements for all wind turbines in the area, and numerical weather predictions of wind speed and wind direction. If necessary, the total region is broken into a number of subareas. The predictions for the total region are then calculated using a two-branch approach: the first model, by using online measurements of power production as well as numerical weather predictions as input, makes calculated predictions of wind power for a number of wind farms. The prediction of the total power production in the area is calculated by upscaling the sum of the predictions for the individual wind farms. The second model, by using offline measurements of area power production to the numerical weather predictions, calculates the area power production. For both model branches the power prediction for the total region is calculated as a sum of the predictions for the subareas. The final prediction of the wind power production for the total region is then calculated as a weighted average of the predictions from the two model branches. A central part of this system is statistical models for short-term predictions of the wind power production in wind farms or areas.

Recent research has indicated that conditional parametric models (nonlinear model formulated as a linear model in which the parameters are replaced by smooth, but otherwise unknown, functions of one or more explanatory variables) present a significant improvement of the prediction performance. For online applications it is preferable to allow the function estimates to be modified as data become available. Furthermore, because the system may change slowly over time, observations should be down-weighted as they become older. For this reason, a time adaptive and recursive estimation method is applied. The time-adaptivity of the estimation is an important property in this application of the method as the total system consisting of wind farm or area, surroundings, and the numerical weather prediction (NWP) model will be subject to changes over time. This is caused by the effects incurred by factors such as wind turbines, changes in the monitors of those turbines, and most important, changes of atmospheric processes and the number of operating wind turbines in the wind farm.

Since 2000 the ISET (Institut für Solare Energieversorgungstechnik) has operatively worked with short-term forecasting, using the DWD model and neural networks. It came out of the German federal monitoring program WMEP (Wissenschaftliches Mess- und EvaluierungsProgramm), where the growth of wind energy in Germany was to be monitored in detail (Durstewitz et al., 2001). Their first customer was E.On, who initially lacked an overview of the current wind power production and therefore wanted a good tool (Ernst et al., 2000). Then their model was called the Advanced Wind Power Prediction Tool (AWPT).

EWind is an U.S.-American model by TrueWind, Inc. (Bailey et al., 1999). Instead of using a once-and-for-all parametrisation for the local effects, like the Ris approach does with WAsP, they run the ForeWind numerical weather model as a mesoscale model using boundary conditions from a regional weather model. This way, more physical processes are captured, and the prediction can be tailored better to the local site.

Finally, the University Carlos III of Madrid developed the Siprelico tool, which was used by the popular electric company in Spain, called Red Eléctrica de Espa (the Spanish TSO). There are nine different models, depending on the availability of data. One is a time series analysis model, which does not use NWP input at all. Three more include increasingly higher terms of the forecasted wind speed, while another three also take the forecasted wind direction into account. The last two are combinations of the other ones, plus a nonparametric prediction of the diurnal cycle. These nine models are recursively estimated with both a recursive least squares (RLS) algorithm and a

Kalman filter. For the RLS algorithm, a novel approach is used to determine an adaptive forgetting factor based on the link between the influence of a new observation, using Cook's distance as a measure, and the probability that the parameters have changed.

This chapter encourages the use of neuro-fuzzy models and offers an open forecasting technique to calculate the daily wind energy production.

23.3 Methodology

Fuzzy logic theory was first formulated by Zadeh (1965) as a new means of characterising no probabilistic uncertainties. In contrast to the Boolean 1–0 logic, fuzzy logic also permits in-between values for any judged statement; i.e., it applies a continuous, multivalued logic between 0 and 1. A fuzzy inference system (FIS) is a computing framework that combines the concepts of fuzzy logic, fuzzy decision rules, and fuzzy reasoning (Jang, 1993). The fuzzy decision rules are the way a FIS relates an input variable x to an output variable y. In the case where more than one variable is involved on the premise side, the structure of the rule takes the form:

If $x1$ is A and $x2$ is B, then y is Z,

where $x1$ and$x2$ are the input variables and A, B, and Z are linguistic values (small or big, etc.) defined as the membership function (MF) in the input and output spaces. The steps to create a fuzzy inference model are as follows:

(a) Fuzzification: the input variables are compared with the MFs on the premise part of the fuzzy rules to obtain the probability of each linguistic label.
(b) Combine (through logic operators) the probability on the premise part to get the weight (fire strength) of each rule.
(c) Application of firing strength to the premise MFs of each rule to generate the qualified consequent of each rule depending on its weight.
(d) Defuzzification: Aggregate the qualified consequents to produce a crisp output.

In early examples of fuzzy modeling, attempts were made to extract the fuzzy rules directly from the expert's knowledge. Later, new methods have been developed that use an automatic process to generate the fuzzy rules, taking advantage of neural network algorithms.

Neural networks try to imitate the function of the brain and for this reason the connections between neurons determine the function of the network. Layers of neurons form a neural network. A layer includes the weight matrix, the summations, the bias vector, the transfer function, and the output vector. A layer whose output is the network output is named the output layer. All the others are called hidden layers. A neuro-fuzzy system is defined as a combination of neural networks and fuzzy inference system. Jang and Sun (1993) introduced an adaptive neuro-fuzzy inference system (ANFIS) where the MF parameters are fitted to a dataset through a hybrid-learning algorithm. The basis of the ANFIS model is the theory of artificial neural networks (ANN). An example of ANFIS consists of a first-order Sugeno type FIS, with two input variables (x and y), one output (z), and two if-and-then rules. Each input space has been characterised by two intuitively labelled bell MFs, drawn separately for clarity and for graphical representation of each rule (Jang and Sun, 1993).

The structure of the ANFIS is considered for simplicity as a fuzzy system with only two inputs and one output of first-order Sugeno type. The output of each layer is the input of the next layer. The first layer is the input layer that is adaptive for the nonlinear parameters and carries out the fuzzification of each numeric input variable and the output is the value of each membership function. The second layer that is nonadaptive makes T-norm operations of each combination of the defined fuzzy sets. Its output is the firing strength value of each T-norm operation. Layer 3 that is nonadaptive carries out the normalisation of all firing strengths and this output is normalised firing strengths. Layer 4 that is adaptive for the linear parameters calculates the product of each normalised firing strength by each of the Sugeno first-order crisp function values. Layer 5, which is not adaptive, derives the summation of all incoming signals (products) and gives as an output the desired prediction.

23.4 Model presentation

The forecasting of wind energy production is carried out by an ANFIS model, which uses a first-order Sugeno-type FIS (Jang, 1993). The model forecasts the daily energy production one step ahead (next day). The method of trial and error is used in order to decide the type of membership function that best describes the model and provides the lowest error. An estimate of the mean square error between observed and modeled values is computed for each trial, and the best structure is determined by considering a trade-off between the mean square error and the number of parameters involved in computation. The trapezoidal membership function derived better results than the Gauss2mf, gbell, triangular, and Gauss membership functions. Finally, two-membership functions of trapezoidal shape are chosen for each input variable, of the following form,

$$trapmf(x,a,b,c,d) = \max\left(\min\left(\frac{x-a}{b-a}, 1, \frac{d-x}{d-c}\right), 0\right)$$

Once the ANFIS structure is identified, the parameters of the trapezoidal MFs and the output constants are fitted by the hybrid learning algorithm (Jang, 1993). ANFIS applies a mixture of the least squares method (for the consequent part of the rules) and the backpropagation gradient descent method (for the premise part of the rules) for training the fuzzy inference system membership function parameters to emulate a given training dataset. Also, it uses a checking dataset for checking the model over fitting.

Four ANFIS models are developed using different input variables for each model. Table 23.1 presents the models and the different input variables. The inputs of each of the models are lagged values (k − 1, k − 2, or k − 3) of the independent variables. In fact, the inputs represent the wind speed value one day before and two or three days before. The output of the models is the next day wind energy production value.

The linguistic labels of each input are ‘low’ and ‘high’. After training the models the rules are automatically created by the ANFIS and they have the following form.

R1:**If** x **is** A_1 **and** y **is** B_1 **then** $f_1 = p_1 \times x + q_1 \times y + r_1$
R2:**If** x **is** A_2 y **is** B_2 **then** $f_2 = p_2 \times x + q_2 \times y + r_2$

Table 23.1. Input variables for each model

Models	Description of inputs	No. of MFs
ANFIS 1	M_S AVG: the average speed of wind at time k − 1	2
	M_S AVG: the average speed of wind at time k − 2	2
ANFIS 2	M_S AVG: the average speed of wind at time k − 1	2
	M_S AVG: the average speed of wind at time k − 3	2
ANFIS 3	M_S MAX: the maximum speed of wind at time k − 1	2
	M_S AVG: the maximum speed of wind at time k − 2	2
ANFIS 4	M_S MAX: the maximum speed of wind at time k − 1	2
	M_S AVG: the maximum speed of wind at time k − 3	2

Table 23.2 reports the four rules that the ANFIS models use to reach a conclusion. ANFIS 2 uses the same rules as ANFIS 1 but with different lagged days (k − 3). ANFIS 3 uses the same rules as ANFIS 1 but with different input variables. ANFIS 4 uses the same rules as ANFIS 3 but with a different lagged day (k − 3).

Table 23.2. The rules of the ANFIS 1 model

Rule	Rule's description
R1:	If M_S AVG (k-1) is low and M_S AVG (k − 2) is low then output is out1mf1
R2:	If M_S AVG (k-1) is low and M_S AVG (k − 2) is high then output is out1mf2
R3:	If M_S AVG (k-2) is high and M_S AVG (k − 1) is low then output is out1mf3
R4:	If M_S AVG (k-2) is high and M_S AVG (k − 1) is high then output isout1mf4

The model is tested many times by using a different number of epochs. Finally, good results are obtained after 300 epochs. Figure 23.1 represents the initial membership functions of each input variable before the training of the model. Figure 23.2 presents the final form of membership functions after the completion of the training process.

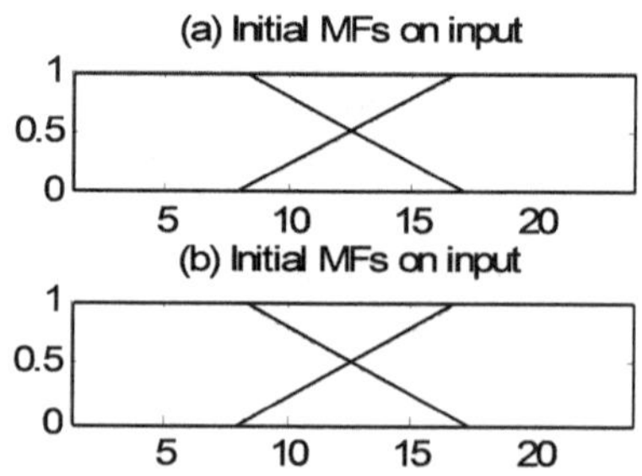

Figure 23.1. MFs before training

Figure 23.3 depicts a graphical representation of the fuzzy reasoning mechanism. The rows represent the rules and the columns represent the membership functions of each input and the output (i.e., if the value of the first input is 12.5 and the value of the second input is 12.5 then the output value is 11.3).

Finally, 21 notes and 28 parameters are created in this version of ANFIS. Twelve parameters are linear and 16 are nonlinear.

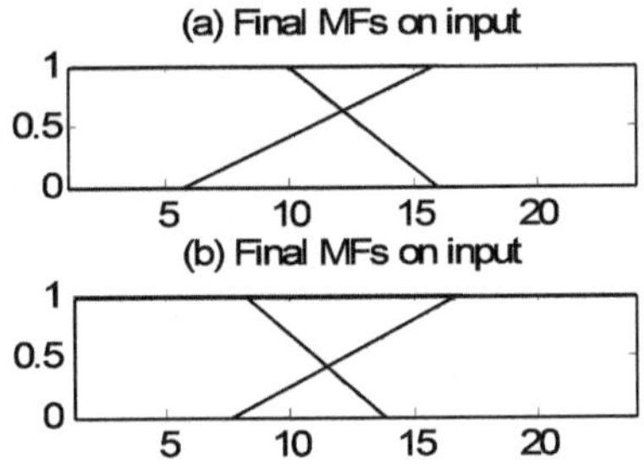

Figure 23.2. MFs after the training

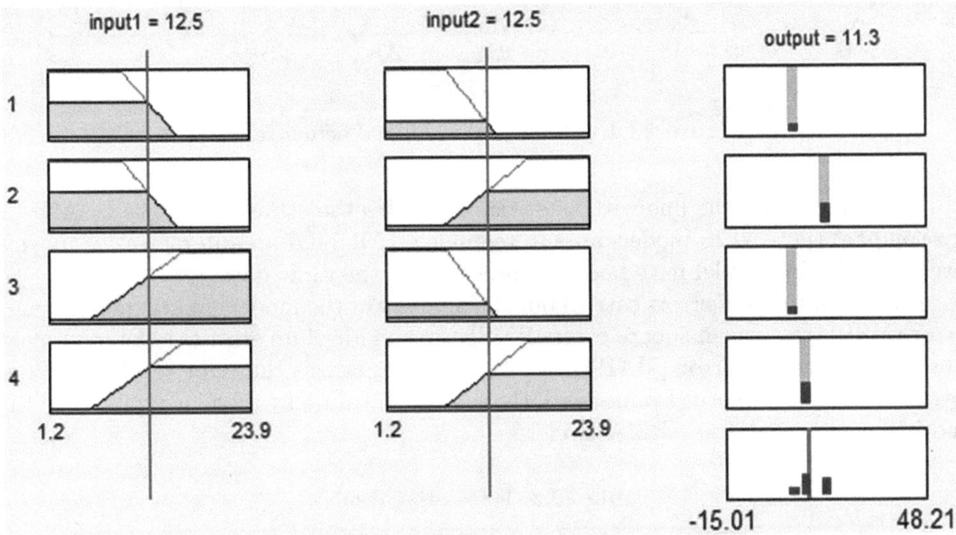

Figure 23.3. A view of the rules and the decision mechanism

23.5 Results

The data are real-world data and have been derived from the wind plant island of Evia. A total of 365 samples was collected from 9 March 2005 until 22 March 2006. The first 68.5% (248 samples) of data are used for training and 31.5% (113 samples) for testing. Four types of ANFIS models have been trained and tested. ANFIS 1 has as input, the values of the average speed of the wind at times $k-1$ and $k-2$, ANFIS 2, has as inputs the values of the average speed of the wind at times $k-1$ and $k-2$. ANFIS 3, has as inputs the values of the maximum speed of the wind at times $k-1$ and $k-2$, and ANFIS 4, has as inputs the values of the maximum speed of the wind at times $k-1$ and $k-3$. The parameter k symbolizes the time (lagged days) of the values of the variables. The disposition of the values is determined by the phases of training and testing data, which have been defined by the algorithm.

A graphical comparison between actual values and ANFIS 1 estimated values (a part of the samples) is illustrated in Figure 23.4. The line with square boxes illustrates the

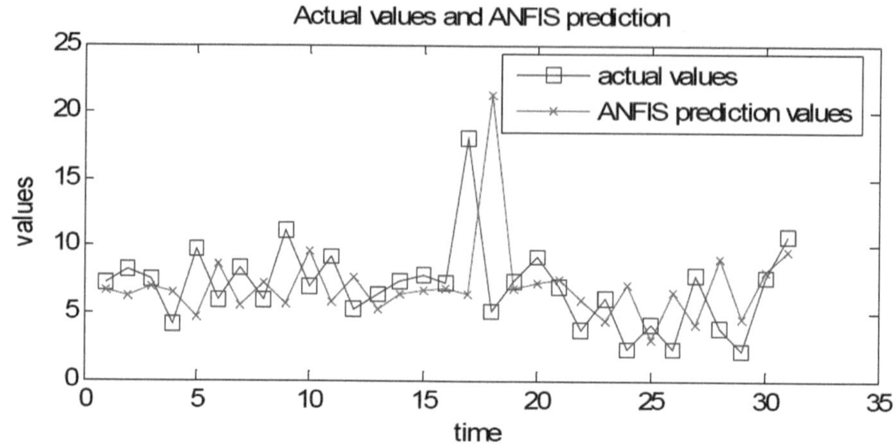

Figure 23.4. ANFIS prediction and actual values

observed values and the line with asterisks illustrates the estimated values by ANFIS. To compare the ANFIS models an autoregressive (AR) and an autoregressive moving average (ARMA) model have been estimated using the same data.

Four main types of errors carried out the analysis of the model quality: mean square error (MSE), root mean square error (RMSE), mean absolute error (MAE), and mean absolute percentage error (MAPE). Table 23.3 summarizes the error results obtained by comparing the observed values with the estimated values by the four types of ANFIS models and the AR and ARMA models.

Table 23.3. Forecasting results

Errors	ANFIS 1	ANFIS 2	ANFIS 3	ANFIS 4	AR	ARMA
MSE	**15.79**	24.95	91.27	65.81	17.07	16.04
RMSE	**3.97**	4.99	9.95	8.11	4.13	4.05
MAE	**2.73**	3.28	6.84	6.13	2.82	2.79
MAPE	**36.04**	41.33	36.98	37.00	37.61	37.17

As the above table of errors indicates, the model with the best results is the ANFIS 1 model, as it has the same average speed of wind as an input with lagged values $k-1$ and $k-2$. It gives the lowest value (in bold numbers) of error in all error measures and reveals the superiority of ANFIS versus to the traditional models.

23.6 Conclusion

This chapter presents an ANFIS model to forecast the next day wind energy production. The results are presented and compared based on four different kinds of errors: MSE, RMSE, MAE, and MAPE. Using various input variables, four types of ANFIS models are estimated and their results are compared with the results that are calculated by the AR and ARMA models. ANFIS 1 outperforms the other models.

ANFIS is a model-free, easy to implement approach. In contrast to traditional time forecasting methods, little training is needed to calculate wind energy predictions. It implements a single-fitting procedure to nonlinear situations, without the need of establishing a formal model for the problem being resolved. Thus, no a priori information is required to determine the empirical relationship between the explanatory and predicted variables, and the method suitability is always tested a posteriori. Moreover, the transparent rule structure of ANFIS allows the researcher to extract information about the empirical relationship between the wind speed and the energy production over time and to provide concise explanations.

Despite the above advantages, the ANFIS must be implemented very carefully. The minimum number of data samples must be at least 150, and the number of model parameters should not exceed one fourth of the number of samples in the training sets, in order to avoid the risk of overfitting and losing generality.

References

Atsalakis, G. (2007). Wind energy production forecasting by neural networks and genetic algorithms. *European Computer Conference*, Athens, www.wseas.org

Atsalakis, G., Ucenic, C. (2006a). Electric load forecasting by neuro-fuzzy approach. *WSEAS International Conference on Energy and Environmental Systems*, Chalkida, Evia, Greece. www.wseas.org

Atsalakis, G., Ucenic, C. (2006b). Forecasting the electricity demand using a neuro-fuzzy approach versus traditional methods. *Journal of WSEAS Transactions on Business and Economics*, 3:9–17.

Atsalakis, G., Ucenic, C., Plokamakis, G. (2005). Forecasting of electricity demand using neuro-fuzzy (ANFIS) approach. *Proceedings of International Conference on NHIBE*, Corfu, Greece, pp. 335–341.

Bailey, B., Brower, M.C., Zack, C. (1999). Short-term wind forecasting. *Proceedings of the European Wind Energy Conference*, Nice, France, pp. 1062–1065.

Beyer, H.G., Heinemann, D., Mellinghoff, H., Monnich, K., Waldl, H.P. (1999). Forecast of regional power output of wind turbines, *Proceedings of the European Wind Energy Conference*, Nice, France, pp. 1070–1073.

Cadenas, E., Rivera, W. (2008). Short term wind speed forecasting in La Venta, Oaxaca, Mexico, using artificial neural networks. *Journal of Renewable Energy*, 34:274–278.

Connor, J., Martin R., Atlas, L. (1994). Recurrent neural networks and robust time series prediction. *IEEE Transactions on Neural Networks*, 5:240–253.

Durstewitz, M., Ensslin, C., Hahn B., Hoppe-Kilpper, M. (2001). Annual evaluation of the scientific measurement and evaluation programme (WMEP). *Working paper*, Kassel.

Ernst, B., Rohrig, K., Regber H., Schorn, P. (2000). Managing 3000 MW wind power in a transmission system operation center. *Proceedings of the European Wind Energy Conference*, Copenhagen, Denmark, pp. 890–893.

Focken, U., Lange, M., Waldl, H.P. (2001). A wind power prediction system with an innovative upscaling algorithm. *Proceedings of the European Wind Energy Conference*, Copenhagen, Denmark, pp. 826–829.

Giebel, G. (2000). On the benefits of distributed generation of wind energy in Europe. *Working paper*, University of Oldenburg, pp. 1–2.

Giebel, G. (1996). Utility wind-modeling planning meeting. *Working paper*, National Renewable Energy Laboratory, Golden, Colorado, USA, pp. 22–23.

Goh, S.L., Chen, M., Popovic, D.H., Aiharab, K., Obradovic, D., Mandic, D.P. (2006). Complex-valued forecasting of wind profile. *Renewable Energy*, 31:1733–1750.

Hush, D., Horne, B. (1993). Progress in supervised neural networks, what is new since Lipmann? *IEEE Signal Processing Magazine*, 10:8–39.

Jang, J.S. (1993). ANFIS: Adaptive –network –based fuzzy inference system. *IEEE Transactions on Systems, Man and Cybernetics*, 23:665–684.

Jang, J.S., Sun, C. (1993). Predicting chaotic time series with fuzzy if-then rules. *Proceedings of the IEEE International Conference on Fuzzy Systems*, pp. 1079–1084.

Jorgensen, J., Moehrlen, C., Gallagoir, B.O., Sattler, K., McKeogh, E. (2002). HIRPOM: Description of an operational numerical wind power prediction model for large scale integration of on- and offshore wind power in Denmark. *Poster on the Global Windpower Conference and Exhibition*, Paris, France, www.anemos.cma.fr

Kalogirou, S.A. (2000). Applications of artificial neural networks for energy systems. *Applied Energy*, 67:17–35.

Landberg, L. (1994). Short-term prediction of local wind conditions. PhD Thesis, Riso-R-702 (EN), Riso National Laboratory, Roskilde, Denmark, www.springerlink.com

Louka, P., Galanis, G., Siebert, N., Kariniotakis, G., Katsafados, P., Pytharoulis, I., Kallos, G. (2008). Improvements in wind speed forecasts for wind power prediction purposes using Kalman filtering. *Journal of Wind Engineering and Industrial Aerodynamics*, 96:2348–2362.

Martí Perez, I. (2002). Wind forecasting activities. *Proceedings of the First IEA Joint Action Symposium on Wind Forecasting Techniques*, Published by FOI–Swedish Defence Research Agency Norrkoping, Sweden, pp. 11–20, www.anemos.cma.fr

Monnich, K. (2000). Vorhersage der Leistungsabgabe netzeinspeisender windkraftanlagen zur Unterstützung der Kraftwerkseinsatzplanung. PhD thesis, Carl von Ossietzky Universitat, Oldenburg, www.docserver.bis.uni-oldenburg.de

Morales, G., Sipreolico, G. (2002). Wind power prediction experience, Talk slides accompanied by the paper: Sánchez, I., Usaola, J., Ravelo, O., Velasco, C., Domínguez, G., Lobo, M.G, González, G., Soto, F., Díaz-Guerra B., Alonso M. A wind power prediction system based on flexible combination of dynamic models, Application to the Spanish power system. *Proceedings of the First IEA Joint Action Symposium on Wind Forecasting Techniques* Published by FOI–Swedish Defence Research Agency, Norrkoping, Sweden, pp. 197–214.

More, A., Deo, M.C. (2003). Forecasting wind with neural networks. *Marine Structures*, 16:35–49.

Nielsen, T.S., Madsen, H., Tofting, G. (1999). Experiences with statistical methods for wind power prediction. *Proceedings of the European Wind Energy Conference*, Nice, France, pp. 1066–1069.

Riahy, G.H., Abedi, M. (2008). Short term wind speed forecasting for wind turbine applications using linear prediction method. *Renewable Energy*, 33:35–41.

Rnaweera, D., Hubele, N., Papalexopoulos, A. (1995). Application of radial basis function network model for short term load forecasting. *IEEE Proceedings Generation Transmission Distribution*, 142(1): 45–50.

Sahin, A.D. (2004). Progress and recent trends in wind energy. *Progress in Energy and Combustion Science*, 30:501–543.

Zadeh, L.A. (1965). Fuzzy sets. *Information and Control*, 8:338–353.

Part VII

Parametric and Nonparametric Statistics

24

Nonparametric Comparison of Several Sequential k-out-of-n Systems

Eric Beutner

Department of Quantitative Economics, Maastricht University, Tongersestraat 53, NL-6200 MD Maastricht, The Netherlands (e-mail: e.beutner@maastrichtuniversity.nl)

Abstract: Sequential order statistics have been introduced to model sequential k-out-of-n systems which, as an extension of k-out-of-n systems, allow the failure of some components of the system to influence the remaining ones. Here, we consider nonparametric hypothesis testing for making the decision whether the baseline distributions of several sequential k-out-of-n systems are equal. The asymptotic distribution of the test statistics are derived for the case of known model parameters and for the case where the model parameters of the systems are unknown.

Keywords and phrases: k-out-of-n systems, sequential k-out-of-n systems, counting processes, nonparametric K-sample tests

24.1 Introduction

An n component system functioning as long as k $(1 \leq k \leq n)$ components work is called a *k-out-of-n system*. Particular cases are parallel and series systems corresponding to $k = 1$ and $k = n$, respectively. The failure times T_i, $1 \leq i \leq n$, of the n components are often assumed to be iid random variables; see, for example, Barlow and Proschan (1981), Meeker and Escobar (1998), and Navarro and Rychlik (2007) for the exchangeable case. The particular probabilistic model in which it is supposed that the failure times T_i, $1 \leq i \leq n$, of the n components are iid random variables is hereinafter referred to as the common k-out-of-n model. Implicit in this assumption is that the failure of any component of the system does not affect the remaining lifetime of the components that are still at work. In many situations, however, the assumption of the failure times being iid random variables may not be reasonable. For example, the failure of a high-voltage transmission line will increase the load put on the remaining high-voltage transmission lines, thus violating the iid assumption.

In this context, extended models have been proposed in the literature; see Kamps (1995a) for the *sequential k-out-of-n model* as well as Hollander and Peña (1995) for an extension using a counting process approach. Both models are flexible in the sense that they allow for the distribution of the residual lifetime of the remaining components,

C.H. Skiadas (ed.), *Advances in Data Analysis*, Statistics for Industry and Technology, DOI 10.1007/978-0-8176-4799-5_24,

after the failure of some component, to change; i.e., the underlying failure rate of the remaining components is adjusted according to the number of preceding failures.

The sequential k-out-of-n model is defined via random variables $X_1^*, \ldots, X_{n-k+1}^*$, which are called sequential order statistics. These random variables describe the failure times of a k-out-of-n system. Thus, in the sequential k-out-of-n model, the life length of a k-out-of-n system is given by X_{n-k+1}^*. In the particular setting of sequential order statistics chosen here, the distribution of the random variables $X_1^*, \ldots, X_{n-k+1}^*$ is determined by a distribution function F, called the baseline distribution, and model parameters $\alpha_2, \ldots, \alpha_{n-k+1}$, which describe the adjustment of the failure rate of the remaining components according to the number of preceding failures. It is worth mentioning that, if we take $\alpha_2 = \cdots = \alpha_{n-k+1} = 1$, the random variables $X_1^*, \ldots, X_{n-k+1}^*$ are distributed as order statistics from a random sample of size n with underlying distribution function F. Hence, the sequential k-out-of-n model comprises the common k-out-of-n model. For general theoretical properties and applications of sequential order statistics, one may refer to Cramer and Kamps (2001b, 2003), and Balakrishnan et al. (2008). Belzunce et al. (2003) consider conditions for certain aging properties of a vector of sequential order statistics (see also Hu and Zhuang, 2006). Comparison results for sequential order statistics can be found in Belzunce et al. (2005) and Zhuang and Hu (2007).

In this chapter, we discuss the problem of testing whether the underlying distribution functions of several sequential k-out-of-n systems are equal. Formally, the problem is to decide if K different sequential k-out-of-n systems have the same underlying distribution function, i.e., to test

$$H_0: \quad F_1 = \cdots = F_K = F_0.$$

There is a large literature on parametric methods for the same testing problem in the common k-out-of-n system; one may refer to Kalbfleisch and Prentice (1980) and Lawless (2003). Parametric statistical inference for sequential order statistics in the K sample case may be found in Cramer and Kamps (2001a). We concentrate here on nonparametric methods. The special case of $K = 2$ and a special weight function nonparametric method for the above problem are treated in Beutner (2008). Here we extend these results to arbitrary K and a much wider class of weight functions.

In Section 24.2, we give a short description of sequential order statistics, derive the test statistics, and state some properties of counting processes based on sequential order statistics. Next, in Section 24.3, we derive the asymptotic distribution of the test statistic in the case where the model parameters $\alpha_2, \ldots, \alpha_{n-k+1}$ are known. Finally, in Section 24.4 the results are extended to the case where the model parameters $\alpha_2, \ldots, \alpha_{n-k+1}$ are unknown.

24.2 Preliminaries and derivation of the test statistics

24.2.1 Sequential order statistics: Introduction and motivation

A definition of sequential order statistics with a view to the motivation given in the introduction can be found in Cramer and Kamps (1996). As shown in Cramer and Kamps (2003) they can also be defined as follows.

Definition 1. *Let $F_1, \ldots, F_n$ be distribution functions with $F_1^{-1}(1) \leq \cdots \leq F_n^{-1}(1)$, and let $V_1, \ldots, V_n$ be independent random variables with $V_r \sim Beta(n-r+1,1)$, $1 \leq r \leq n$. Then the random variables*

$$X_r^* = F_r^{-1}(1 - V_r R_r(X_{r-1}^*)), \qquad 1 \leq r \leq n, \qquad X_0^* = -\infty,$$

are called sequential order statistics (based on $F_1, \ldots, F_n$), where R_r denotes the reliability function $1 - F_r$.

Assumption 1 *In the following we restrict ourselves to a particular choice of the distribution functions $F_1, \ldots, F_n$, namely*

$$F_1(t) = F(t), \quad \text{and} \quad F_i(t) = 1 - (1 - F(t))^{\alpha_i}, \qquad i = 2, \ldots, n \tag{24.1}$$

for positive real numbers $\alpha_2, \ldots, \alpha_n$.

Remark 5.3. The restriction to the choice $F_1(t) = F(t)$, and $F_i(t) = 1 - (1 - F(t))^{\alpha_i}, i = 2, \ldots, n$, has two advantages. The first advantage is that the distribution of $X_1^*, \ldots, X_n^*$ depends only on the distribution function F, called the baseline distribution, and the parameters $\alpha_2, \ldots, \alpha_n$, thus, reducing the model uncertainty. The second advantage is that, in this case, the model of sequential order statistics coincides with the model of generalized order statistics in the distributional theoretical sense. The model of generalized order statistics contains, for example, order statistics and progressively Type-II censored order statistics. For nonparametric estimation and hypothesis testing with progressively Type-II censored order statistics see Bordes (2004), Alvarez-Andrade and Bordes (2004), Alvarez-Andrade et al. (2007), Guilbaud (2001, 2004), Balakrishnan (2007), and Balakrishnan et al. (2007).

From the above definition of sequential order statistics or from Kamps (1995b, p. 4) we can derive the following property. For $t > s$ we have

$$\begin{aligned} P(X_i^* > t | X_{i-1}^* = s) &= \left(\frac{1 - (1 - (1 - F(t))^{\alpha_i})}{1 - (1 - (1 - F(s))^{\alpha_i})} \right)^{n-i+1} \\ &= \left(\frac{1 - F(t)}{1 - F(s)} \right)^{(n-i+1)\alpha_i}. \end{aligned}$$

Hence, the conditional hazard rate function $\tilde{\lambda}_i$ of the ith failure time given that the $(i-1)$th failure occurred at time s is given by

$$\tilde{\lambda}_i(t) = (n - i + 1)\alpha_i \frac{f(t)}{1 - F(t)}, \qquad t > s.$$

The last equation is suitable to explain why the sequential k-out-of-n model is more flexible than the common k-out-of-n model. Recall that in the common k-out-of-n model the conditional hazard rates are given by

$$\tilde{\lambda}_i(t) = (n - i + 1) \frac{f(t)}{1 - F(t)}, \qquad i = 1, \ldots, n - k + 1.$$

Thus, in the common k-out-of-n model a failure does not affect the conditional hazard rate functions of the components still at work, whereas in the sequential k-out-of-n model they jump to

$$\alpha_i \frac{f(t)}{1-F(t)}$$

after a failure. Here, the parameters α_i, $2 \le i \le n-k+1$, allow the researcher to model the adjustment of the load put on the remaining components.

24.2.2 Sequential order statistics and associated counting processes

Our test statistics are based on the following data. For each of the K sequential k_i-out-of-n_i systems we have m_i, $1 \le i \le K$, independent observations during the time interval $[0,T]$. The associated vectors of failure times are denoted by

$$\mathbf{X}^*_{i,j} = (X^*_{1,i,j}, \ldots, X^*_{n_i-k_i+1,i,j}), \qquad 1 \le i \le K, \qquad 1 \le j \le m_i.$$

To facilitate the presentation we introduce the following notations where $0 \le t \le T$.

$$\begin{aligned}
N_{i,j}(t) &= \sum_{\ell=1}^{n_i-k_i+1} I_{\{X^*_{\ell,i,j} \le t\}}, \qquad 1 \le i \le K, 1 \le j \le m_i,\\
\bar{N}_{m_i}(t) &= \sum_{j=1}^{m_i} N_{i,j}(t), \qquad 1 \le i \le K,\\
\bar{N}_m(t) &= \sum_{i=1}^{K} \bar{N}_{m_i}(t),\\
\boldsymbol{\rho}_{i,j}(t) &= (\rho_{1,i,j}(t), \ldots, \rho_{n_i-k_i+1,i,j}(t))\\
&= \left(I_{\{X^*_{0,i,j} < t \le X^*_{1,i,j}\}}, \ldots, I_{\{X^*_{n_i-k_i,i,j} < t \le X^*_{n_i-k_i+1,i,j}\}}\right),\\
&\quad 1 \le i \le K, 1 \le j \le m_i,\\
\mathbf{Y}_{i,j}(t) &= (Y_{1,i,j}(t), \ldots, Y_{n_i-k_i+1,i,j}(t))\\
&= (n_i, n_i-1, \ldots, k_i) * \boldsymbol{\rho}_{i,j}(t), \qquad 1 \le i \le K, 1 \le j \le m_i,\\
\bar{\mathbf{Y}}_{m_i}(t) &= (\bar{Y}_{1,m_i}(t), \ldots, \bar{Y}_{n_i-k_i+1,m_i}(t))\\
&= \sum_{j=1}^{m_i} \mathbf{Y}_{i,j}(t), \qquad 1 \le i \le K,\\
\bar{\mathbf{Y}}_m(t) &= \sum_{i=1}^{K} \bar{\mathbf{Y}}_{m_i}(t),\\
\boldsymbol{\alpha}_i &= (1, \alpha_{2,i}, \ldots, \alpha_{n_i-k_i+1,i}), \qquad 1 \le i \le K,\\
\boldsymbol{\nu}_{m_i}(\boldsymbol{\alpha}_i, t) &= \frac{1}{\boldsymbol{\alpha}_i \bar{\mathbf{Y}}'_{m_i}(t)} \boldsymbol{\alpha}_i * \bar{\mathbf{Y}}_{m_i}(t), \qquad 1 \le i \le K,\\
\mathbf{e}_i(t) &= (e_{1,i}(t), \ldots, e_{n_i-k_i+1,i}(t))\\
&= (E[Y_{1,i}(t)], \ldots, E[Y_{n_i-k_i+1,i}(t)]), \qquad 1 \le i \le K,
\end{aligned}$$

$$\Psi_i(\boldsymbol{\alpha}_i, t) = \int_0^t \left[\mathbf{D}\left(\frac{\boldsymbol{\alpha}_i * \mathbf{e}_i(s)}{\boldsymbol{\alpha}_i \mathbf{e}'_i(s)}\right) - \left(\frac{\boldsymbol{\alpha}_i * \mathbf{e}_i(s)}{\boldsymbol{\alpha}_i \mathbf{e}'_i(s)}\right)' \cdot \left(\frac{\boldsymbol{\alpha}_i * \mathbf{e}_i(s)}{\boldsymbol{\alpha}_i \mathbf{e}'_i(s)}\right)\right] \boldsymbol{\alpha}_i \mathbf{e}'_i(s) \lambda_i(s) ds,$$
$$1 \le i \le K.$$

Here and in the following, I denotes the indicator function, $\star$ represents component-by-component multiplication for vectors, $'$ denotes the transpose, and λ_i is the hazard rate function of F_i. Given a vector $\boldsymbol{\zeta}$ we denote by $\mathbf{D}(\boldsymbol{\zeta})$ a diagonal matrix with diagonal elements $\boldsymbol{\zeta}$.

The processes $N_{i,j}$ describe the number of failures of the jth, $1 \leq j \leq m_i$, observation of the ith, $1 \leq i \leq K$, system up to time t; the processes $\bar{N}_{m_i}$ describe the number of failures of the ith, $1 \leq i \leq K$, system up to time t if we combine all observations of the ith system; and $\bar{N}_m$ gives us the number of failures up to time t if all systems and observations are combined. The processes $\mathbf{Y}_{i,j}$ represent for the jth, $1 \leq j \leq m_i$, observation of the ith, $1 \leq i \leq K$, system the number of the risk set at time t.

We have the following result, a proof of which can be found in Beutner (2008).

Lemma 1. *Let λ_i be the hazard rate function of F_i. Then, for every i, $1 \leq i \leq K$, and every j, $1 \leq j \leq m_i$, the processes*

$$M_{i,j}(t) = N_{i,j}(t) - \int_0^t \boldsymbol{\alpha}_i \mathbf{Y}'_{i,j}(s)\lambda_i(s)ds, \qquad 0 \leq t \leq T,$$

are square-integrable martingales with respect to the natural filtration denoted by $\mathcal{F}_t^{i,j}$.

Hence, we obtain that the processes

$$\bar{M}_{m_i}(t) = \bar{N}_{m_i}(t) - \int_0^t \boldsymbol{\alpha}_i \bar{\mathbf{Y}}'_{m_i}(s)\lambda_i(s)ds, \qquad 0 \leq t \leq T, \qquad i = 1, \ldots, K,$$

are square-integrable martingales with respect to the filtration $\mathcal{F}_t^i = \vee_{j=1}^{m_i} \mathcal{F}_t^{i,j}$ and that

$$\bar{M}_m(t) = \bar{N}_m(t) - \sum_{i=1}^{K} \int_0^t \boldsymbol{\alpha}_i \bar{\mathbf{Y}}'_{m_i}(s)\lambda_i(s)ds, \qquad 0 \leq t \leq T,$$

is a square-integrable martingale with respect to the filtration $\mathcal{F}_t = \vee_{i=1}^{K} \mathcal{F}_t^i$.

Suppose that the parameter vectors $\boldsymbol{\alpha}_i = (1, \alpha_{2,i}, \ldots, \alpha_{n_i-k_i+1,i})$, $1 \leq i \leq K$, are known. Since the $\bar{M}_{m_i}$s, $1 \leq i \leq K$, are martingales, an obvious estimator for the cumulative hazard rate function Λ_i of the underlying baseline distribution F_i of the ith system (based on m_i observations) is given by

$$\hat{\Lambda}_{m_i}(t) = \int_0^t \frac{J_{m_i}(s)}{\boldsymbol{\alpha}_i \bar{\mathbf{Y}}'_{m_i}(s)} d\bar{N}_{m_i}(s), \qquad 0 \leq t \leq T, \tag{24.2}$$

where $J_{m_i}(s) = I_{\{\boldsymbol{\alpha}_i \bar{\mathbf{Y}}'_{m_i}(s) > 0\}}$. Notice that under the hypothesis each $\hat{\Lambda}_{m_i}(t)$, $1 \leq i \leq K$, is also an estimator for the cumulative hazard rate function Λ_0 of the common baseline distribution F_0. Moreover,

$$\hat{\Lambda}_m(t) = \int_0^t \frac{J_m(s)}{\sum_{q=1}^{K} \boldsymbol{\alpha}_q \bar{\mathbf{Y}}'_{m_q}(s)} d\bar{N}_m(s), \qquad 0 \leq t \leq T,$$

where $J_m(s) = I_{\left\{\sum_{q=1}^{K} \boldsymbol{\alpha}_q \bar{\mathbf{Y}}'_{m_q}(s)>0\right\}}$ is also an estimator for Λ_0. Please note that $\hat{\Lambda}_{m_i}$, $1 \leq i \leq K$, and $\hat{\Lambda}_m$ are generalized Nelson – Aalen estimators.

Following the derivation of k-sample tests (cf. Andersen et al., 1993, Chapter V) our test statistics are based on the processes

$$Z_{m_i}(t) = \int_0^t \tilde{W}_{m_i}(s) d(\hat{\Lambda}_{m_i} - \breve{\Lambda}_{m_i})(s), \qquad 0 \leq t \leq T, \qquad 1 \leq i \leq K,$$

where

$$\breve{\Lambda}_{m_i}(t) = \int_0^t \frac{J_{m_i}(s)}{\sum_{q=1}^{K} \boldsymbol{\alpha}_q \bar{\mathbf{Y}}'_{m_q}(s)} d\bar{N}_m(s), \qquad 0 \leq t \leq T.$$

Notice that the processes Z_{m_i}, $1 \leq i \leq K$, are the accumulated weighted differences in the increments of $\hat{\Lambda}_{m_i}$ and $\breve{\Lambda}_{m_i}$. Using weight functions of the form

$$\tilde{W}_{m_i}(t) = W_m(t) \boldsymbol{\alpha}_i \bar{\mathbf{Y}}'_{m_i}(t), \qquad 0 \leq t \leq T, \qquad 1 \leq i \leq K,$$

where W_m is a predictable and locally bounded process, we have under the hypothesis

$$Z_{m_i}(t) = \sum_{\ell=1}^{K} \int_0^t W_m(s) \left(\delta_{i\ell} - \frac{\boldsymbol{\alpha}_i \bar{\mathbf{Y}}'_{m_i}(s)}{\sum_{q=1}^{K} \boldsymbol{\alpha}_q \bar{\mathbf{Y}}'_{m_q}(s)} \right) d\bar{M}_{m_\ell}(s), \qquad 0 \leq t \leq T, \quad (24.3)$$

where δ_{il} denotes a Kronecker delta.

Obviously, the basis of our test statistics depends on the parameter vectors $\boldsymbol{\alpha}_i$, $i = 1, \ldots, K$. In the case where the parameter vectors are unknown we propose to proceed as follows. First notice that Z_{m_i} can be written as

$$Z_{m_i}(t) = \int_0^t W_m(s) d\bar{N}_{m_i}(s) - \int_0^t W_m(s) \frac{\alpha_i \bar{\mathbf{Y}}'_{m_i}(s)}{\sum_{q=1}^{K} \boldsymbol{\alpha}_q \bar{\mathbf{Y}}'_{m_q}(s)} d\bar{N}_m(s), \; 0 \leq t \leq T. \quad (24.4)$$

According to Jacod (1975) for every i, $1 \leq i \leq K$, the full likelihood of the counting processes $N_{i,j}$, $1 \leq j \leq m_i$, is, up to a factor, given by

$$\prod_{j=1}^{m_i} \left[\prod_{0 \leq s \leq T} \left[\boldsymbol{\alpha}_i \mathbf{Y}'_{i,j}(s) d\Lambda_i(s) \right]^{dN_{i,j}(s)} \cdot \exp\left(- \int_0^T \boldsymbol{\alpha}_i \mathbf{Y}'_{i,j}(s) d\Lambda_i(s) \right) \right], \qquad (24.5)$$

which has the same mathematical structure as the likelihood derived by Kvam and Peña (2005) for a dynamic reliability model. Here, the second product in (24.5) denotes the product-integral. Hence, following Kvam and Peña (2005) we may estimate $\boldsymbol{\alpha}_i$, $1 \leq i \leq K$, by solving the $n_i - k_i$ equations

$$U_{m_i}(T, \boldsymbol{\alpha}_i) = 0. \qquad (24.6)$$

Here $U_{m_i}(\cdot, \boldsymbol{\alpha}_i)$, $1 \leq i \leq K$, is the profile (partial) score process of the ith system. We denote by $\hat{\boldsymbol{\alpha}}_{m_i}$ the estimator obtained from (24.6) for $\boldsymbol{\alpha}_i$, $1 \leq i \leq K$. In the case where the parameter vectors $\boldsymbol{\alpha}_i$, $1 \leq i \leq K$, are unknown, we replace them in the representation (24.4) of the processes Z_{m_i} by their estimates $\hat{\boldsymbol{\alpha}}_{m_i}$ to obtain the following processes which will be the basis of our test statistics for unknown $\boldsymbol{\alpha}$'s:

$$\tilde{Z}_{m_i}(t) = \int_0^t W_m(s)d\bar{N}_{m_i}(s) - \int_0^t W_m(s)\frac{\hat{\boldsymbol{\alpha}}_{m_i}\bar{\mathbf{Y}}'_{m_i}(s)}{\sum_{q=1}^K \hat{\boldsymbol{\alpha}}_q \bar{\mathbf{Y}}'_{m_q}(s)} d\bar{N}_m(s), \; 0 \le t \le T. \quad (24.7)$$

24.3 K-sample tests for known $\boldsymbol{\alpha}$'s

In order to derive the asymptotic distribution of the test statistics in this and the next section we assume that the following assumptions are satisfied

Assumption 2 *There exist $\kappa_i \in (0,1)$ such that $\sum_{i=1}^K \kappa_i = 1$, and for every i, $1 \le i \le K$, we have*

$$\lim_{m\to\infty} \frac{m_i}{m} \to \kappa_i, \; \textit{where } m = \sum_{i=1}^K m_i.$$

Assumption 3 $F_0(T) < \infty$.

Assumption 4 *There is a deterministic function w defined on $[0,T]$ such that*

$$\sup_{t\in[0,T]} |W_m(t) - w(t)| \xrightarrow{P} 0, \; \textit{as } m \to \infty.$$

The following lemma from Beutner (2008) is useful when deriving the asymptotic distribution of the basis of the test statistics. By $||.||_0^T$ we denote the supremum norm on $[0,T]$.

Lemma 2. *Under Assumption 3 with probability 1, $(1/m_i)\boldsymbol{\alpha}_i\bar{\mathbf{Y}}'_{m_i}$ converges to $\boldsymbol{\alpha}_i\mathbf{e}'_i$ uniformly on $[0,T]$ as $m_i \to \infty$.*

Therefore, under Assumptions 2 and 3 we have that

$$\frac{1}{m}\sum_{i=1}^K \boldsymbol{\alpha}_i\bar{\mathbf{Y}}'_{m_i}(t)$$

converges with probability 1, uniformly on $[0,T]$ as $m \to \infty$ to $e(\boldsymbol{\alpha},t) = \sum_{i=1}^K \kappa_i\boldsymbol{\alpha}_i\mathbf{e}'_i(t)$. Here, and in the following $\boldsymbol{\alpha} = (\boldsymbol{\alpha}_1, \dots, \boldsymbol{\alpha}_K)$.

In the following, we denote by $D[0,T]^K$ the cadlag functions on $[0,T]^K$, and by $\Rightarrow$ weak convergence.

Theorem 1. *Let Assumptions 2, 3, and 4 be satisfied. Then under the hypothesis $F_1 = \dots = F_K = F_0$:*

(i) For the process $\mathbf{Z}_m(t) = (Z_{m_1}(t), \dots, Z_{m_K}(t))$ we obtain

$$\frac{1}{\sqrt{m}}\mathbf{Z}_m \Rightarrow (G_1, \dots, G_K) \quad \textit{as } m \to \infty,$$

in $D[0,T]^K$, where G_i, $i = 1, \dots, K$, are zero-mean Gaussian martingales with covariance matrix function $\Sigma(\boldsymbol{\alpha},t) = (\sigma_{ij}(\boldsymbol{\alpha},t))_{1\le i,j\le K}$ given by

$$\sigma_{ij}(\boldsymbol{\alpha},t) = \int_0^t w^2(s)\frac{\kappa_i \cdot \boldsymbol{\alpha}_i\mathbf{e}'_i(s)}{e(\boldsymbol{\alpha},s)}\left(\delta_{ij} - \frac{\kappa_j \cdot \boldsymbol{\alpha}_j\mathbf{e}'_j(s)}{e(\boldsymbol{\alpha},s)}\right) e(\boldsymbol{\alpha},s)\lambda_0(s)ds. \quad (24.8)$$

(ii) Moreover, σ_{ij} may be estimated unbiasedly by

$$\hat{\sigma}_{ij}(\boldsymbol{\alpha},t)=\frac{1}{m}\int_0^t W_m^2(s)\frac{\boldsymbol{\alpha}_i\bar{\mathbf{Y}}'_{m_i}(s)}{\sum_{q=1}^K\boldsymbol{\alpha}_q\bar{\mathbf{Y}}'_{m_q}(s)}\cdot\left(\delta_{ij}-\frac{\boldsymbol{\alpha}_j\bar{\mathbf{Y}}'_{m_j}(s)}{\sum_{q=1}^K\boldsymbol{\alpha}_q\bar{\mathbf{Y}}'_{m_q}(s)}\right)d\bar{N}_m(s). \tag{24.9}$$

Proof. In order to apply Rebolledo's theorem (see Andersen et al., 1993, Theorem II. 5.1) to the martingale $(1/\sqrt{m})\mathbf{Z}_m$ we have to show that

(a) $\left\langle\frac{1}{\sqrt{m}}Z_{m_i},\frac{1}{\sqrt{m}}Z_{m_j}\right\rangle(t)\xrightarrow{P}\sigma_{ij}(t)$ for all $1\le i,j\le K$ and $t\in[0,T]$ as $m\to\infty$,
(b) For all $\epsilon>0$

$$\left\langle\frac{1}{\sqrt{m}}Z^{\epsilon}_{m_i}\right\rangle(t)\xrightarrow{P}0 \quad \text{for all } 1\le i\le K \quad \text{and} \quad t\in[0,T] \quad \text{as } m\to\infty$$

where

$$Z^{\epsilon}_{m_i}(t)=\sum_{\ell=1}^K\int_0^t W_m(s)\left(\delta_{i\ell}-\frac{\boldsymbol{\alpha}_i\bar{\mathbf{Y}}'_{m_i}(s)}{\sum_{q=1}^K\boldsymbol{\alpha}_q\bar{\mathbf{Y}}'_{m_q}(s)}\right)\times I_{\left\{\left|\frac{W_m(s)}{\sqrt{m}}\left(\delta_{il}-\frac{\boldsymbol{\alpha}_i\bar{\mathbf{Y}}'_{m_i}(s)}{\sum_{q=1}^K\boldsymbol{\alpha}_q\bar{\mathbf{Y}}'_{m_q}(s)}\right)\right|>\epsilon\right\}}d\bar{M}_\ell(s).$$

To prove (a) notice that under the hypothesis

$$\begin{aligned}\left\langle\frac{1}{\sqrt{m}}Z_{m_i},\frac{1}{\sqrt{m}}Z_{m_j}\right\rangle(t)=&\frac{1}{m}\int_0^t W_m^2(s)\frac{\boldsymbol{\alpha}_i\bar{\mathbf{Y}}'_{m_i}(s)}{\sum_{q=1}^K\boldsymbol{\alpha}_q\bar{\mathbf{Y}}'_{m_q}(s)}\cdot\left(\delta_{ij}-\frac{\boldsymbol{\alpha}_j\bar{\mathbf{Y}}'_{m_j}(s)}{\sum_{q=1}^K\boldsymbol{\alpha}_q\bar{\mathbf{Y}}'_{m_q}(s)}\right)\\&\times\left(\sum_{q=1}^K\boldsymbol{\alpha}_q\bar{\mathbf{Y}}'_{m_q}(s)\right)\lambda_0(s)ds\\=&\int_0^t W_m^2(s)\cdot\frac{m_i}{m}\cdot\frac{\boldsymbol{\alpha}_i\frac{\bar{\mathbf{Y}}'_{m_i}(s)}{m_i}}{\sum_{q=1}^K\boldsymbol{\alpha}_q\frac{\bar{\mathbf{Y}}'_{m_q}(s)}{m}}\\&\left(\delta_{ij}-\frac{m_j}{m}\frac{\boldsymbol{\alpha}_j\frac{\bar{\mathbf{Y}}'_{m_j}(s)}{m_j}}{\sum_{q=1}^K\boldsymbol{\alpha}_q\frac{\bar{\mathbf{Y}}'_{m_q}(s)}{m}}\right)\times\left(\sum_{q=1}^K\boldsymbol{\alpha}_q\frac{\bar{\mathbf{Y}}'_{m_q}(s)}{m}\right)\lambda_0(s)ds.\end{aligned}$$

The result follows now from the fact that the processes

$$\frac{\boldsymbol{\alpha}_i\bar{\mathbf{Y}}'_{m_i}}{\sum_{q=1}^K\boldsymbol{\alpha}_q\bar{\mathbf{Y}}'_{m_q}}, \qquad 1\le i\le K, \qquad \text{and} \quad \sum_{q=1}^K\boldsymbol{\alpha}_q\frac{\bar{\mathbf{Y}}'_{m_q}}{m}$$

are bounded on $[0,T]$, Assumptions 2 and 4, and Lemma 2.

To show that condition (b) is satisfied notice that

$$\left\langle\frac{1}{\sqrt{m}}Z^{\epsilon}_{m_i}\right\rangle(t)=\sum_{\ell=1}^K\int_0^t W_m^2(s)\left(\delta_{il}-\frac{\boldsymbol{\alpha}_i\bar{\mathbf{Y}}'_{m_i}(s)}{\sum_{q=1}^K\boldsymbol{\alpha}_q\bar{\mathbf{Y}}'_{m_q}(s)}\right)^2 I_{\left\{\left|\frac{W_m(s)}{\sqrt{m}}\left(\delta_{il}-\frac{\boldsymbol{\alpha}_i\bar{\mathbf{Y}}'_{m_i}(s)}{\sum_{q=1}^K\boldsymbol{\alpha}_q\bar{\mathbf{Y}}'_{m_q}(s)}\right)\right|>\epsilon\right\}}\times\frac{\boldsymbol{\alpha}_\ell\bar{\mathbf{Y}}'_{m_\ell}(s)}{m}\lambda_0(s)ds.$$

Since for m sufficiently large

$$I_{\left\{\left|\frac{W_m(s)}{\sqrt{m}}\left(\delta_{i\ell}-\frac{\boldsymbol{\alpha}_i\bar{\mathbf{Y}}'_{m_i}(s)}{\sum_{q=1}^K \boldsymbol{\alpha}_q\bar{\mathbf{Y}}'_{m_q}(s)}\right)\right|>\epsilon\right\}}$$

is equal to zero, condition (b) is satisfied and the assertion follows.

(ii) The result follows by an application of Lenglart's inequality and Lemma 2.

□

Remark 5.3. (i) In the above theorem we may allow the weight function W_m to depend on the known parameters $\boldsymbol{\alpha}_i$, $1 \le i \le K$, without changing either the result or the proof.

(ii) Assumption 4 may be replaced by: there is a deterministic function w defined on $[0, T]$ and a sequence of nonnegative numbers $(a_m)_m$ such that

$$\sup_{t\in[0,T]} |a_m \cdot W_m(t) - w(t)| \xrightarrow{P} 0, \qquad \text{as } m \to \infty.$$

In this case the above theorem remains valid if we replace the normalizing constant $1/\sqrt{m}$ by $a_m/\sqrt{m}$.

Since $\sum_{i=1}^K Z_{m_i}(t) = 0$, $0 \le t \le T$, the test statistics are based on $\mathbf{Z}_{m,K-1} = (Z_{m_1}, \ldots, Z_{m_{K-1}})$. By $\hat{\Sigma}(\boldsymbol{\alpha}, \cdot)$ we denote the $K \times K$ matrix with elements given by (24.9), and by $\hat{\Sigma}(\boldsymbol{\alpha}, \cdot)_{K-1}$ the matrix obtained by deleting the last row and the last column of $\hat{\Sigma}(\boldsymbol{\alpha}, \cdot)$. From Theorem 1 we then obtain that for every $0 < t \le T$ the test statistic

$$\mathbf{Z}_{m,K-1}(t) \cdot \hat{\Sigma}^{-1}(\boldsymbol{\alpha}, t)_{K-1} \cdot \mathbf{Z}'_{m,K-1}(t)$$

converges to a χ^2 distribution with $K - 1$ degrees of freedom as $m \to \infty$.

Alternatively, in order to avoid the problems arising from the choice of exactly one time point, one can proceed as follows. Let $\mathbf{t} = \{0 = t_0 < t_1 < \cdots < t_\ell = T\}$ be a partition of the time interval $[0, T]$. Then, it follows from Theorem 1 that $\mathbf{Z}_{m,\ell(K-1)}(\mathbf{t}) = (\mathbf{Z}_{m,K-1}(t_1) - \mathbf{Z}_{m,K-1}(t_0), \mathbf{Z}_{m,K-1}(t_2) - \mathbf{Z}_{m,K-1}(t_1), \ldots, \mathbf{Z}_{m,K-1}(t_\ell) - \mathbf{Z}_{m,K-1}(t_{\ell-1}))$ has, as $m \to \infty$, a normal distribution with covariance matrix $\hat{\Sigma}(\boldsymbol{\alpha}, \mathbf{t})_{\ell(K-1)} = \text{diag}(\hat{\Sigma}(\boldsymbol{\alpha}, t_1)_{K-1} - \hat{\Sigma}(\boldsymbol{\alpha}, t_0)_{K-1}, \ldots, \hat{\Sigma}(\boldsymbol{\alpha}, t_\ell)_{K-1} - \hat{\Sigma}(\boldsymbol{\alpha}, t_{\ell-1})_{K-1})$. Hence,

$$\mathbf{Z}_{m,\ell(K-1)}(\mathbf{t}) \cdot \hat{\Sigma}^{-1}(\boldsymbol{\alpha}, \mathbf{t})_{\ell(K-1)} \cdot \mathbf{Z}'_{m,\ell(K-1)}(\mathbf{t})$$

has asymptotically a χ^2 distribution with $\ell \cdot (K - 1)$ degrees of freedom.

24.4 K-sample tests for unknown α's

In this section, we extend the results to the case where the parameter vectors $\boldsymbol{\alpha}_i$, $1 \le i \le K$, are unknown. In the following, we denote the true parameter vectors by $\boldsymbol{\alpha}_i^0 = (1, \alpha_{2,i}^0, \ldots, \alpha_{n_i-k_i+1,i}^0)$, $1 \le i \le K$. Furthermore, as mentioned in Remark 5.3, part (i) the weight function may depend on the $\boldsymbol{\alpha}_i$s. Since, in this section, they are assumed to be unknown this dependency is made explicitly by denoting the weight function by $W_m(\boldsymbol{\alpha}, s)$, where $\boldsymbol{\alpha} \in \mathbb{R}^{(n_1-k_1)\cdot(n_2-k_2)\cdot\ldots\cdot(n_K-k_K)}$. Finally, Assumption 4 is replaced by

Assumption 5 *The weight function W_m is differentiable with respect to $\boldsymbol{\alpha}$, there is a deterministic function w defined on $[0,T]$ such that in a neighborhood of $\boldsymbol{\alpha}^0 = (\boldsymbol{\alpha}_1^0, \ldots, \boldsymbol{\alpha}_K^0)$*

$$\sup_{t \in [0,T]} |W_m(\boldsymbol{\alpha}, t) - w(\boldsymbol{\alpha}, t)| \xrightarrow{P} 0, \qquad \text{as } m \to \infty,$$

and the derivative of W_m with respect to $\boldsymbol{\alpha}$ is bounded on $[0,T]$ in a neighborhood of $\boldsymbol{\alpha}^0$.

Theorem 2. *Let Assumptions 2, 3, and 5 be satisfied. Then under the hypothesis $F_1 = \cdots = F_K = F_0$: For the process $\tilde{\mathbf{Z}}_m(t) = (\tilde{Z}_{m_1}(t), \ldots, \tilde{Z}_{m_K}(t))$ we obtain*

$$\frac{1}{\sqrt{m}} \tilde{\mathbf{Z}}_m \Rightarrow (\tilde{G}_1, \ldots, \tilde{G}_K) \qquad \text{as } m \to \infty,$$

in $D[0,T]^K$, where $\tilde{G}_i$, $i = 1, \ldots, K$, are zero-mean Gaussian processes with covariance matrix function $\tilde{\Sigma}(\boldsymbol{\alpha}^0, t) = (\tilde{\sigma}_{ij}(\boldsymbol{\alpha}^0, t))_{1 \le i,j \le K}$ given by

$$\tilde{\Sigma}(\boldsymbol{\alpha}^0, t) = \Sigma(\boldsymbol{\alpha}^0, t) + b(\boldsymbol{\alpha}^0, t)\mathbf{D}(\boldsymbol{\alpha}^0)\Psi^{-1}(\boldsymbol{\alpha}^0, T)\mathbf{D}(\boldsymbol{\alpha}^0)b(\boldsymbol{\alpha}^0, t)$$

where

$$\Psi^{-1}(\boldsymbol{\alpha}^0, t) = \begin{pmatrix} \frac{1}{\kappa_1} \cdot \Psi_1^{-1}\left(\boldsymbol{\alpha}_1^0, t\right) & \mathbf{0}_{(n_1-k_1)\times(n_2-k_2)\cdot\ldots\cdot(n_K-k_K)} & \\ \mathbf{0}_{(n_2-k_2)\times(n_1-k_1)} & \frac{1}{\kappa_2} \cdot \Psi_2^{-1}\left(\boldsymbol{\alpha}_2^0, t\right) & \mathbf{0}_{(n_2-k_2)\times(n_3-k_3)\cdot\ldots\cdot(n_K-k_K)} \\ \vdots & \ddots & \vdots \\ & \mathbf{0}_{(n_K-k_K)\times(n_1-k_1)\cdot\ldots\cdot(n_{K-1}-k_{K-1})} & \frac{1}{\kappa_K} \cdot \Psi_K^{-1}(\boldsymbol{\alpha}_K^0, t) \end{pmatrix}$$

with $\mathbf{0}_{\ell \times \mathbf{k}}$ denoting the zero matrix with ℓ rows and k columns, and b is given by

$$b(\boldsymbol{\alpha}^0, t) = \int_0^t w(\boldsymbol{\alpha}^0, s) A'(\boldsymbol{\alpha}^0, s) e(\boldsymbol{\alpha}^0, s) \lambda_0(s) ds,$$

where $A(\boldsymbol{\alpha}^0, s)$ is given by (24.14) (see below).

Proof. Using the above notation for the weight function, the ith component of $\tilde{\mathbf{Z}}_m$ (cf. (24.7)) is given by

$$\tilde{Z}_{m_i}(t) = \int_0^t W_m(\hat{\boldsymbol{\alpha}}_m, s) d\bar{N}_{m_i}(s) - \int_0^t W_m(\hat{\boldsymbol{\alpha}}_m, s) \frac{\hat{\boldsymbol{\alpha}}_{m_i} \bar{\mathbf{Y}}'_{m_i}(s)}{\sum_{q=1}^K \hat{\boldsymbol{\alpha}}_{m_q} \bar{\mathbf{Y}}'_{m_q}(s)} d\bar{N}_m(s),$$

where $\hat{\boldsymbol{\alpha}}_m = (\hat{\boldsymbol{\alpha}}_{m_1}, \ldots, \hat{\boldsymbol{\alpha}}_{m_K}) \in \mathbb{R}^{(n_1-k_1)\cdot(n_2-k_2)\cdot\ldots\cdot(n_K-k_K)}$.

The derivative of

$$\frac{\hat{\boldsymbol{\alpha}}_{m_i} \bar{\mathbf{Y}}'_{m_i}(s)}{\sum_{q=1}^K \hat{\boldsymbol{\alpha}}_{m_q} \bar{\mathbf{Y}}'_{m_q}(s)}$$

with respect to the vector $\boldsymbol{\alpha} = (\boldsymbol{\alpha}_1, \ldots, \boldsymbol{\alpha}_K) = (\alpha_{2,1}, \ldots, \alpha_{n_1-k_1+1,1}, \ldots, \alpha_{2,K}, \ldots, \alpha_{n_K-k_K+1,K}) = (\alpha_{pr})_{2 \le p \le n_r-k_r+1, 1 \le r \le K}$ is given by the vector value process

$$\mathbf{A}_{m_i}(\boldsymbol{\alpha}, s) = \begin{cases} \dfrac{-\boldsymbol{\alpha}_i \bar{\mathbf{Y}}'_{m_i}(s) \bar{Y}_{p,m_r}(s)}{(\sum_{q=1}^K \boldsymbol{\alpha}_q \bar{\mathbf{Y}}'_{m_q}(s))^2}, & \text{for } 2 \le p \le n_r - k_r + 1, r \ne i, \\ \dfrac{\bar{Y}_{p,m_i}(s) \cdot \left(\sum_{q=1}^K \boldsymbol{\alpha}_q \bar{\mathbf{Y}}'_{m_q}(s)\right) - \boldsymbol{\alpha}_i \bar{\mathbf{Y}}'_{m_i}(s) \bar{Y}_{p,m_i}(s)}{(\sum_{q=1}^K \boldsymbol{\alpha}_q \bar{\mathbf{Y}}'_{m_q}(s))^2}, & \text{for } 2 \le p \le n_i - k_i + 1, r = i. \end{cases}$$

Thus, expanding

$$\frac{\hat{\boldsymbol{\alpha}}_{m_i}\bar{\mathbf{Y}}'_{m_i}(s)}{\sum_{q=1}^{K}\hat{\boldsymbol{\alpha}}_{m_q}\bar{\mathbf{Y}}'_{m_q}(s)}$$

in a first-order Taylor series around $\boldsymbol{\alpha}^0$, the ith component of $(1/\sqrt{m})\tilde{\mathbf{Z}}_m$ becomes

$$\frac{1}{\sqrt{m}}\tilde{Z}_{m_i}(t) = \frac{1}{\sqrt{m}}\left[\int_0^t W_m(\hat{\boldsymbol{\alpha}}_m, s)d\bar{N}_{m_i}(s) - \int_0^t W_m(\hat{\boldsymbol{\alpha}}_m, s)\frac{\boldsymbol{\alpha}_i^0\bar{\mathbf{Y}}'_{m_i}(s)}{\sum_{q=1}^{K}\boldsymbol{\alpha}_q^0\bar{\mathbf{Y}}'_{m_q}(s)}d\bar{N}_m(s)\right]$$
$$- \sqrt{m}(\hat{\boldsymbol{\alpha}}_m - \boldsymbol{\alpha}^0)\int_0^t \frac{1}{m}W_m(\hat{\boldsymbol{\alpha}}_m, s)\mathbf{A}'_{m_i}(\tilde{\boldsymbol{\alpha}}_m, s)d\bar{N}_m(s), \qquad (24.10)$$

where $\tilde{\boldsymbol{\alpha}}_m$ lies in the line segment connecting $\boldsymbol{\alpha}^0$ and $\hat{\boldsymbol{\alpha}}_m$.

Expanding W_m in the first line of (24.10) in a first-order Taylor series around $\boldsymbol{\alpha}^0$ we obtain that the first line in (24.10) equals

$$\frac{1}{\sqrt{m}}\left[\int_0^t W_m(\boldsymbol{\alpha}^0, s)d\bar{N}_{m_i}(s) - \int_0^t W_m(\boldsymbol{\alpha}^0, s)\frac{\boldsymbol{\alpha}_i^0\bar{\mathbf{Y}}'_{m_i}(s)}{\sum_{q=1}^{K}\boldsymbol{\alpha}_q^0\bar{\mathbf{Y}}'_{m_q}(s)}d\bar{N}_m(s)\right] \qquad (24.11)$$
$$+(\hat{\boldsymbol{\alpha}}_m - \boldsymbol{\alpha}^0)\frac{1}{\sqrt{m}}\left[\int_0^t \frac{\partial W'_m(\breve{\boldsymbol{\alpha}}_m, s)}{\partial\boldsymbol{\alpha}}d\bar{N}_{m_i}(s)\right.$$
$$\left.- \int_0^t \frac{\partial W'_m(\breve{\boldsymbol{\alpha}}_m, s)}{\partial\boldsymbol{\alpha}}\frac{\boldsymbol{\alpha}_i^0\bar{\mathbf{Y}}'_{m_i}(s)}{\sum_{q=1}^{K}\boldsymbol{\alpha}_q^0\bar{\mathbf{Y}}'_{m_q}(s)}d\bar{N}_m(s)\right],$$

where $\breve{\boldsymbol{\alpha}}_m$ lies in the line segment connecting $\boldsymbol{\alpha}^0$ and $\hat{\boldsymbol{\alpha}}_m$. Notice that the term in brackets in the second and third line of (24.11) are equal to

$$\int_0^t \frac{\partial W'_m(\breve{\boldsymbol{\alpha}}_m, s)}{\partial\boldsymbol{\alpha}}d\bar{M}_{m_i}(s) - \int_0^t \frac{\partial W'_m(\breve{\boldsymbol{\alpha}}_m, s)}{\partial\boldsymbol{\alpha}}\frac{\boldsymbol{\alpha}_i^0\bar{\mathbf{Y}}'_{m_i}(s)}{\sum_{q=1}^{K}\boldsymbol{\alpha}_q^0\bar{\mathbf{Y}}'_{m_q}(s)}d\bar{M}_m(s).$$

Hence, the second and third line in (24.11) are equal to

$$\sqrt{m}(\hat{\boldsymbol{\alpha}}_m - \boldsymbol{\alpha}^0)\left[\int_0^t \frac{1}{m}\frac{\partial W'_m(\breve{\boldsymbol{\alpha}}_m, s)}{\partial\boldsymbol{\alpha}}d\bar{M}_{m_i}(s)\right.$$
$$\left.- \int_0^t \frac{1}{m}\frac{\partial W'_m(\breve{\boldsymbol{\alpha}}_m, s)}{\partial\boldsymbol{\alpha}}\frac{\boldsymbol{\alpha}_i^0\bar{\mathbf{Y}}'_{m_i}(s)}{\sum_{q=1}^{K}\boldsymbol{\alpha}_q^0\bar{\mathbf{Y}}'_{m_q}(s)}d\bar{M}_m(s)\right]. \qquad (24.12)$$

From Kvam and Peña (2005, Theorem 2) we have that $\sqrt{m}(\hat{\boldsymbol{\alpha}}_m - \boldsymbol{\alpha}^0)$ (see also below) converges to a normal distribution, and from Lenglart's inequality, Assumption 5, and Lemma 2 we obtain that the term in brackets in (24.12) converges uniformly on $[0, T]$ to zero in probability. Thus, we have that the first line in (24.10) is equal to

$$\frac{1}{\sqrt{m}}\left[\int_0^t W_m(\boldsymbol{\alpha}^0, s)d\bar{N}_{m_i}(s) - \int_0^t W_m(\boldsymbol{\alpha}^0, s)\frac{\boldsymbol{\alpha}_i^0\bar{\mathbf{Y}}'_{m_i}(s)}{\sum_{q=1}^{K}\boldsymbol{\alpha}_q^0\bar{\mathbf{Y}}'_{m_q}(s)}d\bar{N}_m(s)\right] + o_P(1).$$

In the same way it can be proved that the second line in (24.10) equals

$$\sqrt{m}(\hat{\boldsymbol{\alpha}}_m - \boldsymbol{\alpha}^0)\int_0^t \frac{1}{m}W_m(\boldsymbol{\alpha}^0, s)\mathbf{A}'_{m_i}(\tilde{\boldsymbol{\alpha}}_m, s)d\bar{N}_m(s) + o_P(1).$$

Hence, the ith component of $(1/\sqrt{m})\tilde{\mathbf{Z}}_m$ can be written as

$$\frac{1}{\sqrt{m}}\left[\int_0^t W_m(\boldsymbol{\alpha}^0,s)d\bar{N}_{m_i}(s) - \int_0^t W_m(\boldsymbol{\alpha}^0,s)\frac{\boldsymbol{\alpha}_i^0\bar{\mathbf{Y}}'_{m_i}(s)}{\sum_{q=1}^K \boldsymbol{\alpha}_q^0\bar{\mathbf{Y}}'_{m_q}(s)}d\bar{N}_m(s)\right]$$
$$-\sqrt{m}(\hat{\boldsymbol{\alpha}}_m - \boldsymbol{\alpha}^0)\int_0^t \frac{1}{m}W_m(\boldsymbol{\alpha}^0,s)\mathbf{A}'_{m_i}(\tilde{\boldsymbol{\alpha}}_m,s)d\bar{N}_m(s) + o_P(1).$$

Thus, we have that $(1/\sqrt{m})\tilde{\mathbf{Z}}_m$ is equal to

$$\frac{1}{\sqrt{m}}\mathbf{Z}_m(t) - \sqrt{m}(\hat{\boldsymbol{\alpha}}_m - \boldsymbol{\alpha}^0)\int_0^t \frac{1}{m}W_m(\boldsymbol{\alpha}^0,s)\mathbf{A}'_m(\tilde{\boldsymbol{\alpha}}_m,s)d\bar{N}_m(s) + o_P(1),$$

where the ith row, $1 \le i \le K$, of the matrix $\mathbf{A}_m$ is given by $\mathbf{A}_{m_i}$. According to Theorem 1 we have that $(1/\sqrt{m})\mathbf{Z}_m(t)$ converges to a zero-mean Gaussian martingale with covariance matrix function given by (24.8). Moreover, it follows by an application of Lenglart's inequality, the structure of $\mathbf{A}_m$, Lemma 2, and the fact that $\tilde{\boldsymbol{\alpha}}_m = \boldsymbol{\alpha}^0 + o_P(1)$,

$$\int_0^t \frac{1}{m}W_m(\boldsymbol{\alpha}^0,s)\mathbf{A}'_m(\tilde{\boldsymbol{\alpha}}_m,s)d\bar{N}_m(s) \xrightarrow{P}$$
$$\int_0^t w(\boldsymbol{\alpha}^0,s)\mathbf{A}'(\boldsymbol{\alpha}^0,s)e(\boldsymbol{\alpha}^0,s)\lambda_0(s)ds, \tag{24.13}$$

where the ith row, $1 \le i \le K$, of the matrix value function $\mathbf{A}(\boldsymbol{\alpha}^0,t)$ is given by

$$\frac{1}{e^2(\boldsymbol{\alpha}^0,t)}\left(\kappa_1\kappa_i\boldsymbol{\alpha}_i^0\mathbf{e}_i'(t)e_{1,1}(t),\ldots,\kappa_1\kappa_i\boldsymbol{\alpha}_i^0\mathbf{e}_i'(t)e_{n_1-k_1+1,1}(t),\right.$$
$$\vdots$$
$$\kappa_{i-1}\kappa_i\boldsymbol{\alpha}_i^0\mathbf{e}_i'(t)e_{1,i-1}(t),\ldots,\kappa_{i-1}\kappa_i\boldsymbol{\alpha}_i^0\mathbf{e}_i'(t)e_{n_{i-1}-k_{i-1}+1,i-1}(t),$$
$$\kappa_i e_{1,i}(t)e(\boldsymbol{\alpha}^0,t) - \kappa_i^2\boldsymbol{\alpha}_i^0\mathbf{e}_i'(t)e_{1,i}(t),\ldots,\kappa_i e_{n_i-k_i+1,i}(t)e(\boldsymbol{\alpha}^0,t) - \kappa_i^2\boldsymbol{\alpha}_i^0\mathbf{e}_i'(t)e_{n_i-k_i+1,i}(t),$$
$$\kappa_{i+1}\kappa_i\boldsymbol{\alpha}_i^0\mathbf{e}_i'(t)e_{1,i+1}(t),\ldots,\kappa_{i+1}\kappa_i\boldsymbol{\alpha}_i^0\mathbf{e}_i'(t)e_{n_{i+1}-k_{i+1}+1,i+1}(t),$$
$$\vdots$$
$$\left.\kappa_K\kappa_i\boldsymbol{\alpha}_i^0\mathbf{e}_i'(t)e_{1,K}(t),\ldots,\kappa_K\kappa_i\boldsymbol{\alpha}_i^0\mathbf{e}_i'(t)e_{n_K-k_K+1,K}(t)\right). \tag{24.14}$$

From Kvam and Peña (2005) we have for i, $1 \le i \le K$, that $\sqrt{m_i}(\hat{\boldsymbol{\alpha}}_{m_i} - \boldsymbol{\alpha}_i^0)$ has a representation of the form

$$\mathbf{D}(\boldsymbol{\alpha}_i^0)\left(\frac{1}{m_i}\int_0^t [\mathbf{D}(\nu_{m_i}(\boldsymbol{\alpha}_i^0,s)) - \nu_{m_i}(\boldsymbol{\alpha}_i^0,s)' * \nu_{m_i}(\boldsymbol{\alpha}_i^0,s)]d\bar{N}_{m_i}(s)\right)^{-1}$$
$$\times \frac{1}{\sqrt{m_i}}\sum_{j=1}^{m_i}\int_0^t [\boldsymbol{\rho}_{i,j}(s) - \boldsymbol{\nu}_{m_i}(\boldsymbol{\alpha}_i^0,s)]dM_{i,j}(s) + o_P(1),$$

and that $\sqrt{m_i}(\hat{\boldsymbol{\alpha}}_{m_i} - \boldsymbol{\alpha}_i^0)$ converges to a normal distribution with expectation equal to $\mathbf{0}$ and covariance matrix given by

$$\mathbf{D}(\boldsymbol{\alpha}_i^0)\Psi_i^{-1}(\boldsymbol{\alpha}_i^0, T)\mathbf{D}(\boldsymbol{\alpha}_i^0).$$

From the independence assumption and Assumption 2 it follows that $\sqrt{m}(\hat{\boldsymbol{\alpha}}_m - \boldsymbol{\alpha}^0)$ converges to a normal distribution with expectation equal to $\mathbf{0}$ and covariance matrix given by

$$\mathbf{D}(\boldsymbol{\alpha}^0)\Psi^{-1}(\boldsymbol{\alpha}^0, T)\mathbf{D}(\boldsymbol{\alpha}^0).$$

Therefore, we only have to determine the covariance process between $\mathbf{Z}_m$ and

$$\left(\sum_{j=1}^{m_1}\int_0^t [\boldsymbol{\rho}_{1,j}(s) - \boldsymbol{\nu}_{m_1}(\boldsymbol{\alpha}_1^0, s)]dM_{1,j}(s), \ldots, \sum_{j=1}^{m_K}\int_0^t [\boldsymbol{\rho}_{K,j}(s) - \boldsymbol{\nu}_{m_K}(\boldsymbol{\alpha}_K^0, s)]dM_{K,j}(s)\right).$$

A direct calculation shows (cf. Kvam and Peña, 2005, Lemma 1) that the covariation process is equal to $\mathbf{0}$. This finishes the proof. □

It is worth mentioning that the covariance matrix $\tilde{\Sigma}$ can be consistently estimated by the following steps: plugging in the estimator $\hat{\boldsymbol{\alpha}}_m$ into the estimator $\hat{\Sigma}$, using $\hat{b}(\hat{\boldsymbol{\alpha}}_m, t) = \int_0^t (1/m) W_m(\hat{\boldsymbol{\alpha}}_m, s) A_m'(\hat{\boldsymbol{\alpha}}_m, s) d\bar{N}_m(s)$ as an estimator for $b(\boldsymbol{\alpha}^0, t)$ (cf. (24.13)), and using $(1/m_i)\int_0^t [\mathbf{D}(\nu_{m_i}(\hat{\boldsymbol{\alpha}}_{m_i}, s)) - \nu_{m_i}(\hat{\boldsymbol{\alpha}}_{m_i}, s)' * \nu_{m_i}(\hat{\boldsymbol{\alpha}}_{m_i}, s)] d\bar{N}_{m_i}(s)$ as an estimator for $\Psi_i(\boldsymbol{\alpha}_i^0, t)$. Using Theorem 2 we can then construct test statistics according to the discussion following Theorem 1.

References

Alvarez-Andrade, S. and Bordes, L. (2004). Empirical quantile process under Type-II progressive censoring. *Statistics & Probability Letters*, 68:111–123.

Alvarez-Andrade, S., Balakrishnan, N., and Bordes, L. (2007). Homogeneity tests based on several progressively Type-II censored samples. *Journal of Multivariate Statistics*, 98:1195–1213.

Andersen, P.K., Borgan, O., Gill, R.D., and Keidling, N. (1993). *Statistical Models Based on Counting Processes.* Springer, New York.

Balakrishnan, N. (2007). Progressive censoring methodology: An appraisal. *Test*, 6:211–296 (with discussion) DOI: 10.1007/s11749-008-0133-7.

Balakrishnan, N., Beutner, E., and Cramer, E. (2007). *Exact two-sample non-parametric confidence, prediction, and tolerance intervals based on ordinary and progressively Type-II right censored data.* submitted.

Balakrishnan, N., Beutner, E., and Kamps, U. (2008). Order restricted inference for sequential k-out-of-n systems. *Journal of Multivariate Analysis*, 99:1489–1502.

Barlow, R.E. and Proschan, F. (1981). *Statistical Theory of Reliability and Life Testing*, To Begin With, Silver Spring, MD.

Belzunce, F., Mercader, J.-A., and Ruiz, J.-M. (2003). Multivariate aging properties of epoch times of nonhomogeneous processes. *Journal of Multivariate Analysis*, 84: 335–350.

Belzunce, F., Mercader, J.-A., and Ruiz, J.-M. (2005). Stochastic comparisons of generalized order statistics. *Probability in the Engineering and Informational Sciences*, 19:99–120.

Beutner, E., (2008). Nonparametric inference for sequential k-out-of-n systems. *Annals of the Institute of Statistical Mathematics*, 60:605–626.

Bordes, L. (2004). Nonparametric estimation under progressive censoring. *Journal of Statistical Planning and Inference*, 119:171–189.

Cramer, E. and Kamps, U. (1996). Sequential order statistics and k-out-of-n systems with sequentially adjusted failure rates. *Annals of the Institute of Statistical Mathematics*, 48:535–549.

Cramer, E. and Kamps, U. (2001a). Estimation with sequential order statistics from exponential distributions. *Annals of the Institute of Statistical Mathematics*, 53:307–324.

Cramer, E. and Kamps, U. (2001b). Sequential k-out-of-n systems. In: N. Balakrishnan, C.R. Rao (Eds.), *Handbook of Statistics*, Vol. 20: Advances in Reliability. Elsevier, Amsterdam, pp. 301–372.

Cramer, E. and Kamps, U. (2003). Marginal distributions of sequential and generalized order statistics. *Metrika*, 58:293–310.

Guilbaud, O. (2001). Exact non-parametric confidence intervals for quantiles with progressive Type-II censoring. *Scandinavian Journal of Statistics*, 28:699–713.

Guilbaud, O. (2004). Exact non-parametric confidence, prediction and tolerance intervals with progressive Type-II censoring. *Scandinavian Journal of Statistics*, 31:265–281.

Hollander, M. and Peña, E.A. (1995). Dynamic reliability models with conditional proportional hazards. *Lifetime Data Analysis*, 1:377–401.

Hu, T. and Zhuang, W. (2006). Stochastic orderings between p-spacings of generalized order statistics from two samples. *Probability in the Engineering and Informational Sciences*, 20:465–479.

Jacod, J. (1975). Multivariate point processes: Predictable projection, Radon-Nikodym derivatives representation of martingales. *Zeitschrift für Wahrscheinlichkeitstheorie und Verwandte Gebiete*, 31:235–253.

Kalbfleisch, J.D. and Prentice, R.L. (1980). *The Statistical Analysis of Failure Time Data.* Wiley, New York.

Kamps, U. (1995a). *A Concept of Generalized Order Statistics.* Teubner, Stuttgart

Kamps, U. (1995b). A concept of generalized order statistics. *Journal of Statistical Planning and Inference*, 48:1–23.

Kvam, P.H. and Peña, E.A. (2005). Estimating load-sharing properties in a dynamic reliability system. *Journal of the American Statistical Association*, 100:262–272.

Lawless, J.F. (2003). *Statistical Models and Methods for Lifetime Data.* John Wiley & Sons, Hoboken, NJ.

Meeker, W.O. and Escobar, L.A. (1998). *Statistical Methods for Reliability Data*, Wiley, New York.

Navarro, J. and Rychlik, T. (2007). Reliability and expectation bounds for coherent systems with exchangeable components. *Journal of Multivariate Analysis*, 98: 102–113.

Zhuang, W. and Hu, T. (2007). Stochastic comparison results of sequential order statistics. *Probability in the Engineering and Informational Sciences*, 21:47–66.

25

Adjusting p-Values when n Is Large in the Presence of Nuisance Parameters

Sonia Migliorati and Andrea Ongaro

Department of Statistics, University of Milano-Bicocca, via Bicocca degli Arcimboldi 8, 20126 Milano, Italy (e-mail: sonia.migliorati@unimib.it, andrea.ongaro@unimib.it)

Abstract: A precise null hypothesis formulation (instead of the more realistic interval one) is usually adopted by statistical packages although it generally leads to excessive (and often misleading) rates of rejection whenever the sample size is large. In a previous paper (Migliorati and Ongaro, 2007) we proposed a calibration procedure aimed at adjusting test levels and p-values when testing the mean of a Normal model with known variance. We now address the more complicated calibration issues arising when a nuisance parameter (e.g., the variance) is present. As procedures for testing the interval null hypothesis available in the literature are shown to be unsatisfactory for calibration purposes, this entails, in particular, the construction of suitable new tests.

Keywords and phrases: Precise versus interval null hypothesis, nuisance parameter, calibration, large sample size

25.1 Introduction

Very often the null hypothesis is formulated as a precise one specifying an exact value for the parameter(s) of the model. However, in many practical situations it is more realistic to test an interval hypothesis stating that the parameter is "close" enough to a given value. In other words a distinction between statistical significance and practical (or substantial) significance should be made (Berger and Sellke, 1987; Berger and Delampady, 1987).

The approximation of the true null (interval) hypothesis by a precise one does not lead to strong inconsistencies as long as the sample size n is small; on the contrary, when n is large the use of precise hypotheses can be heavily misleading. This is because in the presence of huge statistical information even a very small departure from the precise null is detected (leading to rejection) due to the high power of the test. Such a problem is well known both in the literature (Hodges and Lehmann, 1954) and in empirical studies where an anomalously high rejection rate is a widely reported experience whenever the available data are extensive.

In Migliorati and Ongaro (2007) a practically oriented calibration procedure has been proposed, which is theoretically well grounded and easily implementable as well.

C.H. Skiadas (ed.), *Advances in Data Analysis*, Statistics for Industry and Technology, DOI 10.1007/978-0-8176-4799-5_25,

More precisely, starting from the interval hypothesis formulation, such a procedure enables us to adjust (precise null hypotheses based) standard package outputs so as to reconcile the two concepts of statistical and practical significance.

The aforementioned procedure has been developed for dealing with the no nuisance parameter case. Here we tackle the calibration problem in the presence of nuisance parameters taking as the motivating example the Normal mean problem when the variance is unknown. In this more interesting setup new issues arise calling for suitable modifications of the calibration method. Such a method is based on the comparison of the (traditional) precise null hypothesis test with a test for the interval null having the same structure. If no nuisance parameters are present, such a comparison is, in principle, simple to be performed as one can use the same type of test. This is not usually the case when the model includes nuisance parameters. In particular, in the Normal setup, the traditional t-test turns out to be inadequate for testing an interval null hypothesis. Besides, other known tests specifically built for the interval null hypothesis framework are shown not fully appropriate for calibration. Consequently we develop alternative new tests and assess their properties.

The chapter is organized as follows. In Section 25.2 the calibration for the Normal model with known variance is briefly summarized and a new in-depth study of the extent of the adjustment needed and of the accuracy of some of its approximations is given. Section 25.3 contains the main results which enable us to tackle the nuisance parameter presence (specifically the Normal model with unknown variance). Furthermore both exact and approximate formulas as well as numerical tables necessary to implement calibration and to assess accuracy of approximations are given and discussed. Finally a brief discussion, including further examples of application of the calibration method, is presented in Section 25.4.

25.2 Normal model with known variance

Let $\mathbf{x}$ be a vector of i.i.d. observations from a Normal random variable (r.v.) with unknown mean μ and known variance σ^2. The UMPU level α^* test for the precise hypothesis $H_0^* : \mu = \mu_0$ versus $H_1^* : \mu \neq \mu_0$ rejects if $|z| \geq z_{1-\alpha^*/2}$ where z is the observed value of $Z = \sqrt{n}(\bar{X} - \mu_0)/\sigma$, $\bar{X}$ is the sample mean, and z_q is the standard Normal quantile of level q.

Suppose now that δ represents the smallest departure from μ_0 which is considered practically significant for the problem under consideration. Then the precise hypothesis should be replaced by the more realistic interval one: $H_0 : |\mu - \mu_0| \leq \delta$ versus $H_1 : |\mu - \mu_0| > \delta$.

If an exact value for δ can be specified, a level α rejection region for the interval hypothesis based on the same statistic used for testing the precise one is given by $\{|z| \geq k_\alpha\}$, where k_α is such that $\alpha = \sup_{|\mu-\mu_0|\leq\delta} P_{\mu,\sigma^2}(|Z| \geq k_\alpha) = 2 - [\Phi(k_\alpha - \sqrt{n}\delta/\sigma) + \Phi(k_\alpha + \sqrt{n}\delta/\sigma)]$ and must be computed numerically. Here Φ denotes the standard Normal distribution function (d.f.) and P_{μ,σ^2} is the distribution of $\mathbf{x}$ when μ and σ^2 are the true values of the parameters.

The relationship between the (nominal) level α^*, that is, the level implicitly considered by standard package outputs, and the (real) level α can be obtained by equating the two rejection thresholds, that is, by imposing $k_\alpha = z_{1-\alpha^*/2}$. Consequently, the nominal level to be used in order to achieve a given real level α is the following,

$$\alpha^* = 2[1 - \Phi(k_\alpha)]. \tag{25.1}$$

Notice that for any given α the nominal level α^* is a decreasing function of $\delta_{st} = \sqrt{n}\delta/\sigma$ ranging from α to zero. This is in agreement with the common intuition that for large sample sizes an adjustment is needed which reduces the rejection rate. Moreover, in general the extent of the required adjustment is substantial even for small values of δ_{st}: it is easy to verify numerically that the nominal level can be considered approximately correct (less than 10% error) only for values of δ_{st} smaller than 0.2 or 0.1. Such an aspect becomes more evident if one just considers the following approximations for the relationship between α and α^*. First of all, if n, and therefore δ_{st}, is large enough one can neglect $1 - \Phi(k_\alpha + \sqrt{n}\delta/\sigma) = \Phi(-k_\alpha - \delta_{st})$ in the computation of the level of the test, so that the threshold k_α can be approximated by

$$k_\alpha \approx z_{1-\alpha} + \delta_{st}. \tag{25.2}$$

The precision of such an approximation can be evaluated purely on the basis of the values of δ_{st}. More precisely, by inspection of the Normal d.f. a precision up to any desired digit can be achieved by asking that $k_\alpha + \delta_{st} > c$ for an appropriate positive constant c. For example, $c = 4$ leads to a precision up to the fourth digit, $c = 5$ to the sixth, and so on. Thus, being $k_\alpha > \delta_{st}$ for any reasonable value of α (i.e., $\alpha < 0.5$), the chosen precision can be reached for $\delta_{st} > c/2$.

The above approximation leads to the following (approximate) calibration formula

$$\alpha^* = 2[1 - \Phi(z_{1-\alpha} + \delta_{st})] \tag{25.3}$$

which corresponds to a real level α test with precision up to the desired digit. For example, suppose one wishes to test at level α with absolute error lower than 0.0001, which is negligible to any practical purpose. This means that $c = 4$, implying that (25.3) can be resorted to whenever $\delta_{st} > 2$. Actually a numerical investigation shows that (25.3) gives a very accurate approximation also for smaller values of δ_{st}. In particular if $\delta_{st} > 0.5$ the relative error on α^* is lower than 10% for any $\alpha \leq 0.1$ and if $\delta_{st} \geq 1$ it is even lower than 1%.

A further approximate value for α^* can be obtained through the asymptotic series relative to the tail values of the d.f. Φ (Johnson et al., 1994). More precisely, if $x >> 1$ then

$$\Phi(x) = 1 - \frac{e^{-x^2/2}}{\sqrt{2\pi}}(x^{-1} - x^{-3} + 3x^{-5} - 15x^{-7} + \cdots). \tag{25.4}$$

Therefore if δ_{st} is large enough, truncating (25.4) and replacing it into (25.3) gives rise to a new expression for α^*. Notice that (25.3) provides an approximation by excess which can be compensated by truncating (25.4) to an even term given that the remainder term in brackets of expansion (25.4) is less in absolute value than the last term taken into account. For example, a truncation to the second term leads to the following expression,

$$\alpha^* = 2e^{-(\delta_{st}+z_{1-\alpha})^2/2}\frac{(\delta_{st} + z_{1-\alpha})^2 - 1}{\sqrt{2\pi}(\delta_{st} + z_{1-\alpha})^3}. \tag{25.5}$$

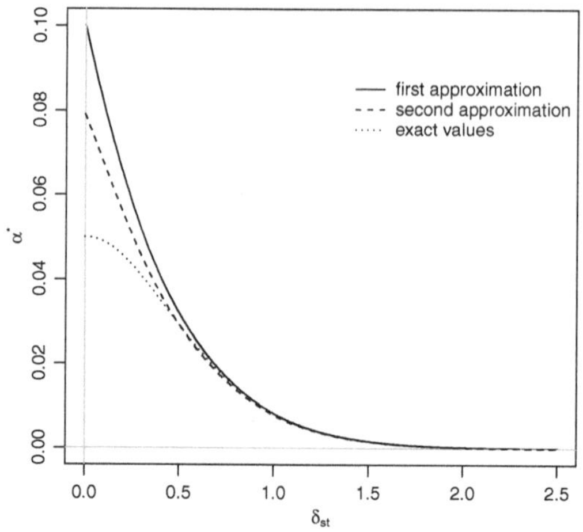

Figure 25.1. α^* as a function of δ_{st} for fixed $\alpha = 0.05$: exact values (*bottom line*), first approximation (equation (25.3), *top line*) and second approximation (equation (25.5), *middle line*)

In Figure 25.1 the two approximations (25.3) and (25.5) together with the exact α^* as a function of δ_{st} for fixed $\alpha = 0.05$ are compared.

The accuracy of the two approximations for large values of δ_{st} is confirmed by graphical inspection and a similar behaviour can be observed for other values of α. Expression (25.5), though less straightforward than (25.3), shows that the nominal level decreases at an exponential rate as a function of δ_{st} – and thus of $\sqrt{n}$ – which confirms the severity of the adjustment required when large sample sizes are taken into consideration.

Let us now focus on the practical application of the calibration procedure. In order to test at a given real level α, it is enough to compare the p-value p^* reported by the package, that is, the one relative to the precise hypothesis, with the nominal level α^* computed according to (25.1) (or to the approximations given by (25.3) or (25.5) for sufficiently large values of δ_{st}). Rejection should then be chosen only if p^* is smaller than α^*.

As far as the p-value calibration is concerned (i.e., the computation of the correct value p which should be used instead of p^*), one can resort to the following

$$p = 2 - [\Phi(z_{1-p^*/2} - \delta_{st}) + \Phi(z_{1-p^*/2} + \delta_{st})]. \qquad (25.6)$$

In many contexts it is difficult to determine an exact value for δ_{st}, only partial information on it being available instead.

Let us now examine, then, how the calibration can be achieved in this new and more widespread setup.

First suppose one is interested in testing at a given real level α. Equation (25.1) gives the nominal level α^* as a function of the desired real level α and of the value of δ_{st}, which we write as $\alpha^* = g(\alpha, \delta_{st})$. Let us denote by δ'_{st} the value of δ_{st} such that the corresponding nominal level equals the reported nominal p-value p^*; that is,

$$p^* = g\left(\alpha, \delta'_{st}\right). \qquad (25.7)$$

Then according to the procedure developed for the δ_{st} known case, for any given δ_{st} one should reject if $g(\alpha, \delta_{st})$ is greater than p^* and accept otherwise. This means that rejection has to be chosen if and only if $\delta_{st} < \delta'_{st}$. It follows that to take a decision it is sufficient (and necessary) to know whether δ'_{st} belongs to H_0. In other words to decide whether to accept we simply need to tell whether δ'_{st} has to be considered a practically significant departure from the precise null.

Finally, let us focus on the question of p-value calibration. If the exact value of δ_{st} is unknown, then it is not possible to determine exactly the p-value. In this case we believe that, from a practical point of view, the relevant objective is to position p with respect to some standard critical values of α (e.g., 1, 5, or 10%). This can be achieved by performing level α tests for increasing values of α greater than p^*. More precisely, for each test level the value δ'_{st} given by (25.7) can be judged practically not significant (corresponding to acceptance), unclassifiable, or practically significant (corresponding to rejection). The lower and upper bounds for p coincide therefore with the last (greatest) value of α leading to acceptance and with the first (smallest) value of α leading to rejection. It is noteworthy that the information required by this simple procedure is quite straightforward to retrieve: everything boils down to classifying a small set of δ_{st} values.

25.3 Normal model with unknown variance

Let us now focus on the more realistic problem of testing the mean of a Normal distribution with unknown variance σ^2. The UMPU level α^* test for the precise hypothesis $H_0^* : \mu = \mu_0$ versus $H_1^* : \mu \neq \mu_0$ rejects for sufficiently large values of the statistic $|T_n| = \left|\sqrt{n}(\bar{X} - \mu_0)/S_n\right|$ where S_n^2 is the sample (unbiased) variance. Therefore the rejection region has the form

$$\left\{|T_n| \geq t_{n-1,1-\alpha^*/2}\right\}$$

where $t_{\nu,q}$ is the q quantile of a Student's t-distribution with ν degrees of freedom.

Consider now the interval hypothesis $H_0 : |\mu - \mu_0| \leq \delta$ versus $H_1 : |\mu - \mu_0| > \delta$. In order to apply the calibration procedure we need a level α test for H_0 which is based on the same type of rejection region; i.e., $\{|T_n| \geq k\}$ where $k \geq 0$. Unfortunately this approach is not feasible as any test of this type has level 1. This can be seen as follows. First notice that $|T_n|$ is stochastically increasing with respect to $|\mu - \mu_0|$. Therefore for any given σ^2 we have

$$\sup_{|\mu-\mu_0|\leq\delta} P_{\mu,\sigma^2}(|T_n| \geq k) = P_{|\mu-\mu_0|=\delta;\ \sigma^2}(|T_n| \geq k) = g(\sigma^2).$$

It is then possible to show (see Result 1 in the Appendix) that $g(\sigma^2)$ is decreasing with respect to σ^2 with limit 1 as σ^2 goes to zero.

In such a context we are therefore forced to look for tests with a different structure. In particular, to our purposes an interesting class of tests is obtained by letting the threshold k depend on data. As this class of tests rejects for large values of $|T_n|$ as well, even in this case calibration formulas can be derived by comparing the two thresholds.

Notice that in general it is quite difficult to build tests for the interval null hypothesis with good minimal properties such as unbiasedness (see Hodges and Lehmann, 1954; Schervish, 1995).

The solution proposed in Hodges and Lehmann (1954) is rather intricate and its form is not suitable for calibration. A standard solution for this problem can be found by combining two separate one-tailed t-tests, i.e., rejecting if $\{T_A \leq t_{n-1,\alpha/2}\}$ or $\{T_B \geq t_{n-1,1-\alpha/2}\}$ where $T_A = \sqrt{n}[\bar{X}_n - (\mu_0 - \delta)]/S_n$ and $T_B = \sqrt{n}[\bar{X}_n - (\mu_0 + \delta)]/S_n$. Such a test can be written in the more compact form

$$\left\{|T_n| \geq t_{n-1,1-\alpha/2} + \sqrt{n}\frac{\delta}{S_n}\right\} \tag{25.8}$$

which is of the type required for calibration. Nonetheless, in practice such a test has some serious drawbacks. The test has exact level α, but such level is obtained by letting $\sigma^2 \to \infty$. Actually, the power of the test can be proved to be equal to $\alpha/2$ for small values of σ^2 (see Schervish, 1995). Furthermore, simulation studies show that its level is close to $\alpha/2$ for a range of σ^2 values including relatively large ones. As a consequence its use for calibration purposes is questionable given that it is not clear what level should be used. Some further remarks about it are given later in this section.

An appealing method to obtain a different data-dependent threshold is to replace the nuisance parameter σ^2 with a suitable estimate in the distribution P_{μ,σ^2} used to compute the level of the test. In other words the threshold is derived by acting as if σ^2 were known and equal to its estimate.

More precisely, let $k'_\alpha(\sigma^2)$ be the level α threshold when σ^2 is known, i.e., such that

$$\alpha = P_{|\mu-\mu_0|=\delta;\sigma^2}\left(|T_n| \geq k'_\alpha(\sigma^2)\right).$$

Then, if the nuisance parameter σ^2 is estimated by the sample variance S_n^2, the new test will have rejection region

$$R'_n = \left\{|T_n| \geq k'_\alpha(S_n^2)\right\}. \tag{25.9}$$

The threshold $k'_\alpha(\sigma^2)$ can be computed numerically by solving the equation

$$\alpha = 1 - F_{n-1,\delta_{st}}(k) + F_{n-1,\delta_{st}}(-k) \tag{25.10}$$

where $F_{\nu,\lambda}$ is a noncentral Student's t-distribution with ν degrees of freedom and non centrality parameter λ.

For large n one can replace the Student's t distribution with the Normal as $F_{n-1,\lambda} \to N(\lambda,1)$. This implies that $k'_\alpha(\sigma^2)$ is approximately equal to the Normal threshold $k_\alpha(\sigma^2)$ derived in the previous section. One can therefore consider the more tractable rejection region

$$R_n = \left\{|T_n| \geq k_\alpha\left(S_n^2\right)\right\}. \tag{25.11}$$

Moreover for sufficiently large values of $\hat{\delta}_{st} = \sqrt{n}\delta/S_n$ an even simpler expression can be given by replacing the threshold $k_\alpha\left(S_n^2\right)$ with

$$k_\alpha\left(S_n^2\right) \approx z_{1-\alpha} + \hat{\delta}_{st} \tag{25.12}$$

(see approximation (25.2)).

Clearly the proposed method leads to a simple and easily interpretable solution: it behaves as if the nuisance parameter were equal to its estimate and it can be implemented employing the formulas relative to the σ^2 known case. Although its level cannot be determined explicitly, both analytical results and simulation studies show that, for large n, it is approximately equal to α.

Proposition 1. *The test with rejection region R_n given by* (25.11) *is consistent with asymptotic level α. More precisely, for any given $\sigma^2 > 0$ we have:*

$$\lim_{n\to\infty} P_{\mu,\sigma^2}(R_n) = \begin{cases} \alpha & |\mu-\mu_0| = \delta \\ 0 & |\mu-\mu_0| < \delta \\ 1 & |\mu-\mu_0| > \delta \end{cases}.$$

The same asymptotic result does hold if in R_n the threshold $k_\alpha\left(S_n^2\right)$ is replaced by (25.12).

For the proof see the Appendix.

In order to investigate the rate of convergence of the R_n test level to α we performed a simulation study for various sample sizes n and δ/σ values. Simulations for the level are obtained by taking 40,000 samples of size n (similar results can be obtained by resorting to the simple approximate formula (25.12), slight differences emerging only when $\delta_{st} = \sqrt{n}\delta/\sigma$ is smaller than 1). Table 25.1 shows that the real (simulated) level is relatively close to α even for moderate sample sizes; in particular it is quite accurate

Table 25.1. Simulated levels (based on 40,000 replications) for different values of n and of δ/σ with $\alpha = 0.05$

δ/σ	$n = 30$	$n = 100$	$n = 500$	$n = 1000$
0.1	0.05545	0.05218	0.05105	0.04963
0.5	0.05423	0.05213	0.05113	0.05020
1	0.05418	0.05230	0.04960	0.04978
3	0.05403	0.05210	0.04980	0.05022

for $n \geq 100$. The different δ/σ values do not appear to influence the goodness of the approximation, only a slight improvement emerging essentially for small sample sizes as δ/σ increases. Moreover, further simulations performed with different α values and larger δ/σ deserve similar remarks.

Notice that if in R_n we use approximation (25.12) the resulting test coincides with test (25.8) provided that we substitute $z_{1-\alpha}$ with $t_{n-1,1-\alpha/2}$. As $t_{n-1,1-\alpha/2}$ converges to $z_{1-\alpha/2}$ as $n \to \infty$, Proposition 1 implies that for large n such a test has level $\alpha/2$ for any given $\sigma^2 > 0$. To sum up, the test (25.8) has level $\alpha/2$ when n diverges and $\sigma^2 > 0$ is fixed or for small σ^2 and fixed n. On the other hand it has level α when $\sigma^2 \to \infty$ and n is fixed. As a consequence when both n and σ^2 are large the choice of the level to be used for calibration is particularly difficult.

On the contrary, the test R_n does not suffer from this limitation as for sufficiently large but fixed n, its level is approximately equal to α both when $\sigma^2 \to 0$ and when $\sigma^2 \to \infty$ as stated in the following proposition.

Proposition 2. *Let V_n have a central Student's t-distribution with n degrees of freedom. Then for any given μ we have:*

$$\lim_{\sigma^2 \to \infty} P_{\mu,\sigma^2}(R_n) = P\left(|V_n| \geq z_{1-\alpha/2}\right).$$

Furthermore it holds that

$$\lim_{\sigma^2 \to 0} P_{\mu,\sigma^2}(R_n) = \begin{cases} P(V_n \leq z_\alpha) & |\mu - \mu_0| = \delta \\ 0 & |\mu - \mu_0| < \delta \\ 1 & |\mu - \mu_0| > \delta \end{cases}$$

For the proof see the Appendix.

As the central Student's t-distribution converges to the standard Normal, Proposition 2 implies that the level of the test R_n is rather close to α both when σ^2 tends to zero and when it diverges to infinity provided that n is large. In particular even for moderate values of n, say larger than 50, a good accuracy is obtained for the usual α levels.

As far as calibration is concerned, for large n it is then legitimate to perform it by equating the two thresholds $t_{n-1,1-\alpha^*/2}$ and $k_\alpha(S_n^2)$, obtaining the following expression

$$\alpha^* = 2\left\{1 - \Phi\left[k_\alpha(S_n^2)\right]\right\}, \tag{25.13}$$

which can be simplified by using approximation (25.12), thus getting

$$\alpha^* = 2\left\{1 - \Phi\left[z_{1-\alpha} + \hat{\delta}_{st}\right]\right\}. \tag{25.14}$$

If we employ the test given by R'_n instead of R_n the calibration becomes

$$\alpha^* = 2\left\{1 - F_{n-1,0}\left[k'_\alpha(S_n^2)\right]\right\}. \tag{25.15}$$

Notice that in expression (25.15) α^* depends both on n and on $\hat{\delta}_{st}$ whereas in (25.13) and (25.14) α^* is given purely in terms of $\hat{\delta}_{st}$.

Table 25.2 compares the nominal levels α^* computed on the basis of R'_n (formula (25.15)), R_n (formula (25.13)), and the approximation (25.14). From a practical point of view all the α^* values reported in the table can be considered quite similar for any given positive $\hat{\delta}_{st}$. In particular, the two tests R'_n and R_n are nearly equivalent when $n \geq 100$ giving very close α^* values. Moreover the approximate formula (25.14) for the test R_n is rather accurate for $\hat{\delta}_{st} \geq 0.5$, perfectly fitting the behaviour of α^* computed with (25.13) for $\hat{\delta}_{st} \geq 1$ as already shown in Figure 25.1.

Expressions (25.13) and (25.15) give rise to the following calibration formulas for the p-value:

$$p = 1 - \Phi(z_{1-p^*/2} - \hat{\delta}_{st}) + \Phi(-z_{1-p^*/2} - \hat{\delta}_{st}) \tag{25.16}$$
$$p = 1 - F_{n-1,\hat{\delta}_{st}}(t_{1-p^*/2}) + F_{n-1,\hat{\delta}_{st}}(-t_{1-p^*/2}). \tag{25.17}$$

Let us now focus on the more common case where one is unable to specify exactly the critical value δ which defines the interval null hypothesis. As we have shown in Section 25.2 any calibration procedure in this case is crucially based on the ability of testing at a given level α, which can be performed by finding the value $\hat{\delta}'_{st}$ of $\hat{\delta}_{st}$ which solves

Table 25.2. Nominal levels α^* for different values of n and of $\hat{\delta}_{st}$ when $\alpha = 0.05$: equation (25.15) (columns 2–5), equation (25.13) (column 6), and equation (25.14) (column 7)

$\hat{\delta}_{st}$	$n = 30$	$n = 100$	$n = 500$	$n = 1000$	Equation (25.13)	Equation (25.14)
0	0.05	0.05	0.05	0.05	0.05	0.1
0.5	0.03236	0.02950	0.02922	0.02918	0.02914	0.03196
1	0.00958	0.00856	0.00823	0.00818	0.00814	0.00817
1.5	0.00244	0.00188	0.00170	0.00168	0.00166	0.00166
2	$564{\cdot}10^{-6}$	$342{\cdot}10^{-6}$	$282{\cdot}10^{-6}$	$274{\cdot}10^{-6}$	$268{\cdot}10^{-6}$	$268{\cdot}10^{-6}$
2.5	$122{\cdot}10^{-6}$	$53{\cdot}10^{-6}$	$37{\cdot}10^{-6}$	$36{\cdot}10^{-6}$	$34{\cdot}10^{-6}$	$34{\cdot}10^{-6}$
3	$253{\cdot}10^{-7}$	$70{\cdot}10^{-7}$	$40{\cdot}10^{-7}$	$37{\cdot}10^{-7}$	$34{\cdot}10^{-7}$	$34{\cdot}10^{-7}$
3.5	$516{\cdot}10^{-8}$	$80{\cdot}10^{-8}$	$34{\cdot}10^{-8}$	$30{\cdot}10^{-8}$	$27{\cdot}10^{-8}$	$27{\cdot}10^{-8}$
4	$1052{\cdot}10^{-9}$	$82{\cdot}10^{-9}$	$24{\cdot}10^{-9}$	$20{\cdot}10^{-9}$	$17{\cdot}10^{-9}$	$17{\cdot}10^{-9}$
4.5	$2171{\cdot}10^{-10}$	$76{\cdot}10^{-10}$	$13{\cdot}10^{-10}$	$10{\cdot}10^{-10}$	$8{\cdot}10^{-10}$	$8{\cdot}10^{-10}$
5	$4581{\cdot}10^{-11}$	$65{\cdot}10^{-11}$	$6{\cdot}10^{-11}$	$4{\cdot}10^{-11}$	$3{\cdot}10^{-11}$	$3{\cdot}10^{-11}$

the analogue of equation (25.7). This can be accomplished by substituting α^* with the (nominal) p-value p^* into equation (25.13) and subsequently solving it with respect to $\hat{\delta}_{st}$ (on which the threshold $k_\alpha(S_n^2)$ does depend). Obviously a similar method can be applied to the test R_n', using equation (25.15) instead of (25.13). Moreover, it is noteworthy that approximation (25.14) relative to the test R_n gives rise to the explicit solution

$$\hat{\delta}'_{st} = z_{1-p^*/2} - z_{1-\alpha}. \tag{25.18}$$

Such a simple solution is virtually perfect when $\hat{\delta}'_{st} > 1$, which is often the case when n is large. The performance for smaller values of $\hat{\delta}'_{st}$ is shown in Table 25.3. Clearly the approximation improves as α decreases, being quite accurate for values bigger than 0.4–0.6 depending on α levels.

Table 25.3. Approximate $\hat{\delta}'_{st}$ values (see formula (25.18)) for different α levels (the first column reporting the exact values)

$\hat{\delta}'_{st}$	$\alpha = 0.10$	$\alpha = 0.05$	$\alpha = 0.01$	$\alpha = 0.001$
0	0.3633	0.3151	0.2495	0.2003
0.1	0.3715	0.3249	0.2623	0.2165
0.2	0.3959	0.3537	0.2993	0.2623
0.3	0.4359	0.4002	0.3571	0.3309
0.4	0.4902	0.4621	0.4310	0.4143
0.5	0.5572	0.5366	0.5159	0.5063
0.6	0.6346	0.6205	0.6077	0.6026
0.7	0.7199	0.7110	0.7036	0.7011
0.8	0.8110	0.8056	0.8016	0.8004
0.9	0.9058	0.9027	0.9007	0.9002
1	1.0029	1.0013	1.0003	1.0000

In practice one can use the test R_n and determine the crucial $\hat{\delta}'_{st}$ values as follows. Use the value obtained by (25.18) if it is bigger than say 0.6, otherwise deduce an approximate value on the basis of Table 25.4. For a given α, one has to look for the α^* value closest to the observed p^* and then choose as approximate $\hat{\delta}'_{st}$ value the one

Table 25.4. Nominal levels α^* derived from equation (25.13) for different values of α and of δ_{st}

δ_{st}	$\alpha = 0.10$	$\alpha = 0.05$	$\alpha = 0.01$	$\alpha = 0.001$
0	0.100000	0.050000	0.010000	0.001000
0.1	0.098318	0.048870	0.009637	0.000944
0.2	0.093449	0.045657	0.008649	0.000801
0.3	0.085897	0.040850	0.007286	0.000624
0.4	0.076429	0.035118	0.005828	0.000457
0.5	0.065952	0.029148	0.004480	0.000323
0.6	0.055350	0.023489	0.003345	0.000222
0.7	0.045341	0.018481	0.002446	0.000150
0.8	0.036392	0.014267	0.001760	0.000100
0.9	0.028719	0.010847	0.001251	0.000066
1	0.022344	0.008141	0.000879	0.000043

which entitles the corresponding row. Such approximation is sufficiently precise to most practical purposes as its absolute error is smaller than 0.1, which implies an absolute error in terms of δ/σ smaller than $(10\sqrt{n})^{-1}$.

25.4 Conclusion

We proposed a simple procedure to calibrate p-values when nuisance parameters are present. Such a calibration is based on the comparison between the traditional t-test and a new test specifically developed for testing interval null hypotheses having the form required to perform calibration. The level of the latter has been investigated both from a theoretical and from a practical perspective and a comparison is made with the standard proposal which combines two one-tailed tests.

The implementation of the procedure depends on the specification, for a given set of standard α levels, of crucial values $\hat{\delta}'_{st}$. The actual calibration of a given p-value can then be simply accomplished by deciding whether a departure from the precise null model of size $|\mu - \mu_0| = \hat{\delta}'_{st} S_n/\sqrt{n}$ has to be considered practically significant (the possibility one is unable to decide is accounted for by the procedure, obviously leading to a less accurate evaluation of the correct p-value).

Exact formulas and tables for the determination of $\hat{\delta}'_{st}$ are given together with the explicit approximation $\hat{\delta}'_{st} = z_{1-p^*/2} - z_{1-\alpha}$, whose simplicity and great accuracy for large sample sizes make it a benchmark value for most practical situations.

Our calibration procedure was conceived for the Normal mean testing problem, but it can be extended to more general frameworks. In particular, the method can be directly applied to the family of Student's t type test statistics, such as those arising in the two-sample problem and in regression coefficient testing within Normal models.

As far as the former is involved, suppose that two i.i.d samples of size n_1 and n_2 are drawn independently from $X_1 \sim N(\mu_1, \sigma^2)$ and, respectively, from $X_2 \sim N(\mu_2, \sigma^2)$ to test the precise null $H_0^* : \mu_1 = \mu_2$. The usual test statistic takes the form

$$|T_n| = \frac{\sqrt{n}\,|\bar{X}_1 - \bar{X}_2|}{S_n\sqrt{n_1^{-1} + n_2^{-1}}}$$

where $S_n^2 = \left[(n_1 - 1)\, S_1^2 + (n_2 - 1)\, S_2^2\right] / (n_1 + n_2 - 2)$ and S_1^2 and S_2^2 are the two unbiased sample variances. The distribution of T_n is noncentral Student's t with $n_1 + n_2 - 2$ degrees of freedom and noncentrality parameter $\sqrt{n_1 n_2/(n_1 + n_2)}\, |\mu_1 - \mu_2| / \sigma$. Even in this case the correct null hypothesis can be more realistically formulated as an interval one; namely $H_0 : |\mu_1 - \mu_2| \leq \delta$. It follows that, in terms of calibration, this setup is equivalent to the one of Section 25.3, and all calibration formulas derived there do apply.

Analogously, the problem of testing the generic ith regression coefficient is formally identical to the Normal mean testing problem provided that in the expression of δ'_{st} the quantity $\sqrt{n}$ is replaced by $c_{ii}^{-1/2}$ where c_{ii} is the ith element of the diagonal of $(X'X)^{-1}$, X being the design matrix.

The calibration procedure proposed here can also be directly applied in the very general case where the test statistic has an asymptotic Normal distribution. If $H_0^* : \theta = \theta_0$ and $\hat{\theta}$ is an asymptotic Normal estimator with standard error se, then the calibration formulas relative to the R_n test still hold if we set $\hat{\delta}_{st} = \delta / se$.

Finally, the logic underlying the present approach can be fruitfully employed for dealing with more general models with nuisance parameters which will be shown in forthcoming work.

25.5 Appendix

Result 1. In the setting of Section 25.3, the function

$$g(\sigma^2) = P_{|\mu-\mu_0|=\delta,\sigma^2}\left(|T_n| \geq k\right)$$

is decreasing with respect to σ^2 with range $(q, 1)$ where q is the probability that the absolute value of a central Student's t-distribution with $n-1$ degrees of freedom is greater than k.

Proof. The distribution of $T_n = \sqrt{n}\bar{X}/S_n$ is noncentral Student's t with $n-1$ degrees of freedom and noncentrality parameter $\delta_{st} = \sqrt{n}\delta/\sigma$, which we denote by $F_{n-1,\delta_{st}}$. Let Y and W be two independent r.v.s where Y is a standard Normal and $W = \sqrt{\chi^2_{n-1}/(n-1)}$. Then

$$g(\sigma^2) = P_{|\mu-\mu_0|=\delta,\sigma^2}\left(\left|\frac{Y + \delta_{st}}{W}\right| \geq k\right) = 1 - G_{Y-kW}(-\delta_{st}) + G_{Y+kW}(-\delta_{st})$$

where G_Z denotes the distribution function of the r.v. Z. Therefore it is easy to show that $\lim_{\sigma^2 \to 0} g(\sigma^2) = 1$ and $\lim_{\sigma^2 \to \infty} g(\sigma^2) = P(|T_{n-1,0}| \geq k)$. Furthermore we have that

$$g(\sigma^2) = E\left\{P_{|\mu-\mu_0|=\delta,\sigma^2}\left[\left(\left|\frac{Y + \delta_{st}}{W}\right| \geq k\right) \mid W\right]\right\}$$

and

$$P_{|\mu-\mu_0|=\delta,\sigma^2}\left[\left(\left|\frac{Y+\delta_{st}}{W}\right|\geq k\right)\mid W\right]=1-\Phi\left(kW-\delta_{st}\right)+\Phi\left(-kW-\delta_{st}\right).$$

By differentiating with respect to σ the latter expression can be shown increasing in σ^2 which implies monotonicity of $g(\sigma^2)$.

Lemma 1. *Let X_i be i.i.d. $N(\mu,\sigma^2)$ r.v.s and let $\bar{X}_n$ and S_n^2 be the sample mean and variance. Then for any given σ^2 we have that*

$$\frac{\sqrt{n}}{S_n}\left(\left|\bar{X}_n-\mu_0\right|-\delta\right)\xrightarrow{d}\begin{cases}N(0,1) & |\mu-\mu_0|=\delta\\ -\infty & |\mu-\mu_0|<\delta\\ +\infty & |\mu-\mu_0|>\delta\end{cases}$$

where $\xrightarrow{d}$ means convergence in distribution.

Proof. Let us first derive the asymptotic distribution of

$$W_n=\frac{\sqrt{n}}{\sigma}\left(\left|\bar{X}_n-\mu_0\right|-\delta\right).$$

We have

$$P_{\mu,\sigma^2}(W_n\leq w)$$

$$=P_{\mu,\sigma^2}\left(-w-\sqrt{n}\frac{\delta+\mu-\mu_0}{\sigma}\leq\sqrt{n}\frac{\bar{X}_n-\mu}{\sigma}\leq w+\sqrt{n}\frac{\delta-\mu+\mu_0}{\sigma}\right)$$

which implies that

$$\lim_{n\to\infty}P_{\mu,\sigma^2}(W_n\leq w)=\begin{cases}\Phi(w) & |\mu-\mu_0|=\delta\\ 1 & |\mu-\mu_0|<\delta\\ 0 & |\mu-\mu_0|>\delta\end{cases}$$

i.e., convergence of W_n to $N(0,1)$ if $|\mu-\mu_0|=\delta$, to $-\infty$ if $|\mu-\mu_0|<\delta$, and to $+\infty$ if $|\mu-\mu_0|>\delta$.

The result then follows by applying the Slutsky theorem to the ratio $W_n/(S_n/\sigma)$.

Proof of Proposition 1. In the notation of the previous lemma, let $p_\sigma(k)$ be defined as

$$p_\sigma(k)=P_{|\mu-\mu_0|=\delta,\sigma^2}\left(\sqrt{n}\frac{\bar{X}_n-\mu_0}{\sigma}\geq k\right)=1-\Phi\left(k-\frac{\sqrt{n}\delta}{\sigma}\right)+\Phi\left(-k-\frac{\sqrt{n}\delta}{\sigma}\right).$$

It is easy to check that for any given $\sigma^2>0$ and α

$$\{k\geq k_\alpha(\sigma^2)\}=\{p_\sigma(k)\leq\alpha\}.$$

It follows that R_n can be equivalently written as

$$R_n=\{p_{S_n}(|T_n|)\leq\alpha\}.$$

Let us now study the asymptotic behaviour of

$$p_{S_n}(|T_n|) = 1 - \Phi\left[\frac{\sqrt{n}}{S_n}\left(\left|\bar{X}_n - \mu_0\right| - \delta\right)\right] + \Phi\left[-\frac{\sqrt{n}}{S_n}\left(\left|\bar{X}_n - \mu_0\right| + \delta\right)\right].$$

Clearly, for any μ and σ^2 we have that $\Phi\left[-(\sqrt{n}/S_n)\left(\left|\bar{X}_n - \mu_0\right| + \delta\right)\right]$ converges to zero. Furthermore by the previous lemma we have

$$1 - \Phi\left[\frac{\sqrt{n}}{S_n}\left(\left|\bar{X}_n - \mu_0\right| - \delta\right)\right] \xrightarrow{d} \begin{cases} U(0,1) & |\mu - \mu_0| = \delta \\ 1 & |\mu - \mu_0| < \delta \\ 0 & |\mu - \mu_0| > \delta \end{cases}$$

(where $U(0,1)$ is a uniform r.v. with range $(0,1)$), which proves the asymptotic result for R_n.

The rejection region obtained by replacing $k_\alpha(S_n^2)$ with $z_{1-\alpha} + \sqrt{n}\delta/S_n$ can be written as

$$\left\{\frac{\sqrt{n}}{S_n}\left(\left|\bar{X}_n - \mu_0\right| - \delta\right) \geq z_{1-\alpha}\right\}.$$

The asymptotic behaviour of such a region is then a direct consequence of the lemma.

Proof of Proposition 2. As shown in the proof of Proposition 1, the rejection region R_n can be written as $R_n = \{p_{S_n}(|T_n|) \leq \alpha\}$ where

$$p_{S_n}(|T_n|) = 1 - \Phi\left[\frac{\sqrt{n}}{S_n}\left(\left|\bar{X}_n - \mu_0\right| - \delta\right)\right] + \Phi\left[-\frac{\sqrt{n}}{S_n}\left(\left|\bar{X}_n - \mu_0\right| + \delta\right)\right].$$

It is then not difficult to check that, for any given μ, as $\sigma^2 \to \infty$ $p_{S_n}(|T_n|)$ converges in distribution to $1 - \Phi(|V_n|) + \Phi(-|V_n|)$. Therefore

$$\lim_{\sigma^2 \to \infty} P_{\mu,\sigma^2}(R_n) = Pr\left[2\left(1 - \Phi(|V_n|)\right) \leq \alpha\right] = Pr\left(|V_n| \geq z_{1-\alpha/2}\right).$$

Let us now consider the case $\sigma^2 \to 0$. The term $\Phi\left[-(\sqrt{n}/S_n)\left(\left|\bar{X}_n - \mu_0\right| + \delta\right)\right]$ in $p_{S_n}(|T_n|)$ is easily seen to converge to 0 for any given μ. Let us then derive the behaviour of

$$U_n = \frac{\sqrt{n}}{S_n}\left(\left|\bar{X}_n - \mu_0\right| - \delta\right).$$

Its distribution function can be written as

$$\begin{aligned} &P_{\mu,\sigma^2}(U_n \leq u) \\ &= P_{\mu,\sigma^2}\left(-u - \sqrt{n}\frac{\delta + \mu - \mu_0}{S_n} \leq \sqrt{n}\frac{\bar{X}_n - \mu}{S_n} \leq u + \sqrt{n}\frac{\delta - \mu + \mu_0}{S_n}\right). \end{aligned}$$

Since $S_n \xrightarrow{d} 0$ for $\sigma^2 \to 0$ and, for any given σ^2, $\sqrt{n}((\bar{X}_n - \mu)/S_n)$ is distributed as a central Student's t with $n-1$ degrees of freedom, it then follows that as $\sigma^2 \to 0$,

$$U_n \xrightarrow{d} \begin{cases} V_n & |\mu - \mu_0| = \delta \\ -\infty & |\mu - \mu_0| < \delta \\ +\infty & |\mu - \mu_0| > \delta \end{cases}.$$

Finally this implies

$$p_{S_n}(|T_n|) \xrightarrow{d} \begin{cases} \Phi(V_n) & |\mu - \mu_0| = \delta \\ 1 & |\mu - \mu_0| < \delta \\ 0 & |\mu - \mu_0| > \delta \end{cases}$$

which proves the result.

References

Berger, J.O. and Delampady, M. (1987). Testing precise hypotheses. *Statistical Science*, 2:317–352.

Berger, J.O. and Sellke, T. (1987). Testing a point null hypothesis: The irreconcilability of P values and evidence (with discussion). *Journal of the American Statistical Association*, 82:112–122.

Hodges, J.L. and Lehmann, E.L. (1954). Testing the approximate validity of statistical hypotheses. *Journal of the Royal Statistical Society B*, 16:261–268.

Johnson, N.L., Kotz, S., and Balakrishnan, N. (1994). Continuous univariate distributions. Wiley, New York.

Migliorati, S. and Ongaro, A. (2007). Standard packages outputs when n is large: When and how should they be adjusted? *Proceedings of the 12th International Conference on Applied Stochastic Models and Data Analysis*, http://www.asmda.com//CDasmda2007a/index.html

Schervish, M. (1995). *Theory of Statistics.* Springer, New York.

Part VIII

Statistical Theory and Methods

26

Fitting Pareto II Distributions on Firm Size: Statistical Methodology and Economic Puzzles

Aldo Corbellini[1], Lisa Crosato[2], Piero Ganugi[1], and Marco Mazzoli[1]

[1] Department of Economics and Social Sciences, Università Cattolica del Sacro Cuore, 29100 Piacenza, Italy
[2] Statistics Department, Università di Milano Bicocca, 20126 Milano, Italy

Abstract: We propose here a new implementation of the forward search, which is a powerful general method usually suitable for detecting extreme observations and for determining their effect on fitted models (Atkinson and Riani, 2000). Through the forward search we iteratively fit the Pareto II distribution to firm size data. In particular, a threshold is fixed to the fit of the Pareto II distribution through a progressive adaptation technique, performing at each iteration the χ^2 test to check for the acceptance of the null hypothesis. Yearly Zipf-plots of the truncated empirical distribution with superimposed theoretical Pareto II distribution highlight the adherence of the estimates to data for different size ranges. Possible economic interpretations of the results are then provided, referring in particular to the role of the stock market in shaping firm size distribution and to the *firm size effect* (Banz, 1981; Reingaum, 1981). More in general, we discuss possible implications of introducing our methodology in macroeconomic models.

Keywords and phrases: Firm size distribution, forward search, Pareto II, stock market, firm size effect

26.1 Introduction

The existence of a recurrent probability model of firm size distribution has been investigated in statistical as well as in economic literature since Pareto (1897) and Gibrat's (1931) influential works (for an exhaustive survey see Kleiber and Kotz, 2003). In early as well as in recent literature firm size has been mostly modeled by means of the lognormal and Pareto I distributions (see for example, Hart and Prais, 1956; Steindl, 1965; Quandt, 1966; Ijiri and Simon, 1977; Stanley et al., 1995; and Axtell, 2001). Both distributions have been derived as the outcome of stochastic models of growth, based on the law of proportionate effect which postulates no effect of size on percentage growth rates (Gibrat, 1931). In particular, the goodness-of-fit of the Pareto I distribution is generally assessed on the right tail only, where the lognormal distribution fit is not satisfactory (Stanley et al., 1995; Hart and Oulton, 1997). In this chapter we start

C.H. Skiadas (ed.), *Advances in Data Analysis*, Statistics for Industry and Technology, DOI 10.1007/978-0-8176-4799-5_26,

from these stylized facts, proposing an alternative Paretian model, the Pareto II distribution. This model incorporates the Pareto I as a particular case, modeling therefore what Pareto I models, but presenting the relevant advantage of extending the potential range of firms to capture, on the same time maintaining a low number of parameters. We fit the Pareto II distribution to the Italian chemical sector (1999–2004) through the forward search, which carries on the estimation and goodness-of-fit procedures on an iterative basis. The methodology proposed in this chapter can be profitably employed as an analytical method in several research fields and "economic puzzles," such as the (lack of) empirics in new industrial economics models (i.e., the class of models postulating endogenous market structure as a result of strategic interaction among firms), the issue of microfoundation and aggregation, the *firm size effect* (Banz, 1981; Reingaum, 1981), and the impact of financial markets on firm size and growth. The chapter is organized as follows: in Section 26.2 we briefly describe the dataset; Section 26.3 regards the application of the forward search technique to the fit of the Pareto II distribution. Empirical results are presented in Section 26.4, in Section 26.5 we discuss economic implications of results, and Section 26.6 gives some conclusions and suggestions for further research.

26.2 Data description

Our analysis is based on the AIDA dataset, processed and managed by Bureau van Dijck Electronic Publishing (http://www.bvdep.com/en/aida.html). AIDA is a large dataset which records company accounts and activities for 500,000 Italian companies with sales greater than 500,000 Euros, plus ownership and management for the top 20,000 companies. The dataset goes back to 1995. We first derive a panel for Italian manufacturing sectors, tracking their accounts from 1999 to 2004. Second, we focus on the chemical sector, which presented one of the largest number of firms listed in the Italian stock market (5), obtaining a panel of 1344 firms. Table 26.1 reports the trend of total assets in the six considered years, both on absolute values and in terms of growth rates.

Table 26.1. Total Asset (TA) trend (1999–2004)

Year	Total	Mean	1st Qrt	Median	3rd Qrt
	Absolute Values (TA, Millions of Euro)				
1999	31,390	23,355	1.722	3.701	11.077
2000	34,986	26,031	1.950	4.427	12514
2001	37,148	27,640	2.021	4.684	13.053
2002	38,613	28.730	2.235	4.964	14.137
2003	36,617	27.245	2.343	5.152	13.826
2004	36,635	27.258	2.447	5.575	14.978
	Growth (Index Numbers)				
2000	11.458	–	13.229	19.590	12.975
2001	6.180	–	3.665	5.797	4.311
2002	3.944	–	10.595	5.978	8.304
2003	−5.169	–	4.834	3.795	−2.199
2004	0.047	–	4.435	8.211	8.328

26.3 Fitting the Pareto II distribution by means of the forward search

The Pareto II distribution is the second model Pareto proposed (Pareto, 1897) to describe empirical income distributions, and was studied in particular by D'Addario (2003). Its distribution function is given by C.D.F.

$$F(x) = 1 - \left(1 + \frac{x-\mu}{\sigma}\right)^{-\alpha}$$

As can be seen, it is defined by three parameters, respectively, of location (μ), of scale (σ), and of shape (α). The Pareto I distribution is the special case of Pareto II where $\mu = \sigma$. Due to the augmented flexibility, the threshold of goodness-of-fit, usually constrained by the Pareto I to fall from the 75th percentile onwards, is pushed further on the left by the Pareto II.

Our goal is precisely to find the smallest leftmost firm size unit starting from which the Pareto II distribution shows a satisfactory fit. Through the forward search, we iterate the fitting procedure composed by parameter estimation and goodness-of-fit tests.

The steps of the algorithm are the following.

(1) The first step consists in reverse-ordering the vector $\mathbf{X} := \{x_1, x_2, \ldots, x_n\}$ which contains the size of each firm.
(2) Then we make the left queue of the data our initial subset: we start by considering only the three rightmost observations, i.e., the three largest firms. This choice is intentional given the nature of the Pareto II distribution, the adaptation of the curve to the largest firms being usually good. Therefore, this becomes the robust subset of firms from where the Forward Search starts.
(3) We start fitting the Pareto II distribution to the data included in the initial subset. In order to estimate the unknown $\hat{\theta}$ vector whose components are $\hat{\mu}$, $\hat{\alpha}$, and $\hat{\sigma}$ parameters, we choose the maximum likelihood method (Arnold, 1983). Given that the likelihood will have maximum at $\hat{\mu} = x_{1:n}$ (where $\hat{\mu} = x_{1:n}$ is the first-order statistic) the $\hat{\alpha}$ and $\hat{\sigma}$ estimates are given by the following equations.

$$\hat{\alpha} = \frac{1}{n}\sum_{i=1}^{n} \log\left(1 + \frac{x_i - x_{1:n}}{\hat{\sigma}}\right); \qquad \hat{\sigma} = \frac{\hat{\alpha}+1}{n}\sum_{i=1}^{n}(x_i - x_{1:n})\left[1 + \frac{x_i - x_{1:n}}{\hat{\sigma}}\right]^{-1}$$

For computational purposes, however, it is better to calculate the maximum log-likelihood of the Pareto II density, in the form of:

$$\ell(\mu, \alpha, \sigma) = -(\alpha+1)\sum_{i=1}^{n} \log\left(1 + \frac{x_i - \mu}{\sigma}\right) - n\log\sigma + n\log\alpha$$

The *R* package provides us with the constrained optimization routine collection `nlminb`, which we can employ with the following syntax.

```
par2.y.mle<-nlminb(c(1, 1), objective=function(x) -par2.llik(x,
data=dati, mu=mu), lower=c(1e-07, 1e-07), upper=c(Inf, Inf))
```

Please note that a starting point $x_0 = \{1, 1\}$, upper $b = \{\infty, \infty\}$, and lower $a = \{-\infty, -\infty\}$ limits are needed.

(4) Using $\hat{\theta}$ and employing the method of inversion we can set up simulated datasets. By using the following inverse C.D.F. equation,

$$F^{-1}(x) = \sigma \left[(1 - F(x))^{-1/\alpha} - 1 \right] + \mu$$

we can obtain as many random draws from the PII(x;$\hat{\theta}$) distribution as we like.
(5) We are ready to test if the distribution is Pareto II-type compliant by means of the χ^2 test (Quandt, 1966). If the p-value threshold of 10% is reached then we accept the H_0 hypothesis and proceed to step (6), otherwise the last x_k firm size is removed and the routine jumps to step (7).
(6) A single observation is added from $\mathbf{X}$ and the routine goes back to step (2).
(7) At this point we have reached the smallest leftmost firm size which does not force the empirical distribution to abandon the Pareto II model.

26.4 Empirical results

Among the graphical outcomes derivable from procedure (1) to (7) we draw the Zipf plots of the truncated empirical firm size distribution with superimposed theoretical Pareto II distribution (see Figure 26.1a). Furthermore, Figure 26.1b tracks the path of the χ^2-square p-values through the forward search procedure. As can be seen, notwithstanding the satisfactory fit assessed-through the χ^2-test, the Zipf plots point out a systematic deviation in largest firm empirical size with respect to the estimated one. Table 26.2 reports the outcome of the forward search corresponding to the final set of firms satisfactorily modeled by the Pareto II distribution.

As can be seen from the second column of Table 26.2, the percentage of firms covered by the model is about 96–97% of the totality of firms. The corresponding threshold in terms of total assets is given by the location parameter μ (see column 3).

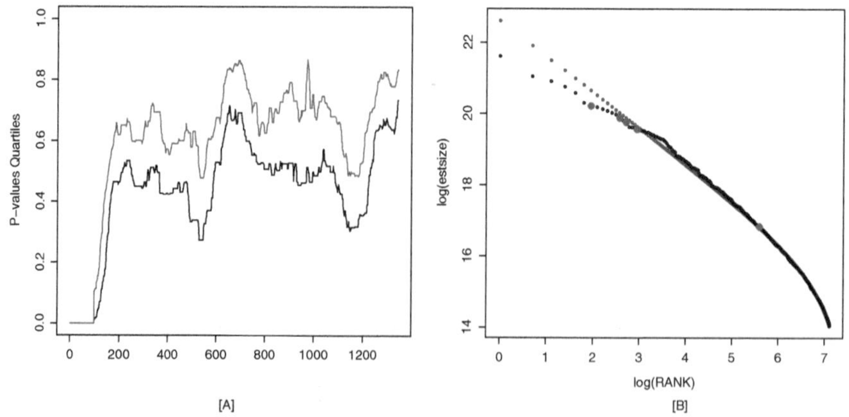

Figure 26.1. (**a**) P-values threshold (Black line: 5th percentile, gray line: 95th percentile) and (**b**) Zipf plot (2004). Gray line: estimated Zipf Plot, black line: empirical Zipf Plot. Large dots represent firms listed in the stock market

Table 26.2. Forward search statistics

Year	Perc	μ	α	σ	χ^2	*df*	*p*-Value
1999	96%	694,105	0.973	3,446,699	42.973	34	0.139
2000	97%	713,494	0.994	4,105,675	41.902	34	0.165
2001	96%	868,204	0.958	3,989,601	43.907	34	0.119
2002	96%	838,834	0.994	4,630,873	43.243	34	0.133
2003	96%	917,986	1.004	4,824,788	43.952	34	0.118
2004	97%	804,792	1.055	5,679,355	44.229	34	0.112

26.5 Economic implications

The first and central result of the analysis is the fact that the firms listed in the stock market (represented by large dots in Figure 26.1) behave in exactly the same way as any other firm of the same size class, whether their observable position lies on or below the Pareto II distribution. Thus the stock market does not seem to play a qualitatively relevant role in affecting firms' behavior or strategies, compared to other elements of financial choice and/or corporate governance.

Second, the methodology proposed here can be fruitfully related to the literature on the so-called *firm size effect*, i.e., the empirical evidence suggesting that risk-adjusted returns are larger for small firms than for large firms. This literature has been associated with finance and, in particular, with the effect of financial market imperfections. The Zipf plot of Figure 26.1b could be interpreted as evidence of the firm *size effect*, to the extent that the Pareto II systematically overestimates the firm size of the largest firm classes. Furthermore, as we said, the firms listed in the stock market behave exactly in the same way as any other firm, for any firm size. Since the stock market is normally regarded as more efficient than informal financial markets this last point may suggest that the *firm size effect* could be rather associated with real phenomena, such as returns to scale (Crosato and Ganugi, 2007) rather than financial market imperfections.

For what concerns the industrial economics implications, the shape parameter (α) of the Pareto II distribution, to the extent that it provides information on how the density varies with firm size, can be interpreted as an observable reduced form of the outcome of firms' strategic interactions (collusion, entry/exit, mergers), given a certain market size. In other words, given, for instance, an oversimplified market demand curve of the kind $P = a - bQ$, where P are prices, Q the output, and a and b parameters, the market size a provides information on how many firms could potentially survive for a given market configuration and market structure, knowing, of course, that there is a multiplicity of equilibria in market configuration, as a result of collusion, entry/exit, and merger decisions of the firms. This point is particularly relevant, to the extent that a well-known criticism of the limits of the new industrial economics lies in the difficulties of implementing and performing empirical analyses based on the implications of the models of strategic interaction. Measuring the changes over time of the slope parameter of the Pareto II distribution (i.e., comparing the different values of its estimates year by year) could provide empirical measures of market structure and market configuration. Furthermore, since in standard frontier macromodels for macroeconomic policy analysis (such as, for instance, the "new Keynesian model for policy analysis" described in Walsh, 2003), the concept of *output gap* is defined as the distance of a given market structure and configuration from the benchmark case of flex prices/perfectly competitive equilibrium, suitable algebraic manipulations of the slope parameter of the Pareto II distribution could be employed in simulations and calibration analyses based on that class of macromodels.

More in general, firm size distribution, giving explicit measures and approximated functional form to changes in the frequencies of heterogeneous firms and agents, also provides a substantial informational contribution to the (lack of) assumptions on aggregation that characterize models based on any kind of representative agent. As we know it, heterogeneity in such models is normally introduced in the behavior function characterizing the agent, either with some random shock extracted from some absolutely naive, and ad hoc distribution function (such as the uniform probability function or some normal distribution a priori assigned parameters), or, in the case of overlapping generation literature, by simply introducing a different time horizon for different classes of agents. Therefore, the logical treatment of the representative agent in standard macroeconomics (which, of course, since the Lucas contributions, belongs to the very basic logical foundation of economic thought) could probably be seen by "hard" scientists, such as mathematicians, physicists, or statisticians, only as a temporary modeling tool, lacking any serious measure analysis and any rigorous statistical study of the frequency distribution of different sized agents, which are likely to be characterized by different features, incentives, and objective functions.

In this regard, even among mainstream economists the amount of relevant methodological contributions is certainly not negligible, and the objections raised on the standard representative agent modeling approach still have not yet received any answer. For instance, Forni and Lippi, in a book containing joint papers with L. Reichlin (Forni and Lippi, 1997), have shown that many statistical features associated with the dynamic structure of a model (such as, for instance, Granger causality and cointegration), when derived from the optimized microeconomic behavior of an agent, do not, in general, survive aggregation of heterogeneous agents. As a consequence, any dynamic macroeconomic simulation, calibration analysis, or even macroeconometric analysis microfounded with some kind of representative agent is very likely to yield biased results. In addition, Blinder (1986) points out that the microfounded "New Econometrics" methodology, by assuming that the observable choices of optimizing individuals are "internal solutions," returns biased estimates when the choices of a relevant portion of individuals are corner solutions. Conversely, by estimating and measuring year by year the parameters of a Pareto II distribution and looking at their changes in time one can obtain a qualitative measure of the bias induced by macroeconomic models microfounded on the basis of a representative agent. In this regard one could argue that any production function of a representative firm or any utility function of a representative consumer (even in their naive and elementary transformation, such as the sum of individual identical behavior function or their integral over a continuum of qualitatively identical individuals or firms) is not a real and proper microfoundation, but rather an ad hoc production or utility function. Of course, any mainstream economist would object to this point by saying that even some kind of "aggregate utility function" or "aggregate production function" would still allow us to model consumers' behavior on the basis of a rigorous and consistent axiom of preferences and firms' behavior on the basis of a set of rigorous and consistent set of assumptions on technology and optimizing behavior. Still, if one really wants to take Lucas' critique seriously, one cannot help observing that naive and ad hoc aggregations of utility functions and production functions fail to account for the endogenous response of different sized agents, which are likely to be characterized by different features, incentives, and objective functions, not to mention the fact that the analytical form of standard utility functions (for instance CRRA) sharply contrast with the actual utility functions measured on the basis of the actual behavior of real individuals such as those emerging from the seminal contributions by Kahneman, Tversky, Diamond, and Shafir and yielding as an analytical form for the utility function the so-called "kinked utility functions," characterized by the so-called "status quo bias" (Kahneman and Tversky, 1979; Kahneman, 1994; Benartzi and Thaler, 1995; Shafir et al., 1997). All that again raises a problem of measure in microfoundations of macroeconomics, and measuring frequency distributions, which is the very issue analyzed by the Pareto II analysis of this chapter, could be a first step in this direction.

26.6 Concluding remarks

In this chapter we propose a forward search (Atkinson and Riani, 2000) approach to fit a suitable probability model to the firm size distribution of the Italian chemical sector, from 1999 to 2004. The iterative nature of the forward search permits us to track the pattern of the χ^2-test, identifying the smallest firm which leads to acceptance of the null distributional hypothesis (in our case, the Pareto II distribution). Our research points out that one cannot reject a distributional hypothesis on the basis of a test performed either only on the whole dataset or on a priori selected subsamples, without first analyzing the impact of each firm on the estimates and goodness-of-fit. The first contribution of the chapter is therefore to avoid the common practice of using an a priori fixed threshold (typically the 75th percentile or over) to model the right tail of the firm size distribution. Empirical results and graphical outputs confirm the satisfactory fit of the Pareto II distribution in all considered years, and highlight some systematic deviations of the largest firms' empirical size from the estimated one. These deviations could be connected with the literature on the *firm size effect*, the causation going both ways. Our results are broadly consistent with Banz' (1981) and Reingaum' (1981) contributions. On the other hand, the literature on the *firm size effect* might suggest, for further research, possible modifications of the Pareto II function that account for the slightly overestimated firm size of the largest sized firms. Moreover, quoted firms show the same behavior of unquoted ones, suggesting first that the stock market does not have a central impact on firms' behavior or strategies; second, the *firm size effect* could be associated with real phenomena rather than financial market imperfections, given that stock markets are generally judged more efficient than informal financial markets. We finally argue that our methodology and results indicate that including empirically grounded distributional hypotheses in new industrial economics models could help to remove part of the bias induced by representative agent-based microfoundation.

References

Arnold, B. (1983). *Pareto Distributions.* International Co-operative, Fairland, ME.

Atkinson, A. C. and Riani, M. (2000). *Robust Diagnostic Regression Analysis.* Springer-Verlag, New York.

Axtell, R. (2001). Zipf distribution of U.S. firm sizes. *Science*, 293: 1818–1820.

Banz, R. (1981). The relationship between return and market value of common stock. *Journal of Financial Economics*, 9: 3–18.

Benartzi, S. and Thaler, R.H. (1995). Myopic loss aversion and the equity premium puzzle. *Quarterly Journal of Economics*, CX: 73–92.

Blinder, A (1986). A skeptical note on the new econometrics, in *Prices, Competition and Equilibrium*, Preston, M.H. and Quandt, R.E., Eds., P. Allan, London, pp. 73–83.

Crosato, L. and Ganugi, P. (2007). Statistical regularity of firm size distribution: The Pareto IV and truncated Yule for Italian SCI manufacturing. *Statistical Methods and Applications*, 16(1): 85–115.

D'Addario (2003). *Monetary Theory and Policy.* MIT Press, Cambridge, MA.

Forni, M. and Lippi, M. (1997). *Aggregation and the microfoundation of Dynamic Macroeconomics.* Oxford University Press, USA.

Gibrat, R. (1931). *Les Inégalités Economiques.* Librairie du Recueil Survey, Paris.

Hart, P.E. and Oulton, N. (1997). Zipf and the size distribution of firms. *Applied Economics Letters*, 4(4): 205–206.

Hart, P.E. and Prais, S.J. (1956). The analysis of business concentration: A statistical approach. *Journal of the Royal Statistical Society, Series A*, 119: 150–191.

Ijiri, Y. and Simon, H.A. (1977). *Skew Distributions and the Sizes of Business Firms.* North-Holland, Amsterdam.

Kahneman, D. (1994). New challenges to the rationality assumption. *Journal of Institutional and Theoretical Economics*, 150: 18–36.

Kahneman, D. and Tversky, A. (1979). Prospect theory: An analysis of decision under risk. *Econometrica*, XLVII: 263–91.

Kleiber, C. and Kotz, S. (2003). *Statistical Size Distributions in Economics and Actuarial Sciences.* Wiley, New York.

Pareto, V. (1897). *Cours d'Economie Politique.* Lausanne, Swisse.

Quandt, R.E. (1966). On the size distribution of firms. *American Economic Review*, 3: 416–432.

Reingaum, J. (1981). Misspecification of capital asset pricing: Empirical anomalies based on earning yields and market value. *Journal of Financial Economics*, 9: 19–46.

Shafir, E., Diamond, P., and Tversky, A. (1997). Money illusion. *Quarterly Journal of Economics*, CXII: 341–374.

Stanley, M.H.R., Buldyrev, S.V., Havlin, S., Mantegna, R.N., Salinger, M.A., and Stanley, H.E. (1995). Zipf plots and the size distribution of firms. *Economics Letters*, 49: 453–457.

Steindl, J. (1965). *Random Processes and the Growth of Firms.* Griffin, London.

Walsh, C. (2003). *Monetary Theory and Policy.* MIT Press, Cambridge, MA.

27

Application of Extreme Value Theory to Economic Capital Estimation

Samit Paul[1] and Andrew Barnes[2]

[1] GE Global Research, John F. Welch Technology Center, EPIP Phase 2, Bangalore, India
[2] GE Global Research, 1 Research Circle, Niskayuna, Niskayuna, NY, USA

Abstract: We are interested in the risk of large losses of certain common financial portfolios (e.g., credit portfolios with default risk). In these cases, we would like to estimate a risk statistic called Value-at-Risk (VaR_α) at an extremely high risk level α (typically 99.99%). This high risk level corresponds to rare events for which it is difficult or impossible to obtain data. We first use Monte Carlo simulation to compute a probability distribution for the portfolio loss. We then use Extreme Value Theory (EVT) to study the tail of this loss distribution. Finally we compute the VaR and associated confidence intervals using bootstrap techniques. For the portfolios under consideration, we have observed that the EVT-based approach results in narrower confidence intervals and hence less sampling uncertainty in computing the VaR. We have also observed that the bootstrap replicate's distribution for the EVT-based method demonstrates a better shape than the empirical method (which is typically very noisy).

Keywords and phrases: Value-at-risk, EVT, economic capital

27.1 Introduction

Modelling financial losses is one of the key technical challenges in risk management. In the past couple of decades we have seen significant instabilities in various financial markets and witnessed several extreme losses (e.g., Barings, Enron, 9/11, Hurricane Katrina). Hence it is very important for financial institutions to look for better and sound methodologies to model extreme losses.

A significant problem in modelling these extreme losses is that in reality we get to see few rare events, and in many cases we don't have any observations at all. Also, standard statistical approaches do not fit very well to model extreme observations because these are typically treated as outliers. Our approach to these issues has been to use Extreme Value Theory (EVT) to aid the analysis of extreme losses.

In this chapter we apply EVT to compute the Value-at-Risk (VaR) for certain credit loss portfolios of commercial loans. These credit portfolios typically have highly skewed loss distributions with potentially large losses occurring with small probabilities.

C.H. Skiadas (ed.), *Advances in Data Analysis*, Statistics for Industry and Technology, DOI 10.1007/978-0-8176-4799-5_27,

Financial institutions typically compute risk measures such as VaR from the tails of these loss distributions. The ultimate use of such VaR calculations is to help set economic capital levels in an appropriate risk-adjusted manner. This economic capital is intended to protect financial institutions from extreme losses, and hence it is based on extreme tail risk measures (such as VaR at a high risk level).

27.2 Background mathematics

27.2.1 Risk measure

We model the loss on a financial portfolio as a random variable X on a probability space $(\Omega, \mathbb{F}, \Pr)$ and let $F(x) = \Pr(X \leq x)$ be its distribution function. For $\alpha \in (0,1)$ we define the Value-at-Risk of X at the risk level α as the smallest value x such that the probability that the loss X exceeds x is less than or equal to $(1-\alpha)$.

$$\begin{aligned} VaR_\alpha(X) &= \inf\{x \in \mathbb{R} : \Pr(X > x) \leq 1 - \alpha\} \\ &= \inf\{x \in \mathbb{R} : F(x) \geq \alpha\}. \end{aligned} \tag{27.1}$$

For a random variable with a continuous distribution VaR is simply a quantile of the loss distribution.

The use of VaR as a risk measure has been widely criticized in the risk literature (see Artzner et al. 1997, 1999; Szego, 2004) because of its failure to satisfy a subadditivity requirement that expresses the notion of diversification. Alternative *coherent* risk measures such as Expected Shortfall (ES) have been proposed as reasonable alternatives to VaR. However, VaR has become engrained in the minds of many risk practitioners and has been also propagated by the Basel Committee on Banking Supervision as a regulatory measure of risk. For brevity in the description below, we exhibit formulae for the use of EVT on VaR estimation, although we have performed the analysis for Expected Shortfall as well.

27.2.2 Extreme value theory

In this section we provide a brief summary of techniques related to extreme value theory. There are two popular approaches for modelling extreme observations. The first approach is called the *block maxima* method. It is based on observations of the maximum (or minimum) of large samples of independent, identically distributed observations. The second approach is known as the *threshold exceedances* method. In this case we consider extreme observations which exceed a given threshold. See Embrechts et al. (2004) and McNeil (2000) for details.

One of the drawbacks of the block maxima method is that it doesn't use a large section of the data except the maximum losses from each block. The threshold exceedances method is a better option for our data, as it uses all the extreme data which exceed a given high threshold.

We define the distribution $F_u(x)$, of excess losses above a threshold u as follows.

$$F_u(x) = \Pr(X - u \leq x \,|\, X > u) = \frac{F(x+u) - F(u)}{1 - F(u)}, \tag{27.2}$$

for $0 \leq x < x_F - u$, where $x_F = \sup\{x \in \mathbb{R} : F(x) < 1\} \leq \infty$ is the right endpoint of F.

The primary distribution function used for modeling exceedances over a threshold is the *Generalized Pareto Distribution* (GPD), defined as

$$G_{\xi,\beta}(x) = \begin{cases} 1 - \left(1 + \xi \frac{x}{\beta}\right)^{-(1/\xi)}, & \text{if } \xi \neq 0; \\ 1 - \exp\left(-\frac{x}{\beta}\right), & \text{if } \xi = 0. \end{cases} \quad (27.3)$$

where $\xi \in \mathbb{R}$, $\beta > 0$. We require $0 \leq x < \infty$ if $\xi \geq 0$, and $0 \leq x \leq -(\beta/\xi)$ if $\xi < 0$.

The fundamental result is a theorem that supports approximating the excess distribution by a generalized Pareto distribution:

Theorem 1 (Pickands, Balkema, de Haan). *For a very rich class of distributions F, there exists a positive function $\beta(u)$ such that,*

$$\lim_{u \to x_F} \sup_{0 \leq x < x_F - u} |F_u(x) - G_{\xi,\beta(u)}(x)| = 0. \quad (27.4)$$

The technical condition that characterizes the class of distributions F, for which the above theorem holds is that F should be in the maximum domain of attraction of a generalized extreme value distribution with parameter ξ. We refer the reader to Embrechts et al. (2004) for details but remark that this class of distributions is extremely rich and includes all commonly used statistical distributions (e.g., Normal, lognormal, Student-t, Beta, uniform, etc.).

27.2.3 Estimating VaR using EVT

Theorem 1 above says that a GPD is a very natural model for the unknown excess distribution given a sufficiently high threshold. In practice, this is tantamount to replacing the excess distribution with a GPD for some ξ and β:

$$F_u(x) = G_{\xi,\beta}(x). \quad (27.5)$$

Writing $y = u + x$ and combining equations (27.2) and (27.5), we obtain:

$$F(y) = (1 - F(u))G_{\xi,\beta}(y - u) + F(u) \quad (27.6)$$

for $y > u$. This gives us a general parametric estimate of the tail of the distribution of X provided we have an estimate of $F(u)$. A simple empirical estimate is:

$$F(u) = \frac{n - N_u}{n}, \quad (27.7)$$

where n is the total number of observations and N_u is the number of observations exceeding the threshold u. See McNeil et al. (2005) for a discussion of this choice of $F(u)$.

In equation(27.7) we have assumed that the point probability distribution of X is specified by a sequence of values $X_1, X_2, \ldots, X_n$ each occurring with the same probability. For weighted observations we replace the empirical estimate in (27.6) by

$$F(u) = \frac{w - W_u}{w}, \quad (27.8)$$

where w is the sum of the weights of all the observations and W_u is the sum of the weights for those observations which exceed the threshold u.

Using the estimate in (27.7) and maximum likelihood estimates of the GPD parameters ξ and β, we obtain the following tail estimate of F for $y > u$ (in the nonweighted case),

$$\hat{F}(y) = 1 - \frac{N_u}{n}\left(1 + \hat{\xi}\frac{y-u}{\hat{\beta}}\right)^{-1/\hat{\xi}}. \tag{27.9}$$

Inverting equation (27.9) we obtain the following estimate for the VaR at risk level α,

$$\widehat{VaR}_\alpha = u + \frac{\hat{\beta}}{\hat{\xi}}\left(\left(\frac{n}{N_u}(1-\alpha)\right)^{-\hat{\xi}} - 1\right). \tag{27.10}$$

For the weighted case, the expression for the VaR is given by:

$$\widehat{VaR}_\alpha^{wgtd} = u + \frac{\hat{\beta}}{\hat{\xi}}\left(\left(\frac{w}{W_u}(1-\alpha)\right)^{-\hat{\xi}} - 1\right). \tag{27.11}$$

27.3 Threshold uncertainty

27.3.1 Tail-data versus accuracy tradeoff

A challenging aspect of the threshold exceedance approach is the optimal choice of the threshold. We need to choose a high enough threshold u, so that equation (27.5) provides a good approximation to the limiting situation where $u \to y_F$ in equation (27.4). But the choice of a very high threshold may lead to a high variance of parameter estimates (due to the lack of data points in the tail). On the other hand, if we choose a low enough threshold we will get sufficient data points for reasonable parameter estimates, but the underlying assumption of Theorem 1 that $u \to y_F$ will be violated.

Therefore the choice of a threshold is a very active area of research. Matthys and Beirlant (2000) have presented a detailed discussion on an adaptive threshold selection method. Here we briefly present two diagnostic methods to pick up the optimal threshold. For detailed discussion about threshold selection see also Embrechts et al. (2004) and McNeil (2000).

27.3.2 Mean residual life plot

The Mean Residual Life (MRL) plot is based on the mean excess function defined as

$$e(u) = E[X - u|X > u], \tag{27.12}$$

where $0 < u < x_F$. In the MRL plot of $e(u)$ versus u, we select the highest threshold where the plot is nearly linear. We also check the width of the confidence band as we move to the extreme right: these bands become wider for large u.

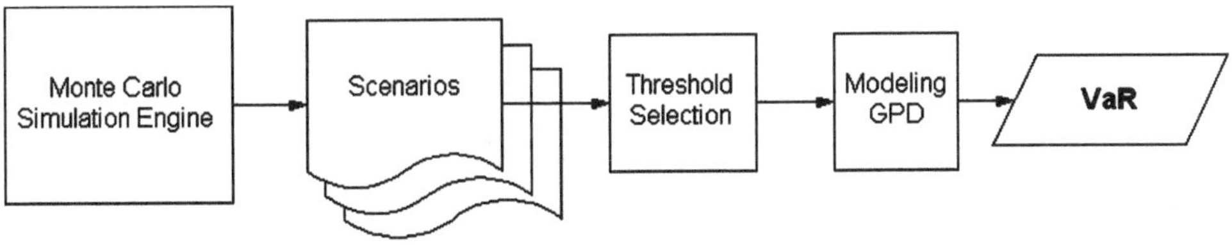

Figure 27.1. Experimental framework

27.3.3 Fit threshold ranges

In the second method we fit the GPD several times, each time using a different threshold and examine the stability of the parameter estimates. This can be done by plotting the scale and shape parameters as a function of the threshold (along with a 95% confidence interval. Ultimately we are interested in VaR estimates given by equations (27.10), (27.11), and analogous expected shortfall estimates. We choose the threshold so that both GPD parameters as well as the VaR and expected shortfall behave in a stable manner (with confidence intervals that are not too wide).

27.4 Experimental framework and results

27.4.1 Data

In this chapter we illustrate the application of EVT to the loss distributions arising from two sample commercial loan portfolios, labelled portfolio A and portfolio B.

27.4.2 Simulation engine

In Figure 27.1 we have presented the experimental framework to calculate VaR using EVT. Since it was not possible for us to obtain high-quality loss data directly, we used Monte Carlo simulations to generate a large number of possible scenarios. The Monte Carlo engine we used is Portfolio Manager®, which is a standard commercial software package from Moody's KMV, for modelling credit risk portfolios. As input it needs portfolio data at a detailed obligor and transaction level. It generates a probability distribution of portfolio losses for a given time horizon (typically one year). It offers two options to generate Monte Carlo loss distributions. The first option is the usual Monte Carlo where each separate loss scenario has equal probability. The second option is an implementation of an importance sampling version of Monte Carlo, which places more emphasis on the tail of the loss distribution. In this second option, the separate loss scenarios do not have equal probabilities, but are appropriately weighted.

27.4.3 Threshold selection

After generating the Monte Carlo loss scenarios the next key step is to select an optimal threshold as we have discussed in Section 27.3. We used a range of 10,000 equally spaced

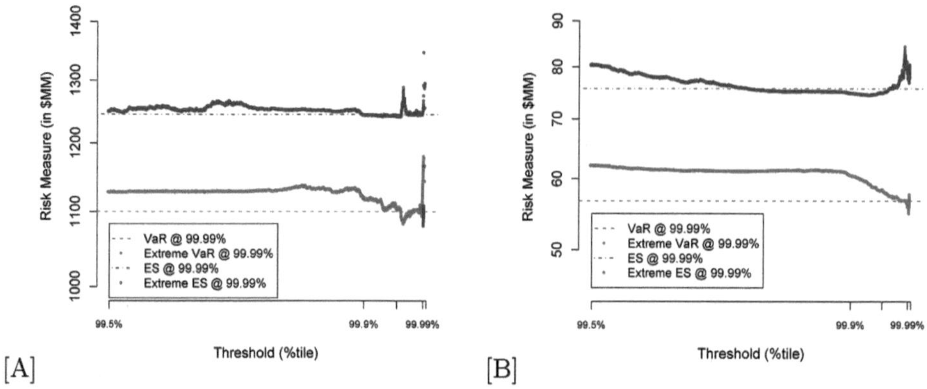

Figure 27.2. Threshold uncertainty analysis for portfolios **A** and **B**

thresholds from the 99.5^{th} percentile to the 99.995^{th} percentile. In Figure 27.2 we have plotted the VaR and ES, based on EVT, as well as the empirical VaR and ES. From these plots it is clear that except for very high thresholds, the EVT, VaR, and ES are fairly stable. Based on these plots, we fixed the threshold at the 99.9^{th} percentile. For this choice of threshold, we have an adequate number of data points in the tail for estimating the shape and scale parameters of the GPD.

27.4.4 Bootstrap results on VaR stability

Here we compare confidence intervals for the VaR based on EVT with VaR confidence intervals based on direct Monte Carlo simulation without EVT. We performed a bootstrap exercise where we drew 2000 bootstrap samples with replacement from the importance sampling Monte Carlo simulation. Based on these bootstrap replicates we have obtained a 95% confidence interval for the directly computed and EVT-based VaR. The bootstrap replicate's distributions are plotted in Figure 27.3 for portfolios A and B. For both the portfolios the left-hand side plot is for the VaR directly computed from the Monte Carlo output, whereas the right-hand side plot is for the EVT-based VaR. The solid lines represent the plug-in estimates and the dashed lines represent the lower and upper 95% confidence bound.

It is clear from these plots that the quality of the VaR distribution is much noisier with the raw Monte Carlo simulation than with the EVT-based VaR. In this sense, the EVT-based VaR is a more robust statistic than the raw Monte Carlo VaR.

27.5 Conclusion

The methods of extreme value theory are a valuable supplement to analyze the tail behaviour of financial loss distributions. We have exhibited such an application of the theory, where conventional methods for VaR estimates in the extreme tail are noisy and unstable. EVT-based methods help cure this difficulty, by providing narrower VaR confidence bands and a more robust estimate of the VaR. This is important if risk metrics such as VaR are to be used for important objectives such as estimation of economic capital for a financial portfolio.

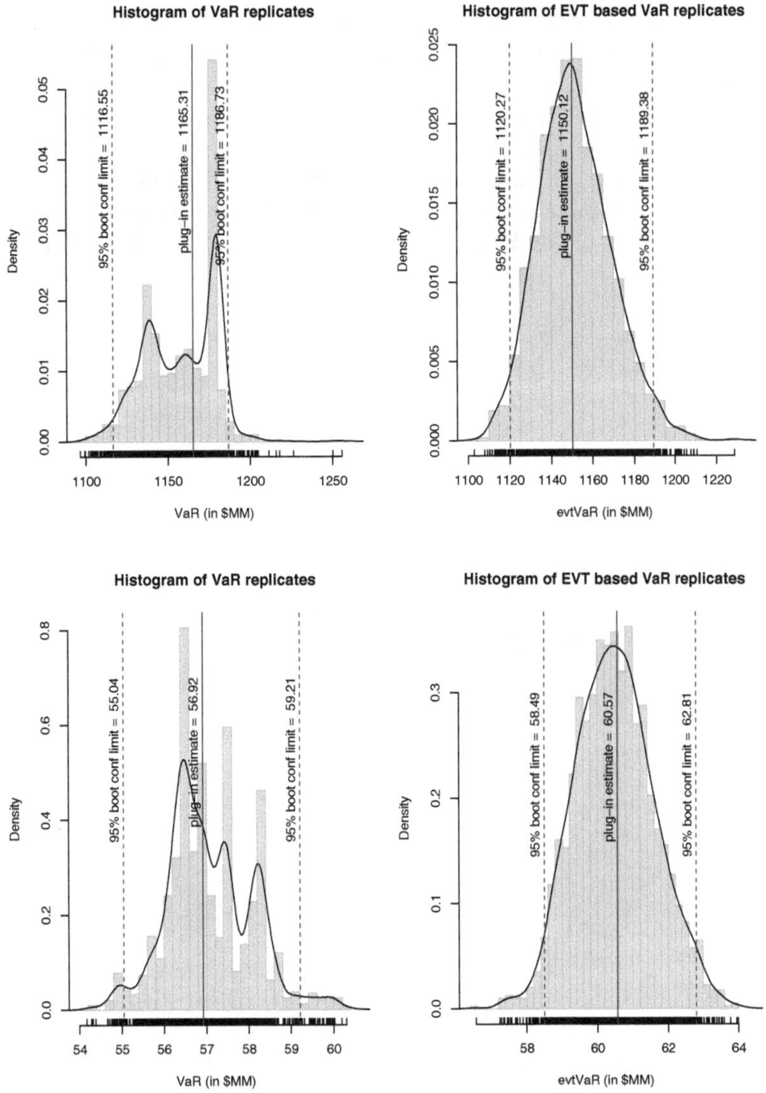

[Portfolio A]

[Portfolio B]

Figure 27.3. Bootstrap results for portfolios **A** and **B**

References

Artzner, P., Delbaen, F., Eber, J.-M., and Heath, D. (1997). Thinking coherently. *Risk*, 10:68–71.

Artzner, P., Delbaen, F., Eber, J.-M., and Heath, D. (1999). Coherent measures of risk. *Mathematical Finance*, 9:203–228.

Embrechts, P., Klppelberg, C., and Mikosch, T (2004). Modelling extremal events for insurance and finance. *Stochastic Modelling and Applied Probability*, Springer, New York.

Matthys, J. and Beirlant, J. (2000). Adaptive threshold selection in tail index estimation. In Paul Embrechts, editor, *Extremes and Integrated Risk Management*, pages 37–49.

McNeil, A.J. (2000). Extreme value theory for risk manager. In *Extremes and Integrated Risk Management*, pages 3–18.

McNeil, A.J., Frey, R., and Embrechts, P. (2005). *Quantitative Risk Management: Concepts, Techniques, and Tools.* Princeton University Press, Princeton, NJ.

Szego, G. (2004). *Risk Measures for the 21st Century.* John Wiley and Sons, New York.

28

Multiresponse Robust Engineering: Industrial Experiment Parameter Estimation

Elena G. Koleva[1] and Ivan N. Vuchkov[2]

[1] Institute of Electronics, Bulgarian Academy of Sciences, Sofia, Bulgaria (e-mail: kolevae@ie.bas.bg)
[2] European Quality Center, University of Chemical Technology and Metallurgy, Sofia, Bulgaria (e-mail: qstat@dir.bg)

Abstract: A new method for estimation of the mean and variance models of the responses, taking into account the correlation between the multiple responses, together with the heteroscedasticity of the observations due to errors in the factor levels is proposed. The application of this method gives us the possibility to use raw industrial data for mean and variance model estimation and leads to reduction of the predicted variance of the responses in production conditions. Recommendations for the experimental design sequential generation, based on the proposed method are made. The introduced new combined method is applied to an electron beam welding experiment.

Keywords and phrases: Multiresponse robust engineering, heteroscedasticity, errors in factor levels, combined method, parameter estimation, experimental design

28.1 Introduction

The Robust Parameter Design (RPD) has been an issue of numerous papers in the literature since 1990 (Box and Jones, 1990; Vining and Myers, 1990), but there are fewer of them in the area of application of RPD (Myers et al., 2004; Vuchkov and Boyadjieva, 2001) for multiple responses. Some of these articles consider the multiresponse case, when replicated observations are available (Chiao and Hamada, 2001), while others are focused on formulation of appropriate optimisation criteria. Examples for such criteria are: (i) a criterion, representing an appropriate compromise between both the process economics and the correlation structure among responses (Vining, 1998); and (ii) a new loss function, incorporating small bias, high robustness, and high quality of predictions (Ko et al., 2005).

The model-based robust approach for improving the quality of the process (Vuchkov and Boyadjieva, 2001) can be successfully applied to different industrial processes. For each of the quality performance characteristics, using their regression models, two other models are estimated for their mean values and their variances. The quality

C.H. Skiadas (ed.), *Advances in Data Analysis*, Statistics for Industry and Technology, DOI 10.1007/978-0-8176-4799-5_28,

improvement is performed using some overall criterion or simply by the performance characteristic variance minimisation, while keeping the mean values close to their target values.

The model of the mean value of the performance characteristic, which is a function of process parameters that are subject to errors during the industrial production process is (Vuchkov and Boyadjieva, 2001)

$$\tilde{y}(\mathbf{p}) = [y(\mathbf{z})] = \eta(\mathbf{p}) + \hat{\theta}^T E(\mathbf{g}) \tag{28.1}$$

where $\eta(\mathbf{p})$ is a model of the quality performance characteristic, for example a polynomial regression model, obtained by the response surface methodology. The second term takes into account the bias caused by the errors transmitted from the process parameters $\mathbf{p}$ to the performance characteristic $\tilde{y}(\mathbf{p})$, where $\hat{\theta}^T$ is the vector of the estimates of the regression coefficients in the model $\eta(\mathbf{p})$. $E(\mathbf{g})$ stands for the mathematical expectation of $\mathbf{g} = \mathbf{h} - \mathbf{f}$, $\mathbf{h}$ is a vector of the regressors $\mathbf{z}$ in the regression model, considered as containing errors $\mathbf{e}$ (for any process parameter, $z_i = p_i + e_i$), and $\mathbf{f}$ is the regressor vector of the process parameters $\mathbf{p}$.

The model for the variance of the quality performance characteristic that is due to errors in factor levels, if the bias that comes from the precision of the estimation of the regression model (negligible in many cases) is taken into account (by the second term), is:

$$\hat{s}^2 = \tilde{s}^2 - tr[\mathbf{\Psi V}(\hat{\theta})] = \hat{\theta}^{\mathbf{T}}\mathbf{\Psi}\hat{\theta} + \mathbf{s}_\varepsilon^2\left(\mathbf{1} - \sum_{\mathbf{i=1}}^{\mathbf{k}} \psi_{\mathbf{ii}}\mathbf{c}_{\mathbf{ii}} - \mathbf{2}\sum_{\mathbf{i=1}}^{\mathbf{k-1}}\sum_{\mathbf{j=i+1}}^{\mathbf{k}} \psi_{\mathbf{ij}}\mathbf{c}_{\mathbf{ij}}\right) \tag{28.2}$$

where $\psi = \mathbf{g} - \mathbf{E}(\mathbf{g})$ is defined on the basis of the variances for each process parameter $\mathbf{p}$, which can be calculated using the tolerance limits of the process parameters or on the base of replicated observations; $\mathbf{\Psi} = \mathbf{E}(\psi\psi^{\mathbf{T}})$ depends on the structure of the regression model and the experimental design; s_ε^2 is the estimate of the random error of the performance characteristic; ψ_{ii} and ψ_{ij} are correspondingly the diagonal and nondiagonal elements of $\mathbf{\Psi}$; c_{ii} and c_{ij} are the diagonal and nondiagonal elements of the variance – covariance matrix, which become smaller with the growth of the number N of the experiments (observations); and k is the number of terms in the regression model. With $\tilde{s}^2$ is denoted the variance of the quality performance characteristic, which is due to the errors in factor levels ($\hat{\theta}^T\mathbf{\Psi}\hat{\theta}$) and the random error s_ε^2 . For a large number of observations or small values of s_ε^2 the bias is negligible.

This chapter considers the mean and variance model parameter estimation in the typical for an industrial process case when there is a correlation between the multiple responses (Khuri and Cornell, 1996; Khuri, 1990) and there are errors in the factor levels in the production stage (there is heteroscedasticity (Vuchkov and Boyadjieva, 2001)). Both the correlation and the heteroscedasticity should be taken into account in order to improve the accuracy of the estimated models. The purpose of this chapter is to present a new combined method for parameter estimation, which will give us the possibility to consider the correlation between the multiple responses. Another big advantage is the possibility to use raw industrial experimental data instead of the necessary very precise parameter estimation of the regression models without errors in the factor levels, done, for example, in laboratory conditions. This new method is applicable in both cases: when there are replicated observations at each experimental run and when there are no such replications. If there are no replicated observations the variance estimation for each experimental run can be done through the tolerance intervals of the factors in the industrial (or laboratory) production process.

28.2 Combined method for regression parameter estimation

The multiresponse approach (Khuri and Cornell, 1996) gives as a result estimates of the regression coefficients that take into consideration the correlation between the responses, which is usually the case. They can be estimated through the equation (Zellner (1962)) referred as two-stage Aitken estimator. The heteroscedasticity of observations can be considered through the application of the weighted least square estimates, (Vuchkov and Boyadjieva, 2001). The two approaches are combined and the overall algorithm of the new combined method for regression parameter estimation ($\hat{\theta}$ used in equations (28.1) and (28.2)) is:

Step 1. The ordinary least squares estimates (OLSE) $\mathbf{b}_0$ are found for each of the responses:

$$\mathbf{b}_{0,i} = (\mathbf{X}_i^T\ \mathbf{X}_i)^{-1}\mathbf{X}_i^T\mathbf{Y}_i.$$

where $\mathbf{X}_i$ are the matrices of known functions $\mathbf{f}$ (regressor vectors) of the process parameters $\mathbf{p}$, defined by the regression models and the performed experiments for each of the responses $\mathbf{Y}_i$, $i = 1, 2, \ldots, r$.

Step 2. An estimate of the random error can be found by:

$$s^2_{\varepsilon,i} = \frac{1}{N - k_i} \sum_{u=1}^{N} (y_{u,i} - \hat{y}_{u,i})^2 .$$

Step 3. The models for the mean equation (28.1) and the variance equation (28.2) are estimated for each of the performance characteristics.

Step 4. The matrix $\tilde{\Sigma}_{\mathrm{h}}$ is estimated:

$$\tilde{\Sigma}_{\mathrm{h}} = \begin{bmatrix} \tilde{\sigma}_1^2 & \cdots & 0 \\ \vdots & \ddots & \vdots \\ 0 & \cdots & \tilde{\sigma}_N^2 \end{bmatrix} \tag{28.3}$$

by calculation of the estimates of the variances $\tilde{\sigma}^2_{\mathrm{u}}$ at each experimental run $u = 1, \ldots, N$ for each of the responses, using equation (28.2). The matrix $\tilde{\Sigma}_{\mathrm{h}}$ estimates the heteroscedasticity of the observations.

Step 5. The variance – covariance matrix $\tilde{\Sigma}_{\mathrm{m}}$ of the random error of all performance characteristics also takes into account their correlation. If it is unknown, its elements ($\hat{\sigma}_{ij}$) can be estimated by:

$$\hat{\sigma}_{ij} = \mathbf{Y}'_i[\mathbf{I}_N - \mathbf{X}_i(\mathbf{X}'_i\mathbf{X}_i)^{-1}\mathbf{X}'_i][\mathbf{I}_N - \mathbf{X}_j(\mathbf{X}'_j\mathbf{X}_j)^{-1}\mathbf{X}'_j]\mathbf{Y}_j/N,$$

for $i, j = 1, 2, \ldots, r$.

Step 6. The combined method variance – covariance matrix is:

$$\tilde{\Delta}* = \begin{bmatrix} \tilde{\hat{\sigma}}_{11,1} & \cdots & 0 & & \tilde{\hat{\sigma}}_{1r,1} & \cdots & 0 \\ \vdots & \ddots & \vdots & \cdots & \vdots & \ddots & \vdots \\ 0 & \cdots & \tilde{\hat{\sigma}}_{11,N} & & 0 & \cdots & \tilde{\hat{\sigma}}_{1r,N} \\ & \cdots & & \ddots & & \cdots & \\ \tilde{\hat{\sigma}}_{r1,1} & \cdots & 0 & & \tilde{\hat{\sigma}}_{rr,1} & \cdots & 0 \\ \vdots & \ddots & \vdots & \cdots & \vdots & \ddots & \vdots \\ 0 & \cdots & \tilde{\hat{\sigma}}_{r1,N} & & 0 & \cdots & \tilde{\hat{\sigma}}_{rr,N} \end{bmatrix}, \tag{28.4}$$

where $\tilde{\Delta}*$ is calculated with elements:

$$\begin{cases} \tilde{\hat{\sigma}}_{ii,u} = \hat{\sigma}_{ii}\tilde{\sigma}^2_{i,u} \\ \tilde{\hat{\sigma}}_{ij,u} = \hat{\sigma}_{ij}\tilde{\sigma}_{i,u}\tilde{\sigma}_{j,u} \end{cases}.$$

Step 7. Combined method parameter estimates are calculated by:

$$\tilde{\theta}* = (\mathbf{Z}'\tilde{\mathbf{\Delta}}*^{-1}\mathbf{Z})^{-1}\mathbf{Z}'\tilde{\mathbf{\Delta}}*^{-1}\mathbf{Y}, \tag{28.5}$$

where $\mathbf{Z} = \text{diag}(\mathbf{X}_1, \ldots, \mathbf{X}_r)$ is a diagonal matrix with diagonal elements – the matrices $\mathbf{X}_i$; and $\mathbf{Y} = [\mathbf{Y}_1', \ldots, \mathbf{Y}_r']'$ is a vector, consisting of the observations for each of the r responses.

Step 8. The criterion Cr is calculated by:

$$Cr_j = \sum_{i=1}^{k_1+\cdots+k_r} \frac{\left(\tilde{\theta} *_{j,i} - \tilde{\theta}*_{j-1,i}\right)^2}{\tilde{\theta}*^2_{j,i}}, \tag{28.6}$$

where $k_1, \ldots, k_r$ are correspondingly the numbers of the regression coefficients in the regression models for each of the responses $i = 1, \ldots, r$ for jth iteration.

Step 9. The procedure continues from step 2, until $\text{Cr} \leq \delta$, where δ is a small positive number.

The proposed procedure for regression parameter estimation takes into account both the heteroscedasticity of the observations and the correlation between the multiple responses. This is an iteration procedure, the convergence of which depends on the accuracy of the initial regression OLSE parameter estimates, nonlinearity of the estimated model, as well as the magnitude of the errors in factor variances, transmitted to the performance characteristics.

When the responses have different dimensions, in order to avoid this influencing the obtained results, the responses should be normalised before the regression models estimation. L_p norms can be used for this purpose. For example, if an L_2 norm is used, the normalized values $Y_{1n,u}$ of Y_1 are:

$$Y_{1n,u} = \frac{Y_{1,u}}{\left(\sum_{u=1}^{N} Y^2_{1,u}\right)^{1/2}}.$$

28.3 Experimental designs

A procedure for the sequential generation of experimental designs, applicable to industrial production processes and based on the proposed combined method and D-optimality criterion, will provide new experimental runs at factor levels coinciding with the most inaccurate regression model predictions and with the largest variance of the observations. The method can also be applied when there is information about the heteroscedasticity of the experiments from repeated observations or previous runs. The procedure begins with the choice of an initial experimental design and the parameter estimation according the combined method parameter estimation procedure, using equation (28.5). Then, for the sequential generation of additional experiments, a D-optimal combined criterion that can be applied together with some optimisation method for finding the maximum of the variance – covariance matrix is:

$$\varphi *'_{\mathbf{N+1}} \left(\mathbf{Z}'_{\mathbf{N}}\tilde{\mathbf{\Delta}}*^{-1}\mathbf{Z}_{\mathbf{N}}\right)^{-1} \varphi *_{\mathbf{N+1}} = \max_{\mathbf{x}} \varphi'_{\mathbf{N+1}} \left(\mathbf{Z}'_{\mathbf{N}}\tilde{\mathbf{\Delta}}*^{-1}\mathbf{Z}_{\mathbf{N}}\right)^{-1} \varphi_{\mathbf{N+1}}, \tag{28.7}$$

where the matrix $\tilde{\mathbf{\Delta}}*$ is defined by equation (28.4), $\varphi_{\mathbf{N+1}}$ is a vector with elements: $[\mathbf{f}_{1,N+1}; \cdots ;\ \mathbf{f}_{r,N+1}]$, where $\mathbf{f}_{i,N+1}$ is the regressors' vector of the candidate experimental point $N+1$ and the response $i=1, \ldots, r$.

28.4 Experimental application

In the last few decades, electron beam welding (EBW) of the refractory metals and alloys, of heterogeneous metal junctions, and of heavy engineering components was widespread. The high joining rate, the deep and narrow weld, and the minimal heat-affected zone are basic advantages leading to the most frequent use of this process. The focused electron beam is one of the highest power density sources and that is why high processing speed are possible; narrow welds with a very narrow heat-affected zone can be produced accurately.

An investigation of the relationships between geometry parameters (weld depth H and mean weld width B) and process parameters (electron beam power (P), welding velocity (v), the difference (dz) between the distances (i) between the magnetic lens of the gun and the focus of the electron beam and (ii) between the magnetic lens of the gun and the surface of the sample) is done for stainless steel type 1H18NT and accelerating voltage 70 kV (Koleva, 2001). Eighty-one experimental runs were performed, forming a design of experiment, which is not statistically designed. The notations of the factors as well as their limits (during the experiments), tolerances, and their basic levels p'_{io} and variation intervals ω_i, are given in Table 28.1.

The influence of the process parameters on the estimated geometrical characteristics of the seam $\hat{H}$ and $\hat{B}$, is investigated by applying the multiresponse regression model estimation approach and conclusions about these dependencies, as well as some optimisation procedures can be found in Koleva (2001). In Koleva and Vuchkov (2005) the estimation of the mean and variance models are estimated for the geometrical characteristics of the welds and optimisation is performed in terms of variance minimization meeting concrete production requirements.

Table 28.1. Experimental conditions

Factors	Dimensions	Coded factors	Min	Max	p'_{io}	ω_i	Tolerance limits
$P - \tilde{p}_1$	kW	p_1	4.2	8.4	6.3	2.1	P ± 2%
$v - \tilde{p}_2$	cm/min	p_2	20	80	50	30	v ± 3%
$dz - \tilde{p}_3$	mm	p_3	−78	62	−8	70	dz ± 2

The calculated coded variances for the process parameters, using the tolerance intervals (Koleva and Vuchkov, 2005) are shown in Table 28.2.

Table 28.2. Coded variances

Factors	Coded variances σ_i^2
p_1	$0.0004 + 0.00026667\ p_1 + 0.00004444\ p_1^2$
p_2	$0.00027778 + 0.00033333\ p_2 + 0.0001\ p_2^2$
p_3	0.0000907

Table 28.3. Regression coefficient estimates for the weld depth H and the weld width B; (i) ordinary least squares estimates (OLSE), (ii) weighted LSE (WLSE), (iii) multiresponse estimates (MRE), (iv) combined method estimates (CME)

H	OLSE	WLSE	MRE	CME	**B**	OLSE	WLSE	MRE	CME
bo	22.80	23.0174	22.9982	23.1942	bo	1.550	1.5691	1.5897	1.5684
b1	4.69	3.5368	4.0933	3.9660	b1	0.265	0.3101	0.2102	0.3061
b2	−6.38	−6.8046	−6.7660	−6.8171	b2	−0.665	−0.6197	−0.6441	−0.6208
b3	−12.60	−12.4014	−11.1480	−9.2928	b3	1.230	1.2819	1.2407	1.2825
b12	−3.16	−2.5419	−2.5153	−2.4707	b12	−0.0785	−0.1196	−0.1139	−0.1201
b13	−1.89	−2.1639	−1.3700	0.5925	b23	−0.1690	−0.0955	−0.1120	−0.0977
b11	−2.48	−1.7729	−1.9951	−1.8504	b11	0.2120	0.1565	0.1720	0.1559
b22	3.90	3.9401	3.7285	3.7374	b22	0.3790	0.4054	0.3959	0.4068
b33	−6.85	−8.3338	−7.2950	−8.3316	b33	1.2800	1.2252	1.1355	1.2264
b133	6.27	8.2416	6.6954	7.1240	b122	0.3470	0.2714	0.3798	0.2770
b113	5.07	4.6156	4.9563	4.9138	b133	−1.1000	−0.9770	−0.9065	−0.9772
b333	7.14	7.6706	4.6672	0.6048	b113	−0.5530	−0.5874	−0.5409	−0.5874

Table 28.3 presents the regression coefficient estimates for the weld depth H and the weld width B, obtained by ordinary least squares, weighted least squares, multiresponse, and the new combined methods.

Figure 28.1 shows the convergence of the combined method presented by the criterion Cr value, equation (28.6).

The mean and the variance models for the two responses are estimated, using all considered regression coefficient estimation methods in order to perform a visual comparison. As an example, Figure 28.2 presents the contour lines of the mean value of the weld width mean $\tilde{y}_H(P, v)$ (solid lines) and its variance $\hat{s}_H^2$ (dashed lines). The methods applied for the regression coefficient estimation are, correspondingly, the combined and the weighted least squares methods. The models equations (28.1) and (28.2) differ by the regression parameter estimates $\hat{\theta}$ and by $tr(\mathbf{\Psi V}(\hat{\theta}))$. It can be seen that the

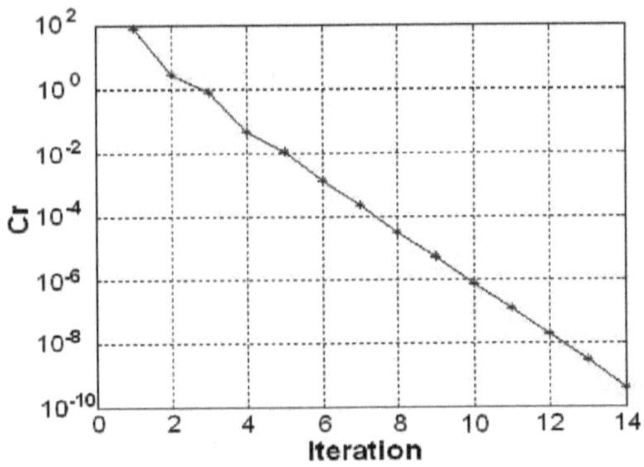

Figure 28.1. Convergence of the combined method

variance of the performance characteristic, when the combined method for parameter estimation is applied, has lower values. In this case the random error $s^2_{\varepsilon,H}$ estimate has a comparatively large value and should be reduced, for example, by performing additional experiments, applying the criterion in equation (28.7).

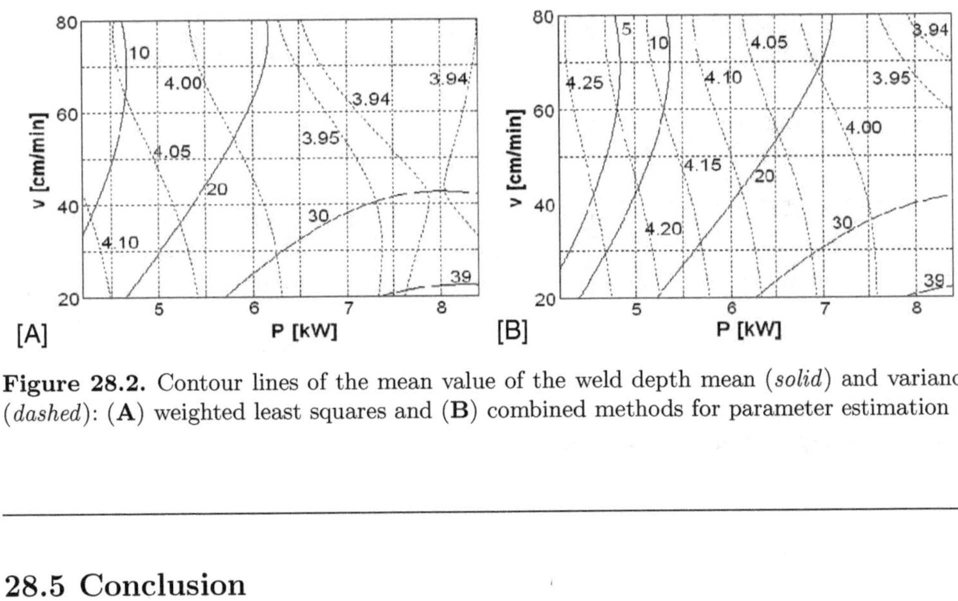

Figure 28.2. Contour lines of the mean value of the weld depth mean (*solid*) and variance (*dashed*): (**A**) weighted least squares and (**B**) combined methods for parameter estimation

28.5 Conclusion

The proposed new method for estimation of regression coefficients takes into account both the correlation and the heteroscedasticity of the performed experiments in order to improve the accuracy of the estimated regression models, as well as the models for the means and variances of the multiple responses. This combined approach can be implemented for the sequential generation of industrial experimental designs.

The application of the proposed approach gives the possibility to use raw industrial experimental data, instead of the necessary very precise regression model estimations without errors in the factor levels, done usually in laboratory conditions.

Acknowledgements. The authors thank the financial support from Bulgarian National Fund "Science Investigations" under contract VUI 307/2007.

References

Box, G. and Jones, S. (1990). Designing products that are robust to the environment. *Quality and Productivity Report 56 Univ. Wisconsin*, CPQI.

Chiao, C. and Hamada, M. (2001). Analyzing experiments with correlated multiple responses. *J. Qual. Technol.*, 33:451–465.

Khuri, A. (1990). Analysis of multiresponse experiments: A review. *Statist. Des. Anal. Indust. Exper.*, 231–246.

Khuri, A. and Cornell, J. (1996). *Response Surfaces*, 37(1):50–59.

Ko, Y., Kim, K., and Jun, Ch. (2005). A new loss function-based method for multiresponse optimization. *J. Qual. Technol.*, Marcel Dekker, Inc.: New York, Basel, Hong Kong, 251–300.

Koleva, E. (2001). Statistical model building and computer programs for optimisation of the EBW of stainless steel. *Vacuum*, 62:151–157.

Koleva, E. and Vuchkov, I. (2005). Model-based approach for quality improvement of EBW applications in mass production. *Vacuum*, 77:423–428.

Myers, R., Montgomery, D., Vining, G., Borror, C., and Kowalski, S. (2004). Response surface methodology: A retrospective and literature survey. *J. Qual. Technol.*, 36(1):53–77.

Vining, G. (1998). A compromise approach to multiresponse optimization. *J. Qual. Technol.*, 30(4):309–313.

Vining, G. and Myers, R. (1990). Combining Taguchi and response surface philosophies: A dual-response approach. *J. Qual. Technol.*, 22:38–45.

Vuchkov, I. and Boyadjieva, L. (2001). *Quality Improvement with Design of Experiment.* Kluwer Academic: The Netherlands.

Zellner, A. (1962). An efficient method of estimating seemingly unrelated regressions and tests aggregation bias. *J. Am. Statist. Assoc.*, 57:348–368.

29

Inference for Binomial Change Point Data

James M. Freeman

Manchester Business School, University of Manchester, Manchester, UK

Abstract: In this chapter we describe a procedure for detecting a systematic change in parameter for a sequence of binomial variables. The procedure is based on a goodness-of-fit argument. Tests for an unknown change point are given. The procedure is found to be appropriate to problems in which the data series has been subject to a single discrete change in binomial parameter or where there have been cumulative changes in binomial parameter, before or after an unknown point.

Keywords and phrases: Binomial parameter, change point, goodness-of-fit

29.1 Introduction

A sequence of independent binomial variables is subject to a change in distribution after an unknown point. Formally, we can describe this situation as follows. $R_1, R_2, R_3, \ldots, R_k$ are independent random variables, such that, for a value τ,

R_i is distributed as

$$\begin{aligned} &B(n_i, \theta_0) \quad (1 \le i \le \tau) \\ &B(n_i, \theta_1) \quad (i \le \tau+1, \tau+2, \ldots, k) \end{aligned} \tag{29.1}$$

Previous work, with this type of model, has been directed toward (i) estimating the change-point, τ and (ii) testing the hypothesis that no change in distribution has occurred.

Most analytical approaches, developed for dealing with binomial change-point data, assume the θ_1 and θ_2 parameters, such as τ, to be unknown. Particular attention has been devoted to the case of the R_i being (Bernouilli) zero-one variables, i.e., with $n_i = 1$ for all i.

The problem has been analysed from a variety of perspectives: Hinkley and Hinkley's (1970) likelihood work has been extended by Pettitt (1980) and Worsley (1983), the latter authors also considering alternative CUSUM-based procedures (see, as well, Page, 1955). In contrast, Smith (1975) and Pettitt (1979) together with Pettitt (1981) offer,

C.H. Skiadas (ed.), *Advances in Data Analysis*, Statistics for Industry and Technology, DOI 10.1007/978-0-8176-4799-5_29,

respectively, Bayesian and nonparametric methodologies. More general techniques, such as those of Worsley (1986), and Kander and Zacks (1966) provide further scope for analysis.

The chapter introduces a new procedure for analysing binomial change-point data. The procedure is more general than many of the techniques developed in this area: it is not only capable of monitoring problems involving a single change in parameter level but also those where the change in parameter level has been cumulative after some unknown point. The technique, based on an unusual 'goodness-of-fit' argument is compared with the maximum likelihood estimation approach of Hinkley and Hinkley [1970]. Finally, the technique is illustrated on a number of relevant datasets mostly from the literature.

29.2 Analysis

Referring to model (29.1), we distinguish between the rival sets of assumptions:

H_1 : θ_θ and θ_1 are fixed with $\theta_0 \neq \theta_1$

$\mathrm{H}_1^{'}$: θ_0 is fixed and θ_1 is a linear function of prechosen scores s_i which we write as $\theta_1 = \theta_1(s_i) = \theta_0 + \beta(s_i - \overline{s})$

where β is fixed and $\overline{s} = \sum_{i=1}^{k} n_i\, s_i \;/\; N$

The choice of scores s_i can be somewhat arbitrary. Options that have been considered (Williams, 1988) include in the case of assay experiments, the dose or logarithm of the dose being tested. Alternatively, s_i may be taken to equate with its index i or defined, for example, as

$$s_i = n_1 + \cdots + n_{i-1} + 0.5(n_i + 1)$$

(the latter choice providing a measure of rank correlation between doses and binary responses.)

Under the null hypothesis H_0 we assume no change in θ_0 has taken place and can therefore write

$$\mathrm{H}_0 : \tau = k \text{ or } \mathrm{H}_0 : \theta_1 = \theta_0$$

Operationally, we assume the data, for which model (29.1) is being considered, arises from an experiment involving the comparison of k groups. In the ith group, there are n_i independent binary responses, comprising r_i 'successes' and $n_i - r_i$ 'failures' ($i = 1, 2, \ldots, k$).

Under H_0, the probability of a successful experimental trial is θ_0.

Correspondingly, the probability of a trial resulting in failure is $1 - \theta_0$.

The r_i can be regarded as realisations of the random variables R_i, described by model (29.1). These data are conveniently arranged in the form of a $2 \times k$ contingency table (Cochran, 1954) as shown in Table 29.1.

Table 29.1. Contingency table formulation

	Group					
	1	**2**	**3**	**....**	**k**	**Total**
Successes	r_1	r_2	r_3	....	r_k	R
Failures	$n_1 - r_1$	$n_2 - r_2$	$n_3 - r_3$	....	$n_k - r_k$	$N - R$
Total trials	n_1	n_2	n_3	....	n_k	N

where $R = \sum_{i=1}^{k} r_i$ and $N = \sum_{i=1}^{k} n_i$

Furthermore, we write $p_i = r_i/n_i$ $(i = 1, 2, 3, \ldots, k)$.

When either H_1 or H' holds, it is often found from a plot of p_i against i (particularly in the case where not all $n_i = 1$) that there appears to be a linear association between the two variables. Such an association would normally be investigated using the procedures, for example, of Cochran (1954) and Armitage (1955). The Cochran – Armitage statistic (Williams, 1988) provides a measure of the strength of this apparent relationship. An alternative, which has received much less attention, is the R square ratio, computed directly from Cochran's (1954) analysis of variance summary.

As we now show, the latter R square statistic can be adapted to suit the specific circumstances of the change-point problem.

Let

$$R_t^2 = (S_{1t} / v_{1t} + S_{2t} / v_{2t}) \,/\, (T_{1t} / v_{1t} + T_{2t} / v_{2t}) \qquad (t = 2, 3, \ldots, k-2) \quad (29.2)$$

where S_{1t} and S_{2t} correspond with the sums of squares from the regression of the p_i ratios on their index i for the first t and last $(k-t)$ observations, respectively, and T_{1t} and T_{2t} are the corresponding corrected sums of squares on p_i. The quantities v_{1t} and v_{2t}, given, respectively, by

$$v_{1t} = N_{1t}\ p_{1t}\ q_{1t} / (N_{1t} - 1), \quad v_{2t} = N_{2t}\ p_{2t}\ q_{2t} / (N_{2t} - 1) \qquad (29.3)$$

are used to convert sums of squares quantities here to corresponding χ^2 values Chapman and Nam (1968).

Note that

$$N_{1t} = \sum_{i=1}^{t} n_i, \ R_{1t} = \sum_{i=1}^{t} r_i, \ p_{1t} = R_{1t} / N_{1t}, \ q_{1t} = 1 - p_{1t} \qquad (29.4)$$

Similarly,

$$N_{2t} = \sum_{t+1}^{k} n_i, \ R_{2t} = \sum_{t+1}^{k} r_i, \ p_{2t} = R_{2t} / N_{2t}, \ q_{2t}, = I - p_{2t} \qquad (29.5)$$

In addition, it can be shown

$$S_{1t} = \left(\sum_{i=1}^{t} r_i (i - \bar{i}_{1t}) \right)^2 \Big/ \left(\sum_{i=1}^{t} n_i (i - \bar{i}_{1t})^2 \right) \qquad (29.6)$$

$$S_{2t} = \left(\sum_{t+1}^{k} r_i (i - \bar{i}_{2t}) \right)^2 \Big/ \left(\sum_{t+1}^{k} n_i (i - \bar{i}_{2t})^2 \right) \qquad (29.7)$$

$$T_{it} = \sum_{i=1}^{t} n_i(p_i - p_{1t})^2 \quad \text{and} \quad \bar{i}_{it} = \left(\sum_{i=1}^{t} i\, n_i\right) \Big/ \left(\sum_{i=1}^{t} n_i\right) \tag{29.8}$$

$$T_{2t} = \sum_{t+1}^{k} n_i(p_i - p_{2t})^2 \quad \text{and} \quad \bar{i}_{2t} = \left(\sum_{t+1}^{k} i\, n_i\right) \Big/ \left(\sum_{t+1}^{k} n_i\right) \tag{29.9}$$

The R_t^2 statistic is analogous to that used by Freeman (1986) in an analysis of normal change-point data. Straightforward application of Freeman's methodology to model (29.1) confirms the following estimation procedure to be appropriate.

Under hypothesis H_1 estimate the change-point τ as the value of t at which R_t^2 is *minimised.*

Under H_1' estimate τ as the value of t at which R_t^2 is *maximised.*

The distribution of R_t^2, which is discrete, can be determined in relation to the multivariate hypergeometric distribution for the R_i, conditioned on R, the number of successes across all experiments. (Note that R is sufficient for θ_0, under the no-change hypothesis (Pettitt, 1979).)

Under H_0, the latter distribution can be shown to tend asymptotically to

$$B\left(R_t^2 | 1, (k-4)/2\right) = \frac{\Gamma((k-2)/2)}{\Gamma 1 \Gamma((k-4)/2)} \left(1 - R_t^{2(k-6)/2}\right) \tag{29.10}$$

but this result is known only to be valid if expected frequencies $n_i p_i$, and $n_i(1 - p_i)$ are sufficiently large, e.g., at least 5 (though this may be conservative (Copas, 1989)). In many applications, expected frequencies are often too small to allow computation of accurate critical values from the asymptotic distribution: zero-one data are an obvious case in hand. Unfortunately, the method usually adopted for overcoming this kind of technical difficulty, that of amalgamating groups, has little to recommend it in the context of a change-point analysis (Connor, 1972).

In addition, the R_t^2 variates themselves are highly correlated.

Notwithstanding this fact, a Type 1 extreme value distribution can be used as an approximation to the distribution of the maximum value of R_t^2. See Freeman [1986] for details and corresponding critical values, since confirmed using computer simulation methods by So [1998]. By default, the distribution of the minimum value of R_t^2 can also be determined (Kendall and Stuart, 1976).

29.3 Applications

29.3.1 Page's data

Forty observations were simulated by Page (1955), the first twenty arising from the $N(5, 1)$ distribution and the remainder from the $N(6, 1)$ distribution. The data were subsequently converted to Bernouilli observations by subtracting 5 from each normal variate and coding the resultant value as 1 if greater than zero, 0 otherwise.

The minimum value of the R_t^2 statistic for these data is $1.713E - 3$ which occurs for $t = 18$ but is not significant.

The corresponding Cochran–Armitage statistic for the entire dataset takes the value 2.471. This is a significant result under H_0 and the sign here points to an *abrupt increase* in probability.

In contrast to the estimated change-point of 18 found here, Page (1955) and Pettitt (1979) independently suggest the value 17 for the change-point. For a one-sided test of the hypothesis H_0 of no change against change, Page obtained results significant at the 1% level. Unlike our procedure, however, Page's relied on the initial mean value 0 being known.

Pettitt's one-sided nonparametric test, which assumed θ_0 unknown, indicated significance just short of the 1% level.

Taking a quite different viewpoint, Smith (1975) deduces the odds on a change having occurred, are about 100 to 1. From a table of posterior probabilities for τ, he derives the two estimates of τ of 18 (posterior median) and 19.24 (posterior mean).

29.3.2 Lindisfarne Scribes' data

The Lindisfarne Scribes' data (Pettitt 1979) refer to the number of occurrences of present indicative third person singular endings '-s' and '- δ' for different sections of Lindisfarne. It is believed different scribes used the endings '-s' and '- δ' in different proportions. A plot of the R_t^2 statistic against t is shown in Figure 29.1. From this, it can be seen the R_t^2 statistic assumes its minimum value of 6.175E-4 at $t = 7$. We therefore deduce an abrupt change in binomial probability occurred after the seventh section. The p-value associated with this result is close to (and slightly greater than) 5% from a corresponding simulation analysis. Pettitt's results suggest the change occurred after the *sixth* section with a (conservative) significance probability of 0.25%. Note that the maximum value of R_t^2 (of 3.657E-1) occurs for $t = 5$, indicating that an incremental change in probability may have occurred beforehand. This view that the

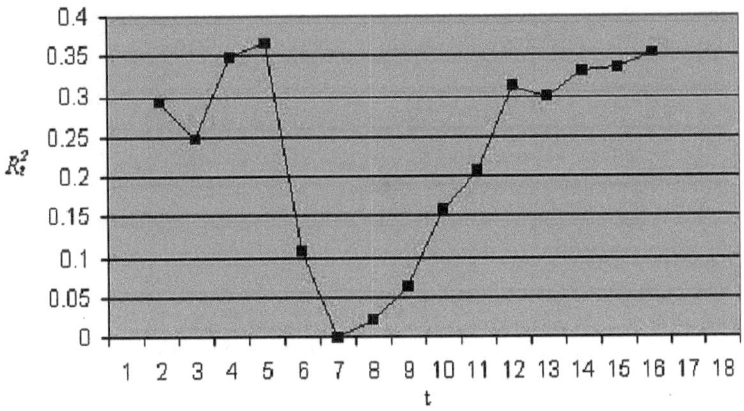

Figure 29.1. R_t^2 plot Lindisfarne Scribes data

data were subject to two change-points is one shared by Smith (Pettitt, 1979) who believes changes occurred after the sixth and seventh sections.

The Cochran–Armitage statistic for these data, using the index t as correlate, yields the significant value of 3.091. From the sign of the coefficient here, we deduce there was an *abrupt increase* in the proportion of '$-s$' endings after the seventh section, preceded by a cumulative increase before the fifth.

29.3.3 Club foot data

Worsley (1983) presents data on the number of cases of birth deformity talipes or club foot in the first month of gestation for the years 1960–1976 in a region of northern New Zealand. It is believed that a change in probability occurred after 1965. Worsley's procedure showed that this was indeed the case, confirming the no-change hypothesis should be rejected at the 5% level.

The R_t^2 plot for these data is shown in Figure 29.2. The minimum value of R_t^2 = 1.029E-2 occurs at t = 6. This coincides with Worsley's estimate of the year an abrupt change occurred.

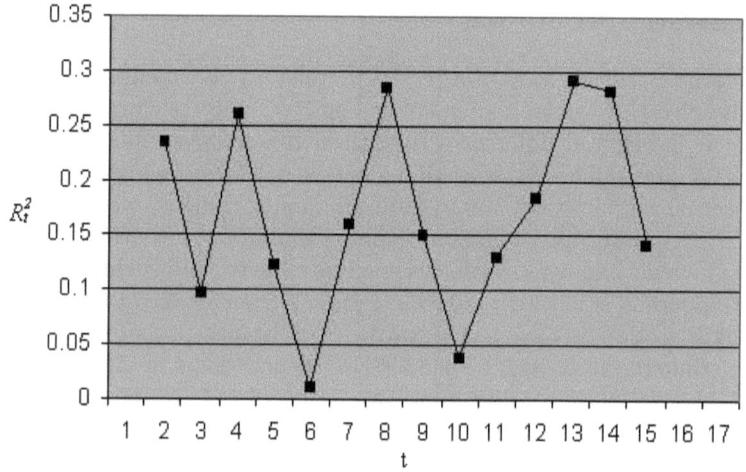

Figure 29.2. R_t^2 plot club foot data

The value of the appropriate Cochran–Armitage statistic for the full dataset is 2.603. This significant result leads to the conclusion that there was an *abrupt increase* in the rate of club foot incidence after 1965.

29.3.4 Simulated data

Data were simulated by the author as follows: (k =) 20 pairs of random digits were drawn from simple random number tables. These were adopted as the n_i values. Assuming a change-point to hold at $\tau = 7$, θ_0 was taken as

$$0.5 \; (i = 1, 2, \ldots, 7)$$

$$0.5 + 0.02i \; (i = 8, 9, \ldots, 20)$$

Applying these probabilities in turn to each of the binary trials within each experimental group, we obtained the r_i data, shown in Table 29.2.

Table 29.2. Simulated data

Group	Successes	Total trials	Group	Successes	Total trials
1	8	15	11	47	85
2	44	82	12	27	37
3	47	82	13	33	44
4	39	77	14	64	96
5	24	38	15	41	76
6	38	81	16	18	26
7	3	12	17	61	91
8	51	97	18	32	47
9	16	33	19	33	41
10	43	75	20	24	35

The R_t^2 for these data is shown in Figure 29.3. The maximum value of $R_t^2 (= 0.556)$ occurs at $t = 5$ which is our estimate of τ under model H_1. The p-value for this result from simulation $< 2.5\%$. The associated Cochran – Armitage statistic for the set is calculated as 4.361. This is highly significant and we deduce a *cumulative increase* in 'success' probability took place after the fifth group.

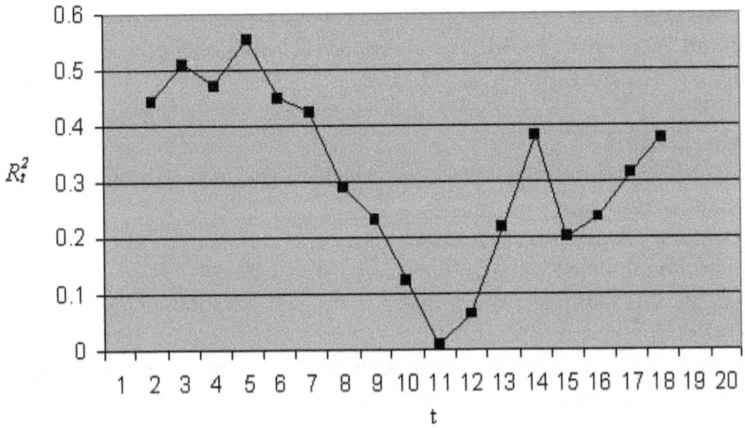

Figure 29.3. R_t^2 plot simulated data

29.4 Conclusion

A novel approach to the identification and testing of an unknown change-point for a series of binomial variates has been introduced. The approach has been demonstrated to have the particular advantage of distinguishing between situations where a binomial parameter undergoes an abrupt as opposed to a cumulative value change. Results obtained using the procedure have been found to compare well with established alternatives across a range of archetypal datasets. However, power characteristics may contrast less favourably – particularly in the special case of Bernouilli observations – and this is an area for future research.

References

P. Armitage (1955). Tests for linear trends in proportions and frequencies. *Biometrics*, 11:375–386.

D.G. Chapman and J. Nam (1968). Asymptotic power of chi square tests for linear trends in proportions. *Biometrics* 24:315–27; correction 25 777.

W.G. Cochran (1954). Some methods for strengthening common χ^2 tests. *Biometrics* 10:417–451.

R.J. Connor (1972). Grouping for testing trends in categorical data. *JASA* 67 339 *Theory and Methods* 601–604.

J.B. Copas (1989). Unweighted sums of squares test for proportions. *Applied Statistics*, 38(1):71–80.

J.M. Freeman (1986). An unknown change point and goodness of fit. *Statistician* 35(3):335–344.

D.V. Hinkley and E.A. Hinkley (1970). Inference about the change point in a sequence of binomial variables. *Biometrika* 57(3):477–488.

Z. Kander and S. Zacks (1966). Test procedures for possible changes in parameters of statistical distributions occurring at unknown time points. *Annals of Mathematical Statistics* 37:1196–1210.

M. Kendall and A. Stuart (1976). *The Advanced Theory of Statistics.* Volume 1, Distribution theory, 4th Edition. Griffin, London.

E.S. Page (1955). A test for a change in a parameter occurring at an unknown point. *Biometrika* 42:523–527.

A.N. Pettitt (1979). A non-parametric approach to the change point problem. *Applied Statistics* 28(2):126–135.

A.N. Pettitt (1980). A simple cumulative sum type statistic for the change point problem with zero-one observations. *Biometrika* 67(1):79–84.

A.N. Pettitt (1981). Posterior probabilities for a change point using ranks. *Biometrika* 68(2):443–450.

A.F.M. Smith (1975). A Bayesian approach to inference about a change point in a sequence of random variables. *Biometrika* 62(2):407–416.

S.-S. So (1998). Test methodology for the Binomial change point problem. Unpublished MSc dissertation. UMIST.

D.A. Williams (1988). Tests for differences between several small proportions. *Applied Statistics* 37(3):421–434.

K.J. Worsley (1983). The power of the likelihood ratio and cumulative sum tests for a change in binomial probability. *Biometrika* 70:455–464.

K.J. Worsley (1986). Confidence regions and tests for a change-point in a sequence of exponential family random variables. *Biometrika* 73(1):91–104.

Index

Note: The letters 'f' and 't' following the locators refers to figure and table respectively

C.H. Skiadas (ed.), *Advances in Data Analysis*, Statistics for Industry and Technology, DOI 10.1007/978-0-8176-4799-5,

Kolibri 360 GmbH
Ferdinand-Jühlke-Straße 7
99095 Erfurt, Deutschland
produktsicherheit@kolibri360.de